OEG-0-73-3823

THE BIOLOGY
OF
ESTUARINE ANIMALS

BIOLOGY SERIES

General Editor: R. Phillips Dales
Reader in Zoology in the University of London at Bedford College

U.S. Editor: Arthur W. Martin
Professor of Physiology, Department of Zoology, University of Washington

---

Practical Invertebrate Zoology
F. E. G. Cox, R. P. Dales, J. Green, J. E. Morton,
D. Nichols, D. Wakelin

Animal Mechanics
R. McNeill Alexander

The Biology of Estuarine Animals
J. Green

Structure and Habit in Vertebrate Evolution
G. S. Carter

Marine Biology
Hermann Freidrich

Molecular Biology
Clyde Manwell and C. M. Ann Baker

IN PREPARATION:

Principles of Histochemistry
W. G. Bruce Casselman

The Investigation of Natural Pigments
G. Y. Kennedy

Description and Classification of Vegetation
David W. Shimwell

Developmental Genetics and Animal Patterns
K. C. Sondhi

The Basis of Animal Distribution
Gustaf de Lattin

Aspects of the Host-Parasite Relationship
W. D. G. Haynes, D. Wakelin, A. Jeffries

The Biology of Estuarine Animals

J. GREEN, D.Sc.

Reader in Zoology
in the
University of London at Westfield College

UNIVERSITY OF WASHINGTON
SEATTLE and LONDON

CONTENTS

PREFACE

The aim of this book is to provide for undergraduates an account of the biology of animals living in estuaries. A broad concept of an estuary is adopted so that the animals dwelling in brackish seas can also be considered. The first part deals with the physico-chemical characteristics of the environment, and gives a concise account of the vegetation which provides shelter and food for many of the animals. The account of the animals is written from the viewpoint of the ways in which they cope with the estuarine environment. In this account the processes of osmotic and ionic regulation occupy a prominent place, but they are not treated in isolation, and have been interrelated with other aspects of the biology of the animals. In this way an attempt is made to present a balanced account of the lives of estuarine animals. The treatment is not exhaustive, but uses examples of different groups to illustrate differing adaptations to the estuarine environment. Many of the examples are of necessity taken from well-studied areas in Europe and North America, but comparisons are also made with examples from many other parts of the world. The final chapters deals with the food webs of estuaries, and considers the feeding habits of a wide range of animals in relation to the general economy of estuaries.

The material on which this book is based has been collected from a wide range of published sources, but no claim is made to complete coverage, and not all the papers that have been consulted have been quoted. The form and content of the book have to some extent been governed by my own field experience. Over a period of about thirteen years I have visited the Gwendraeth estuary at least once a year, sometimes for periods of six or seven weeks. On each occasion some special topic was selected for study, sometimes it was group, such as the oligochaetes

PREFACE

THE aim of this book is to provide for undergraduates an account of the biology of animals living in estuaries. A broad concept of an estuary is adopted so that the animals dwelling in brackish seas can also be considered. The first part deals with the physico-chemical characteristics of the environment, and gives a concise account of the vegetation which provides shelter and food for many of the animals. The account of the animals is written from the viewpoint of the ways in which they cope with the estuarine environment. In this account the processes of osmotic and ionic regulation occupy a prominent place, but they are not treated in isolation, and have been integrated with other aspects of the biology of the animals. In this way an attempt is made to present a balanced account of the lives of estuarine animals. The treatment is not exhaustive, but uses examples of different groups to illustrate differing adaptations to the estuarine environment. Many of the examples are of necessity taken from well-studied areas in Europe and North America, but comparisons are also made with examples from many other parts of the world. The final chapter deals with the food webs of estuaries, and considers the feeding habits of a wide range of animals in relation to the general economy of estuaries.

The material on which this book is based has been collected from a wide range of published sources. But no claim is made to complete coverage, and not all the papers that have been consulted have been quoted. The form and content of the book have to some extent been governed by my own field experience. Over a period of about thirteen years I have visited the Gwendraeth Estuary at least once a year, sometimes for periods of six or seven weeks. On each occasion some special topic was selected for study: sometimes it was group, such as the ciliates, and sometimes a habitat such as the salt marsh pools. In this way a picture of the animal life of an estuary has been slowly assembled. The striking contrast to an estuary provided by a brackish sea was experienced during the summer of 1956 when I worked at the Zoological station at Tvärminne, on the Gulf of Finland.

In the preparation of this book I have been helped by the generosity of authors and publishers who have granted permission for the reproduction of figures. It gives me pleasure to acknowledge the sources as listed on page 377.

CHAPTER 1

THE STRUCTURE AND DYNAMICS
OF ESTUARIES

THE STRUCTURE OF AN ESTUARY

THE configuration of land and water at the meeting of a river
and the sea is to some extent a matter of geological chance.
The detailed shape of an estuary varies with the size and age
of the river. Numerous other factors, such as the nature of the
bedrock, the extent of the tides, the vigour of wave action and
the strength and direction of the longshore drift, are also
important in determining estuarine form. In some circumstances
it may be difficult to define the limits of an estuary. From a
biological point of view it is convenient to regard an estuary as
the region of a river with a variable salinity due to the sea
(Day, 1951). From an oceanographer's point of view an estuary
is often regarded as an inlet where sea water is diluted by inflow
of fresh water. The two views are complementary, and adopting
both of them allows us to consider a wide range of animal
habitats. The absence of any reference to tides in these
definitions is deliberate, because certain estuaries become cut
off from the influence of tides for variable periods during the
dry season when wave action throws sand-banks across their
mouths. In the biological definition the absence of any reference
to dilution of sea water enables us to consider lagoons in which
the salinity may rise above that of sea water when evaporation
exceeds inflow.

The ideal river of the geographer rises among mountains and
flows down to the sea, passing through three tracts: the torrent,
valley and plain. The torrent cuts a narrow gorge, and descends
rapidly. In the valley the river flows more gently and is wider,
while in the plain the channel is even broader and a large
flood plain lies on either side. This ideal river develops in an

area that has been free from large scale movements of the earth's crust for some considerable time. But the earth's crust is not always static. Coastlines may emerge from the sea at a rate of a metre in a century—as certain parts of the Gulf of Bothnia are doing at present, or they may sink slowly into the sea. Movements of this type modify the mouth of a river, and

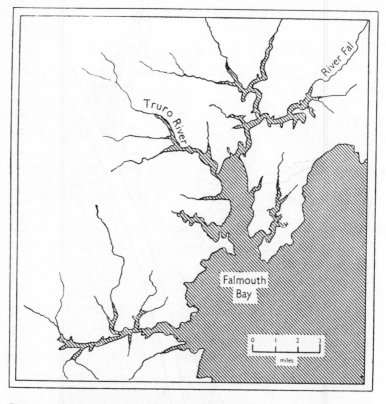

Figure 1. Map of part of the coast of South Cornwall to show drowned river valleys with extensive invasion by the sea

when a coastline sinks the whole of the plain and a large part of the valley tract may be invaded by the sea. A coastline submerged in this way may have many drowned valleys with numerous creeks and sheltered bays. Good examples are found

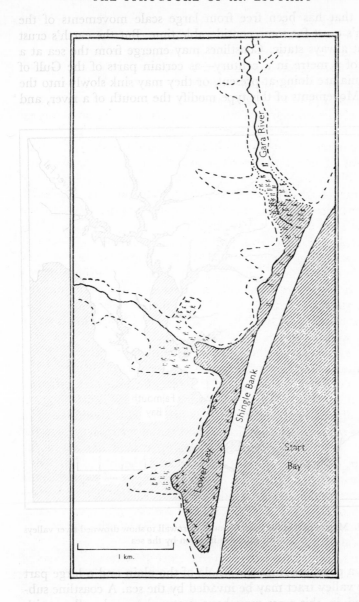

Figure 2. Map of Slapton Ley, Devon. The upper Ley merges gradually with marsh at the northern end

in England on the coasts of South Devon and Cornwall. The most important biological feature of such an area is that marine conditions are carried well within the mouths of the rivers, and the sheltered conditions enjoyed by estuarine animals are made available to truly marine animals.

The mouth of an estuary is often narrowed by the development of shingle spits and sand-banks. The foundation materials for spits and banks are transported by longshore drift, and the degree of narrowing of the estuarine mouth will depend on the amount of material deposited in relation to the flow of the river. When the river is small the drifted shingle may completely close the mouth of the estuary, to form a lagoon which gradually freshens. Provided that the shingle spit is not breached, either by the river or by a storm surge from the sea, the lagoon may eventually become transformed into a lake. In England, Slapton Ley, in Devon, is a classical example of this process (Fig. 2). The shingle spit is large enough and sufficiently stable to carry a motor road. The small River Gara flows into the northern end of the Ley, and is in the process of filling this part with silt. The southern part or lower Ley has open water and a well developed fresh-water fauna.

In regions where there are large seasonal variations in the flow rates of the rivers the mouth may be closed for part of the year and open for the rest of the year. Several such estuaries have been studied in South Africa where the tidal range is generally low, but wave action is heavy. This combination of factors reduces the tidal flow in and out of the estuary, and the violent wave action quickly throws up a sand bar across the mouth of the estuary. In the wet season the flow of the river is sufficient to keep an open channel through the bar, but in the dry season the river flow diminishes and the bar closes the mouth of the estuary. The enclosed lagoon may then freshen, or it may increase in salinity according to whether or not the freshwater inflow exceeds evaporation.

Most rivers carry some material in suspension. The rivers of the world have been estimated to wear away their beds at an average rate of 10 cm in 3,000 years, and to transport 8,000 million tons of finely ground rock down to the sea each year (Holmes, 1944). Some of this material is deposited on the flood plain, but the rate of deposition is increased when the river meets the sea. The salts in the sea water cause the fine suspended particles to group together, or flocculate, and so settle more quickly. In follows from this that fine flocculent mineral material, the basis of mud, is a common feature near the mouth of a river. The detailed distribution of this deposited material will depend on the pattern of currents created by the flow of the river and the movements of the tides. If the sea has a small tidal range and little wave action the river will carry its suspended material out into the sea and form a delta. The Mediterranean has very small tides, and large rivers like the Nile and the Rhône have built deltas which extend several miles into the sea.

The deposition of material in a delta quickly renders a single channel inadequate to take the full flow of the river, so that the banks of the original channel break and new channels are formed. A common feature of deltas is the presence of numerous channels or distributaries. The channels of the Mississippi delta are usually deep and have narrow banks of clay. They extend as narrow projections into the Gulf of Mexico, and individual channels have been observed to increase in length at a rate of 80 metres in a year.

The finger-like development of channels is restricted in many deltas by wave action which prevents the processes from projecting any great distance into the sea, and throws up sand banks to connect the short processes. When a sand bank connects two processes in this way it may enclose a lagoon. Such lagoons are found in the deltas of the Danube, Rhône, Nile and the Volga.

Where the tidal range is great the formation of a delta may be prevented, and the tide will sweep in and out of the estuary, invading the seaward end of the flood plain. The rising tide will slow the rate of flow of the river and so encourage the deposition of mud on the flood plain. This mud may remain bare, or it may become colonised by a variety of plants to form a salt marsh (Chapter 3).

TIDES

Tidal movements of the sea are caused by the gravitational pulls of the sun and of the moon. Of the two heavenly bodies the latter is the more effective, due to its proximity, exerting a little over twice the tide generating force of the sun. The extent of the alternate rising and falling of the surface of the sea varies from place to place and from day to day at any one place. These variations reflect the degree to which the sun and moon are pulling together. When a tide rises to a high level the succeeding low tide is correspondingly low. The reason for this is easy to realise if one considers the rotation of the earth in relation to the gravitational pulls of the sun and moon. Extensive tidal movements will occur when the tide raising forces of the sun and moon are most nearly in line. Such tides are termed spring tides, in contrast to neap tides, which result when the pulls of the sun and moon are exerted at right angles to each other. On a neap tide high water may be only one half the height of a spring tide, and the low neap tide will uncover much less of the shore than the spring tides.

The last statement is not always true of estuaries. On spring tides more water may enter an estuary on the flood than can escape on the ebb. When this occurs the low water of a spring tide may not fall to the low level of neap tides. This phenomenon is found in England in the Severn Estuary. At Sharpness the level of low water varies by less than a metre during the course of a spring-neap cycle, and is lowest on neap tides. The corresponding level of high water varies by nearly five metres and is highest on the spring tides.

On most Atlantic shores the spring tides build up to a peak once a fortnight, about two days after the new and full moons. The tides in between gradually decrease in magnitude and then build up again. Superimposed on this fortnightly cycle there is a seasonal cycle which results in the largest spring tides occurring at the vernal and autumnal equinoxes, when the day and night are of equal length and the sun and the moon are most nearly in line.

Variations in the water level in estuaries are not always caused by the interaction of tides and river flow. The studies of Spencer (1956) on the estuary of the Swan River in Western

Australia have shown that variations in water level are strongly influenced by barometric pressure. The normal tides in this estuarine basin are only about a foot in vertical range, but during cyclonic depressions the onshore gales have the effect of enhancing the flood tide and reducing the ebb tide so that average levels of the water in the estuarine basin are raised, sometimes by as much as three feet. When higher barometric pressures return the ebb tide becomes stronger than the flood tide, and the water level in the basin falls.

If the pattern of tides over the whole world is considered there are many complications. Parts of the China Sea have only one high tide each day. In other areas, where large islands impede the tidal flow there can be four tides per day. The best known example of this is found in Southampton Water on the English South Coast. Here the tide coming in from the Atlantic can enter not only from the west, but also from the east after flowing round the Isle of Wight. This results in the tidal level rising four times during the course of a day.

Tides may be considered as waves of very great length. The height of the tide in the open ocean is only a little over a metre. When this metre-high wave encounters the continental shelf the wave shortens and increases its height. This is brought about by the frictional drag of the bottom slowing the front of the wave and allowing the back of the wave to catch up and increase the height. The actual height of the tidal wave is influenced by the extent and slope of the continental shelf. Further increase in the height of the wave can be induced if the advancing front is gradually constricted, as happens between the coast of S.W. England and S. Wales as the tide flows in towards the Severn.

If the bed of the sea and the mouth of an estuary have an appropriate slope the tide may be projected into the estuary and travel up the river as a 'bore'. The essential structural features of an estuary which give rise to a bore are a funnel shape and a bed which rises steadily as it progresses inland. The rising bed causes the tidal wave to increase in height by retarding the front of the wave. A bore produced in this way may travel far inland, well beyond the limits of the true estuary, and well beyond the region of variable salinity.

The highest tidal bore is found on the Chiang tang kiang, in eastern China. On the highest spring tides the wave reaches a height of eight metres. In Britain the bore on the Severn reaches a height of three metres and travels as far inland as Gloucester. A stimulating account of the formation and characteristics of this bore has recently been given by Rowbotham (1964).

WAVES AND CURRENTS IN ESTUARIES

When waves encounter the mouth of an estuary they may suffer various fates. If the wave enters the mouth squarely it may proceed some distance into the river, but, as the depth decreases, the frictional drag and the flow of the river slow down the wave. The front of the wave becomes steeper and it eventually breaks and dies. If the direction of the wave is inclined at an angle to the mouth of the estuary it may suffer types of deformation similar to those suffered by light rays when they encounter surfaces. Reflection reverses the direction of a wave when it encounters an obstacle; diffraction is the result of passing the end of an obstacle. Fig. 3 shows how the waves change direction and quickly die as they pass the sand spits at the mouth of an estuary.

Refraction of waves is another phenomenon which occurs in shallow water. As the wave approaches the shore it may encounter slopes in which the isobaths are not parallel to the wave front. This encounter results in a change of direction which tends to bring the wave front more nearly parallel to the isobaths.

The overall result of these modifications of waves in estuaries is a diminution in wave action, so that estuarine waters are calm in comparison with the open sea.

In contrast to the general reduction in wave action there is great variation in the currents in estuaries. Major estuarine currents are generated in two ways: by the flow of the river and by the tides. The tidal currents may be modified or re-inforced by wind action. River flow often varies seasonally, and the speed of the current varies with volume of discharge and the width of the estuary. Where the estuary is broad the current will be slower, and where the channel is narrow the speed of the current will increase.

If the river opens into a broad shallow lagoon the river currents diminish, due to the frictional drag of the lagoon floor. Further, if the mouth of the lagoon is restricted by shingle spits or sand banks the tidal currents will also be reduced and the whole lagoon may be free from strong currents.

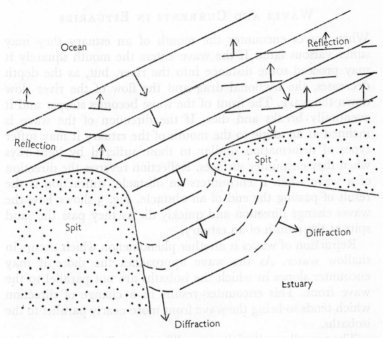

Figure 3. Diagram to show diffraction and reflection of waves at the mouth of an estuary. The reflection of waves from either side of a narrow mouth may interfere with the incoming waves and reduce the effect of the waves entering the mouth

When the tide begins to flow into an estuary it encounters resistance from the river. In many estuaries the inflowing tide moves to one side while the outflowing river moves over to the other, at least during the early part of the tidal cycle. Later the rising tide may push back the river water over the whole width of the estuary, or it may penetrate as a tongue of sea water while the fresh water continues to flow out over the top. There will of course be a certain amount of mixing during the tidal cycle and this factor varies from one estuary to another.

Pritchard (1955) has described a series of estuarine types with differing density stratifications and circulation patterns. Type A is highly stratified, with a wedge of saline water lying beneath outflowing low salinity water. Mixing is reduced to a minimum. Such estuaries are most frequently found in areas with a small tidal range and large river flow.

When the tidal range is increased and the tidal flow involves volumes in excess of the fresh-water outflow, the resulting estuary is less strongly stratified. The rotation of the earth causes the outflowing water to move over to the right-hand side of an estuary in the Northern Hemisphere as it flows towards the sea. The inflowing saline water will then come nearer to the surface on the left-hand side of the estuary. In effect, the boundary between the lower saline water and the upper fresher layers is tilted. This is the condition found in Pritchard's type B estuary. There is still some vertical stratification, but it is not as strong as in type A.

In a type C estuary vertical stratification is not present, but there is still a lateral gradient caused by the rotation of the earth. Such estuaries have relatively large tidal velocities and usually are relatively wide. If the estuary is narrow the lateral gradient may be destroyed by lateral mixing and the estuary then lacks both vertical and lateral salinity gradients. This is Pritchard's type D estuary, which retains only the longitudinal gradient from the sea to fresh water. The four estuarine types are shown in Fig. 4.

The sequence of estuarine types tends to shift from type A through type B to types C and D with changes of the following parameters: (i) decreasing river flow, (ii) increasing tidal velocity, (iii) increasing width, (iv) decreasing depth.

It must not be thought that these categories of estuaries are rigid. Intermediates occur, and it is possible for one estuary to change from type A to type B according to seasonal variations in river flow.

Tidal currents in estuaries may be classified into three basic types (Caldwell, 1955). Much depends upon the nature of the

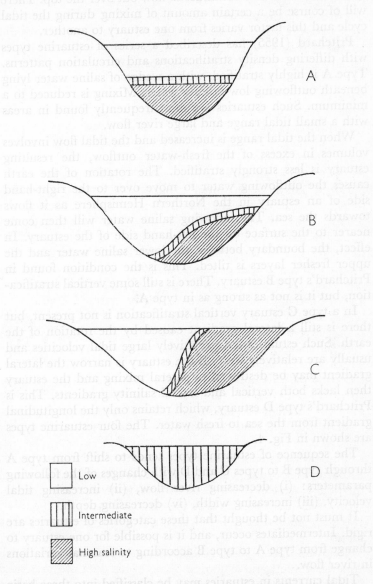

Figure 4. Diagrammatic cross sections of the middle reaches of the four estuarine types defined by Pritchard (1955)

entrance to the estuary. Where the entrance is large enough not to impede the progress of the oceanic tidal wave the current strength and timing in relation to the tides will depend upon the length of the tidal wave. The length of the tidal wave in an estuary can be estimated from the formula

$$L = 48\cdot1\sqrt{d}$$

where 'L' is the length in miles of a 12-hour 25-minutes tide from high tide to high tide, and 'd' is the mean depth in feet of the estuary. The estimate is of necessity only approximate, mainly because the mean depth is not easy to estimate over the whole length of an estuary.

If the length of the tidal reach of an estuary exceeds a quarter of tidal wave-length and there is no restriction at the mouth, the inflowing tide will flow unimpeded into the estuarine basin. The fastest currents can then be expected at high tide and slack waters at mid tide. But if the length of the tidal reach is less than one quarter of the tidal wave-length the inflowing water will pile up against the head of the tidal reach and the strength of the currents will be reduced before high tide. The fastest currents will then occur at about mid tide and slack water will occur at high tide.

A third category is found in estuaries with narrow mouths which restrict the tidal inflow so that the tidal range in the estuarine basin is much lower than the range in the open sea. In such a situation the tidal current depends on the hydraulic gradient through the mouth, which will normally be greatest at high tide. The fastest currents can thus be expected at high tide.

There will of course be intermediates between these categories. Estuaries with a tidal reach which is just about equal to a quarter of the tidal wave-length will be intermediate between the first two types, and intermediates between the third type and the other two may be expected with varying degrees of closure of the mouth.

Caldwell (1955) has studied a large number of North American inlets and estuaries. Some of his data are presented in Table 1, where examples of the three main types and an intermediate are given.

TABLE 1

RELATIONSHIP BETWEEN ESTUARINE STRUCTURE AND TIDAL CURRENTS
(From Caldwell, 1955)

	Estimated tidal reach (miles)	Estimated tidal wave-length (miles)	Estimated depth (feet)	Tidal range inside estuary (feet)	Interval between fastest current and high tide (minutes)	Current speed ft/sec
Hudson River	150	216	20	4·4	40	2·7
St John's River	100	216	20	4·5	11	3·5
St Mary River	9	184	15	6·1	122	4·1
Indian River	10	108	5	0·9	50	3·9
Delaware						
Savannah River	22	184	15	6·7	104	3·6

The strength of estuarine currents also varies with the spring-neap cycle. This fact is well known to fishermen who set nets in estuaries. Nets set low on the shore may be quite safe during a neap tide, but on a spring tide the nets are moved to higher ground because the currents lower on the shore would sweep them away.

Interaction of the river and the tides can result in the development of partly separated inflow and outflow channels. The inflow channels may remain as arms or pools alongside the main outflow at low tide, but serve to conduct the inflowing tide in the initial stages of the flood. The pattern of channels and pools can become very complicated, and varies greatly with the local geological conditions. Separate flood and ebb channels are liable to develop particularly in areas where the estuarine bed consists of sand. The flood channels are relatively straight, because the tidal wave has a relatively high velocity. The ebb channels usually meander because the velocity of the ebb tide, being generated by gravity alone, is less than that of the flood tide, and the flow swings from bank to bank, providing the main erosion forces.

The fastest currents are found at the surface in mid-stream. Towards the bottom and the sides of the channel the current speed is reduced by the frictional drag of the bottom and the banks. Any irregularities in the shape of the estuary are liable to alter the currents and may cause eddy currents which dissipate the energy of flow.

When current speeds are reduced, or even stopped as at high water in some estuaries, any material in suspension is given a chance to settle. The greater the reduction in current speed the finer the particles which will be able to settle. The reduction in current speed over the upper parts of the tidal range at high water means that the finest deposits will be laid down in the upper zone. In the lower zones, where current speeds are generally greater, only the larger heavier particles will settle giving rise to a coarser deposit.

TRANSPORT AND DEPOSITION OF SILT AND SAND IN ESTUARIES

The specific gravity of silt does not vary much from 2·65, and the particles vary in diameter from 4 to 60μ. Larger particles are regarded as fine sand and smaller particles as clay. Because the specific gravity of silt is greater than that of water the particles will sink at a rate which is proportional to their size. A particle with a diameter of 60μ sinks through still fresh water at a rate of about 1 cm in 4 seconds, while a particle with a diameter of 2μ sinks more slowly, at a rate of approximately 1 cm per hour. In sea water, which has a higher specific gravity than fresh water, the rate of sinking will be slower. But the presence of dissolved salts in sea water leads to complications. Small particles frequently clump together in the presence of dissolved salts and so form larger aggregates which settle more quickly than individual small particles. A further complication is brought about by the increased rate of settling at high concentrations of silt. It has been found that two particles close together, even though not actually in contact, will fall more quickly than one particle on its own. A large particle falling through the water may also drag smaller particles with it.

In an estuary the suspended silt, whether carried down by the river or brought in by the sea, will tend to concentrate near the bed. Now if the waters of an estuary are to remain within the normal limits of a spring tide the overall outflow must exceed the tidal inflow by an amount equal to the discharge of the river. But the depth at which most of the water is moved is important. The current velocity near the bed is much lower than that near the surface, both on the flood and on the ebb. Most of the water movement will thus take place in the upper layers. The concentration of silt is highest on the bed of the estuary, so that it is necessary to consider both the current velocities and the concentrations of silt at various depths throughout the complete tidal cycle to understand how the silt is distributed in an estuary.

Inglis and Allen (1957) have made such a study of the transport of suspended material in a reach of the Thames Estuary. Their results show how complicated the movement of silt can be. Fig. 5 shows the variation in current velocity and

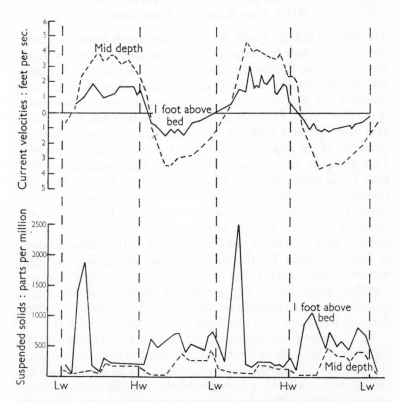

Figure 5. Variation in current velocities and concentrations of suspended solids throughout two tidal cycles in the Thames Estuary at a point 19 miles from London Bridge (after Inglis and Allen. 1957)

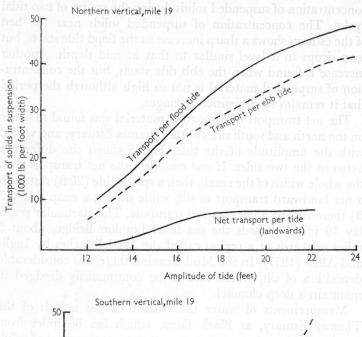

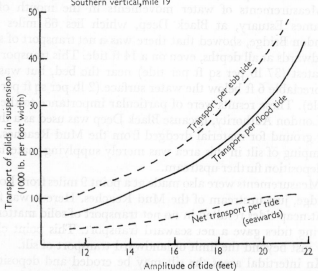

Figure 6. Transport of suspended solids at mile 19 in the Thames Estuary. Landward transport is shown by a continuous line and seaward transport by a broken line (based on the data of Inglis and Allen 1957)

concentration of suspended solids during the course of two tidal cycles. The concentration of suspended solids near the bed of the estuary shows a sharp increase as the flood tide starts, but soon drops to a level similar to that at mid depth. Another increase is found when the ebb tide starts, but the concentration of suspended material is not so high although the period that it remains in suspension is longer.

The net transport of suspended material was found to differ on the north and south sides of the Thames Estuary, and varied with the amplitude of the tide. Fig. 6 shows the difference between the two sides. If one considers the net transport over the whole width of the reach, then a spring tide (20 ft) results in a net landward transport of silt, while during a neap tide (13 ft) there will be a net seaward transport. This particular reach lay 19 miles towards the sea from London Bridge, about 5 miles seawards of a region called the Mud Reaches by Inglis and Allen (1957). In the Mud Reaches there was considerable deposition of silt, which had to be continuously dredged to maintain a deep channel.

Measurements of water movements in the mouth of the Thames Estuary, at Black Deep, which lies 68 miles from London Bridge, showed that there was a net transport of solids landwards at all depths, even on a 14 ft tide. This transport was greatest (37 lb per sq ft per tide) near the bed, but was also appreciable 6 ft below the water surface (2 lb per sq ft per tidal cycle). These results were of particular importance to the Port of London Authority, because Black Deep was used as a dumping ground for material dredged from the Mud Reaches. The dumping of silt in this area was merely supplying material for redeposition further upstream.

Measurements were also made at a point 9 miles from London Bridge, just upstream of the Mud Reaches. Here it was found that neap tides resulted in no net transport of solid matter, but spring tides gave a net seaward transport. This point clearly lay just beyond the limit of landward transport of silt.

In intertidal areas the silt may be eroded and deposited in alternating cycles. On a spring tide material is picked up from the lower parts of a mud flat and carried seawards by the ebbing tide. When the tide floods again it carries the silt inshore and redistributes it all over the mud flat. In a natural estuary, lacking any man-made embankments, the cycle of erosion and deposition is more or less in balance. When embankments and sea walls are built they may reduce the energy of flow by reflection and diffraction so that less energy is available for the transport of solid materials in suspension. The result of this is an increased deposition of solids. A barrage built across the estuary of the River Eider in 1935 caused a change in the relation between the current velocities of the ebb and

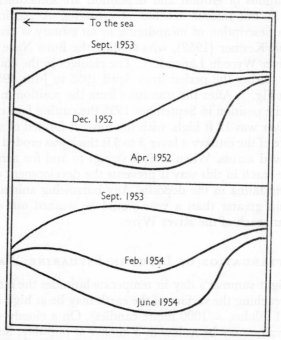

Figure 7. Successive positions of the low water channel in the Burn Naze Reach of the Wyre Estuary (based on data given by Inglis and Kestner, 1958)

flood, which made the ebb too weak to carry away all the material brought in by the flood. The result was a greatly increased rate of silt deposition.

It is clear that the amount of material in suspension in an estuary is influenced by the velocities of the tidal and river currents. The faster the current the larger the size of the particles that it can carry. But once the particles have settled the relation between current speed and erosion is not so simple. Hjulstrom (1935) has shown that sand particles with a diameter ranging from 200 to 500μ can be eroded and moved by a current velocity of about 20 cm per sec. Larger particles required faster currents to move them, and surprisingly so do smaller particles. A uniform sediment of particles with a diameter of 2μ may require a current velocity of over 100 cm per sec to cause erosion once the particles have settled.

Alternations of erosion and deposition are sometimes very marked in the middle reaches of estuaries. A particularly thorough description of meandering in an estuary is given by Inglis and Kestner (1958), who studied the Burn Naze reach of the River Wyre in Lancashire. The changes in the course of the river during the period from April 1952 to June 1954 are shown in Fig. 7. After the transition from the position in April 1952 to the position in September 1953 the cutting face eroded by the river was 12 ft high, with the top 7 ft formed of silt. In the centre of the estuary a layer 3 to 5 ft thick was eroded as the river moved across. When a river swings to and fro across an estuarine reach in this way it prevents the development of any permanent fauna in the deposits. Any burrowing animal with a life span greater than a year would be washed out of this particular reach of the River Wyre.

THE ATTENUATION OF LIGHT IN ESTUARINE WATERS

On a bright summer's day in temperate latitudes the intensity of light reaching the surface of the earth may be as high as 150 kilolux (1 kilolux = 1000 metre candles). On a cloudy winter day the value may be as low as 5 kilolux at mid-day. Some of this light is reflected from the surface of the water; the actual amount varies with the smoothness of the water surface and the angle of inclination of the sun. As much as 25 per cent may be reflected from a choppy sea on a bright day.

The light that penetrates into the water is absorbed, both by the water and by the particles suspended in it. Any increase in the number of particles in suspension increases the absorption of light. With simple suspensions there is a linear relationship between the amount of light absorbed and the concentration of suspended solids (Jones and Wills, 1956).

In many estuaries the water contains sufficient material in suspension to absorb nearly all the light in the top metre or two. Particles in suspension also scatter light, and scatter blue light more than red. This means that red light has the potential of penetrating further than blue into estuarine waters. But at wave-lengths above 650 mμ it has been found that molecules of water absorb light strongly, and as the wave-length increases beyond 700 mμ so the absorption increases. The potential penetration of the long wave-length red light is thus reduced. It is notable that Cooper and Milne (1939) found that a wave-length of about 600 mμ gave the maximum transmission through the turbid waters of the Tamar estuary.

The points made above are all relevant to the possibility of photosynthesis by algae in estuarine waters. Phytoplankton will be able to photosynthesise only in the surface layer of a turbid estuary, and the net production of phytoplankton may be negligible. In such an estuary the production of organic matter is dependent largely on the activities of shore plants, which can photosynthesise when uncovered by the turbid water. Salt marshes assume great importance in the economy of an estuary when the production of phytoplankton is reduced by the turbidity of the water. A quantitative estimate of the contribution made by a salt marsh to an estuarine community is given in Chapter 13.

PROPERTIES OF ESTUARINE DEPOSITS

Analysis of estuarine deposits takes the form of a mechanical separation of the different sizes of particles, and a consideration of the effects of particle size on permeability and other properties. The usual technique is to employ sieves for the coarser particles and sedimentation through water for the finer particles. Sieving is best done after the fine clay and silt have been washed out and the deposit dried. If the whole sample is dried the smallest particles tend to clog together and coat the coarser particles. The terminology of particle sizes most frequently employed is the Wentworth Scale, which is given in Table 2.

TABLE 2

Wentworth Classification Name	Particle size range mm	Phi Scale
Boulder	256	−8
Cobble	64–256	−6, −7
Pebble	32–64	−5
	16–32	−4
	8–16	−3
	4–8	−2
Granule	2–4	−1
Very coarse sand	1–2	0
Coarse sand	0·5–1	1
Medium sand	0·25–0·5	2
Fine sand	0·125–0·25	3
Very fine sand	0·0625–0·125	4
Silt	0·0039–0·0625	5, 6, 7, 8
Clay	0·0039	9

The phi scale of Krumbein (1936), where $\varphi = -\log_2$ of the particle diameter in millimetres, has certain advantages when comparing samples from different localities. It has been found that the size frequency distribution curves of sediments become more symmetrical when the logarithm of the diameter is plotted instead of the diameter. The frequency curves can then be analysed by methods similar to those used in statistics. Mean and median particle size can be calculated and the shape of the frequency distribution curves can be defined by standard deviations and measures of skewness. A detailed discussion of the mathematical techniques involved is given by Inman (1952).

The sizes of the particles in a deposit influence a number of properties which are of biological importance. If the particles are small the total surface area in a given volume of soil will be greater than if they are large. Surfaces are important for the attachment and growth of bacteria. Zobell and Anderson (1936) have shown that the concentration of bacteria in stored sea water was greater in small bottles than in large vessels, and increased greatly in water between sand grains. The concentration of bacteria in natural deposits is much higher in the fine grades than in the coarser (Zobell, 1938). Measurements of organic carbon and nitrogen in the deposits near the mouth of the Thames Estuary (Newell, 1965), showed that the finer deposits were richer than the coarser. A further point is that in the presence of relatively small amounts of organic debris the bacterial population is governed mainly by the fineness of the deposit and not by the amount of organic debris.

Another property related to particle size is the capillary lift of a soil. If dry sand is placed in a glass tube with the base covered in fine gauze and then placed upright in water there is an upward movement of water into the column of sand above the level of the water outside the column. The height that the water rises is determined by the size of the interstices between the sand grains. In fine sand the water may rise ten times as high as in coarse sand. This property will clearly be of importance to animals living in the intertidal zone of estuaries, where retention of water at low tide may be critical.

A related phenomenon is the retention of water by surface forces. An estimate of this can be obtained by studying the rate of evaporation from sands saturated with water. At first the water evaporates at a rate similar to that observed at a free water surface. As the sands approach dryness the rate of evaporation decreases. The decrease commences earlier and is greater in the finer sands. Webb (1958) found that coarse sand, with a mean particle diameter of 0·9 mm retained little or no water, but sand with a mean particle diameter of 0·15 mm retained about 14 per cent of its water against evaporation. This percentage was estimated by noting the point at which there was a sharp decrease in the rate of evaporation.

Variations in temperature and wind speed influence the evaporation of water from natural deposits. In the upper parts of the intertidal zone, where the surface is covered by water only two or three times in a fortnight, the surface may become completely dry. In a sandy area the surface is then liable to wind erosion. In muddy areas the surface becomes hard and may then crack, not only sealing parts of the surface, but exposing animals deeper down to the danger of desiccation.

The porosity of a soil is the volume of spaces between the particles expressed as a percentage of the total soil volume. If a system of uniform spheres is packed as closely as possible the volume of the interstitial spaces is 26 per cent of the total volume irrespective of the size of the spheres (Bruce, 1928). But estuarine deposits are not systems of uniform spheres; they contain particles of varying sizes and shapes, including many very small particles.

The factors influencing variation in porosity have been discussed in detail by Frazer (1935), who gives the following list:

1. Absolute size of grain
2. Non uniformity of grain size
3. Proportions of various sizes of grain
4. Shape of grain
5. Method of deposition
6. Compaction during and following deposition
7. Solidification

In theory the absolute grain size should not influence the porosity. In practice the smaller grain sizes show increased porosity which is caused by friction, adhesion and bridging, which interfere with the packing.

The porosity of graded sands, which have been sieved to give particles with a restricted range of size, may be as high as 46 per cent, but natural sands are often much less porous with porosities ranging from 20 to 38 per cent. Muds, with a high proportion of silt and clay, are much less porous because the smaller particles tend to fill the spaces between the larger particles. Somewhat paradoxically it has been found that newly deposited clays may have porosities as high as 45 to 50 per cent. But these clays are liable to considerable compaction, particularly if exposed intertidally, and then porosity may fall to below 20 per cent.

The effect of particle shape on porosity has been discussed by Tickell and Hiatt (1938). Angularity may either increase or decrease porosity, most often it increases porosity, apparently by bridging and interference with packing. A decrease in porosity is found when the grains are slightly disc shaped; this allows close packing in one dimension which reduces the

B

interstitial space when compared with spheres. Extremely high porosities are found in artificially crushed mica, which forms irregular flat plates. The porosity may exceed 90 per cent, but such high porosities are not found in natural deposits.

The relationship between porosity and mixtures of particle sizes is complex (King, 1898; Frazer, 1935). Small particles may either occupy the interstitial spaces thus reducing porosity, or if the smaller particles exceed a certain critical size at which they will not fit into the interstices they may hold the larger grains apart and tend to increase the porosity. In general it is found that simple mixtures of two sizes of particles have lower porosities than pure samples of a single particle size. The changes in porosity which take place when varying proportions of small particles are added to coarser particles are shown in Fig. 8.

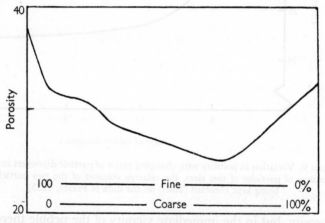

Figure 8. Change in porosity in mixture of coarse (mean diam. 0·483 mm) and fine sand (maximum diam. 0·096 mm) (after King, 1898)

If the relative proportions of two sizes of particles are kept constant, say 50 per cent by volume of each, then the porosity of the mixture will vary with the ratio of the diameters of the particles. Fig. 9 shows the form of the relationship. The decline in porosity is most marked over the range of size ratios from 2:1 to 6:1.

In more complex mixtures, with three or more particle sizes, it is impossible to make any accurate prediction of the porosity.

Local variations in porosity can be induced by the presence of very large particles such as pebbles. If one considers the whole deposit, these large particles cause a reduction in overall

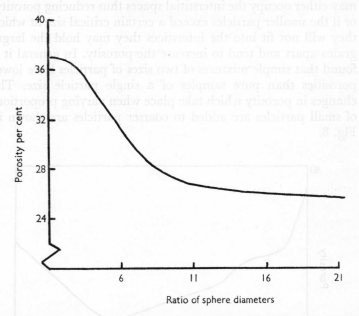

Figure 9. Variation in porosity with changing ratios of particle diameters in simple mixtures of particles of two sizes, the relative volumes of the two particle sizes being kept constant (based on the data of Frazer, 1935)

porosity, but in the immediate vicinity of the pebble there will be an increase in the porosity of the sand. Such an increase is caused by an increase in the average size of the spaces between the sand grains, because the sand cannot pack so tightly against the pebble as against pebble-free sand.

A conglomerate of pebbles and sand will have an average porosity lower than that of pebble-free sand, but the porosity of the sand in the conglomerate will be higher than that of pebble-free sand.

In ordinary hydraulic usage a substance is said to be permeable when it permits the passage of a measurable quantity of fluid in a finite period of time. The term impermeable is applied when the rate at which a substance transmits fluid is so slow as to be negligible (Frazer, 1935).

The permeability of an estuarine deposit is influenced greatly by the size of the particles. For a system of uniform spheres the rate of flow through a standard column is directly proportional to the square of the diameter of the spheres. Pebbles with an average diameter of 50 mm give a permeability over 600 times as great as clean sand with an average grain diameter of 2 mm.

Sand which has passed through a 90-mesh to the inch sieve allows water to drain through much more slowly than sand retained by the same sieve, but which has passed through a 60-mesh sieve. Webb (1958) found that the time taken to drain a standard column of water through a standard 10 cm column of sand which had passed through a 90-mesh sieve was 150 times as long as the time taken to pass through a column of sand retained by the sieve.

Small changes in porosity can produce large changes in permeability; an approximate relation is given by the Fair-Hatch formula:

$$P \propto \frac{n^3}{(1-n)^2}$$

where 'P' is the permeability, and 'n' is the porosity (Franzini, 1951). Such a relationship only applies when all other factors are constant, particularly the grain size.

Addition of silt to a sand sample decreases the permeability. If only 2·5 per cent of silt is present in certain deposits the drainage time for a standard column of water may be four times as long as in the absence of silt. Larger percentages of silt increase the drainage time even more. Clays are characterised by very small particles and hence by minute interstices. The total surface area of clay particles can be enormous. A cubic foot of clay may have a total particle surface area of over 200,000 sq ft (Frazer, 1935). This vast area will tend to retain water by adsorption and will restrict the rate of flow through the minute pores. It follows from this that many estuarine deposits have a low permeability. This may be countered to some extent by the activities of burrowing animals, particularly those that live in U-shaped burrows through which they cause currents of water to flow.

Seepage of water through soil might be expected to increase with increasing temperature. The viscosity of water decreases as the temperature rises, and this might be expected to facilitate the flow of water through soil. But this is not always so. Robinson and Rohwer (1955) found that seepage through a variety of soils increased with decreasing temperatures. They concluded that this effect was caused by variations in the volume and vapour pressure of the air in the soils. These results may be applicable to loose sandy intertidal soils which drain to some distance below the surface when the tide retreats.

Porosity and permeability are factors in determining the water content of a soil. This is important on intertidal shores, particularly in relation to the resistance offered by the soil to the activities of burrowing animals. Chapman (1949) has shown that *Arenicola* can burrow faster into soils with a high water content than into those with a low water content. Under certain conditions a soil will behave in a manner similar to a fluid, and the extent to which this occurs is dependent on the density, viscosity and amount of water present in the soil. These three factors in conjunction with the size of the grains will determine the percentage of grains which are floating, or have no direct contact with other grains. In a quicksand a high proportion of the grains are floating, and such a sand offers little or no resistance to penetration.

The percentage of floating grains can be increased by agitation. This is readily observed on many sandy shores. If the surface of the sand is tapped with a foot the surface becomes wet and the sand becomes soft, offering little resistance to penetration.

Thixotropy has been defined as a reduction in resistance with increasing application of shear. Quicksands are thixotropic. This contrasts with the property of dilatancy, in which the resistance increases as the sheer force increases. A wet dilatant sand whitens underfoot, and is not easily penetrated. Many marine and estuarine sands show a mixture of dilatancy and thixotropy. A sand which whitens underfoot may also soften with agitation and, as Chapman (1949) has shown, the tendency away from dilatancy towards thixotropy increases with increasing water content.

The hardness of the sand in the intertidal zone near the mouth of the Thames Estuary at Whitstable has been found to vary during the course of a year (Perkins, 1958a). In late October or early November the soil suddenly became much softer. In the spring and summer it gradually became harder again. Differences between the surface ridges and troughs were also found. The ridges were softer than the troughs. It appeared that the troughs represented a stable hard layer over which the ridge system moved according to the direction of the wind. Although the troughs were usually covered with water their interstitial water content was low. The ridges, although they projected out of the water were more porous and had a higher interstitial water content, accounting for their softness.

A further complication was found in the form of variation in hardness during the course of a tidal cycle. The sand suddenly became softer a few minutes before being covered by the advancing tide. When actually covered with water the sand became harder. Perkins noted this as a general rule: when the sand was submerged it was harder than when exposed.

CHAPTER 2

SALINITY AND OTHER CHEMICAL FACTORS

AN essential feature of estuaries is that the salinity of the water varies. In the biological definition (p. 1) this was regarded as the basis for determining the limits of an estuary.

For many practical purposes the salts in estuarine waters can be regarded as being of marine origin. Sea water contains about 35 g of salt per litre (35‰), with small local variations above and below this figure. Most rivers have relatively small amounts of salt in solution, usually less than 0·2‰.

MEASURES OF SALT CONCENTRATION

The variations of salinity in estuaries are so large that a relatively crude scale of measurement is all that is necessary. The number of grams of salt per litre (‰) is an adequate measure for most ecological purposes. This figure can be estimated in various ways:

1. Titration with a silver nitrate solution containing 27·25 g/1 using a 10 ml sample of water gives direct values in terms of the number of ml of silver nitrate solution used. For critical oceanographical work the silver nitrate method has been highly refined (cf. Johnston, 1964), but for estuaries the simple method is adequate.

2. Measurement of density with a hydrometer. The temperature has to be measured at the same time and the readings converted to salinity using a graph of the type given by Harvey (1960, p. 128). This is a crude method, and under field conditions with a simple hydrometer is not accurate to more than the nearest 1 or 2‰.

3. Measurement of electrical conductivity. This can give accurate figures with a good modern instrument specially

constructed for the purpose. Some commercial models (for instance, the Dionic Water Tester) can measure conductivity over the whole range from distilled water to twice the strength of sea water, using a logarithmic scale so that the more dilute solutions can be measured with better accuracy. Temperature compensation is necessary, and the instrument should be calibrated against known salinities. This is a highly convenient field method.

4. Measurement of freezing point depression. This is a laboratory method which is not convenient to use in the field, but it can be scaled down to use minute volumes of water so that it can be used for body fluids of small animals.

When animal body fluids are measured the results may be expressed in several different forms. Freezing point depression is given in degrees of the centigrade scale. For instance, the freezing point depression of sea water is 1·98°C, while that of the blood of *Ligia oceanica* varies between 1·71 and 3·48°C (Parry, 1953).

The concentrations of different solutes in animal body fluids are usually given in millimoles (mM) per litre of solution. A millimole is the molecular weight in milligrams (1 milligram = 1/1000 gram). A solution with one millimole in a litre can be termed a millimolar solution. Physiologists studying osmotic and ionic regulation also apply the term millimole to ions instead of the more strictly correct term milligram ion. Molal and milli-molal solutions are measured in terms of moles or millimoles per kilogram of water. In this book where such measures are used they will be given as milligrams per kg water to avoid confusion.

When the concentration of a fluid has been determined by measuring the freezing point depression, or by measuring the osmotic pressure, the results can be calculated in terms of milliosmoles (mOsm), either per litre or per kilogram of water. A milliosmole is the mass of $6·023 \times 10^{2}$ ° (Avogadro's number) osmotically active particles in aqueous solution. For an ideal non-electrolyte this would be the same as a millimole, but salts in sea water and animal body fluids do not behave as ideal non-electrolytes, and the miliosmolar concentration is lower than the concentration in millimoles. Thus, average figures for sea water would be 1000 mOsm/Kg water and 1129 mM/Kg water (Potts and Parry, 1964).

VARIATION IN SALINITY

On a flood tide the salinity will increase along the length of an estuary, and the salinity at any one point will depend on the extent to which sea water travels upstream. On a spring tide the sea will penetrate further into the estuary than on a neap tide, and so will cause a more widespread increase in salinity. There will of course be interaction between the spring-neap cycle and the wet and dry seasons. Thus a spring tide in the dry season, when the river flow is feeble, will cause an increase in salinity further upstream than a corresponding tide in the wet season when the river flows strongly (Fig. 10). When a river is in flood after heavy rain the salinity along the whole length of the estuary will be reduced. If the river is very large it may prevent the entry of sea water and a region of fresh water may be pushed out into the sea, as at the mouth of the Amazon.

When the tide ebbs the estuarine basin may drain, leaving only a narrow channel with fresh water flowing through. But in many large estuaries this does not happen. If the estuary is of the drowned valley type with a deep inlet from the sea there may be a fairly stable salinity gradient from the head to the mouth. This gradient will move backwards and forwards with the tides, but will not exhibit the drastic changes in salinity that are found in small estuaries which drain on every tide.

If the river water mixes thoroughly with the saline waters of the estuary and there is no marked development of a vertical salinity gradient the fresh water entering a large estuary may take several days to reach the sea from the head. The time taken for a unit inflow of fresh water to reach the sea is known as the flushing time. The three main factors which influence the flushing time are:

(1) the discharge rate of the river
(2) the volume of estuarine water at low tide
(3) the tidal volume which flows in and out of the estuary.

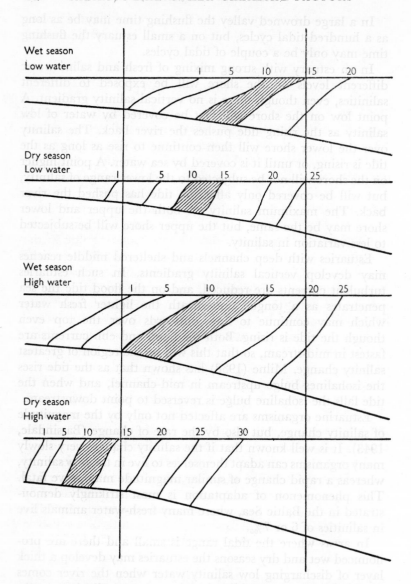

Figure 10. Salinity variation in a hypothetical estuary with a stable salinity gradient and large tidal range. The two vertical lines represent the extreme movements of surface water with a salinity of 15‰.

In a large drowned valley the flushing time may be as long as a hundred tidal cycles, but on a small estuary the flushing time may only be a couple of tidal cycles.

In an estuary with strong mixing of fresh and saline water different levels on the shore will be exposed to different salinities, even though there is no vertical salinity gradient. A point low on the shore will first be covered by water of low salinity as the rising tide pushes the river back. The salinity over the lower shore will then continue to rise as long as the tide is rising, or until it is covered by sea water. A point higher on the shore will not be subjected to the lower range of salinity, but will be covered only after the tide has pushed the river back. The maximum salinity on both the upper and lower shore may be the same, but the upper shore will be subjected to less variation in salinity.

Estuaries with deep channels and sheltered middle reaches may develop vertical salinity gradients. In such estuaries turbulent currents are reduced, and on the flood tide the sea penetrates as a tongue underneath the lighter fresh water which may continue to flow outwards over the top even though the tide is rising. Both the flood and ebb currents are fastest in mid-stream, so that this will be the region of greatest salinity change. Milne (1938) has shown that as the tide rises the isohalines bulge upstream in mid-channel, and when the tide falls the isohaline bulge is reversed to point downstream.

Estuarine organisms are affected not only by the magnitude of salinity change, but also by the rate of change (Bassindale, 1943). It is well known that if the salinity changes very slowly many organisms can adapt themselves to live in the new salinity, whereas a rapid change of similar magnitude may prove fatal. This phenomenon of adaptation is most strikingly demonstrated in the Baltic Sea, where many fresh-water animals live in salinities of 6 or 7‰.

In areas where the tidal range is small and there are pronounced wet and dry seasons the estuaries may develop a thick layer of discharging low salinity water when the river comes into flood. The Swan River in Western Australia is a much-studied example (Spencer, 1956). In this estuary during the dry season the main basin is virtually an arm of the sea with a negligible fresh-water inflow. During the wet season, from May to August, the top 3 metres or more is made up of outflowing low salinity water. There is not much mixing of this layer with the more saline deeper layers because the tidal range is small, rarely exceeding a foot. The main basin is separated from the sea by a sill at a depth of about 5 metres. If the outflowing water is of sufficient volume to reach the depth of the sill it effectively separates the deeper parts of the estuarine basin from the open sea. This deeper water, which extends down to 20 metres in some places, may then stagnate, and because it does not mix with the overflowing low salinity water it gradually becomes depleted of oxygen. When the fresh-water flow declines and rises above the sill there is the possibility of sea water penetrating into the basin again. At first there is some mixing of sea water and fresh water near the sill, so that 'sill water' of intermediate salinity flows inwards under the surface discharge of low salinity water. As the discharge declines there is an increased penetration of sea water, which cascades down over the sill and restores aerobic conditions to the deeper parts of the basin.

Where the tidal range is somewhat greater the duration of the seasonal salinity stratification is reduced. The Hawkesbury River in Eastern Australia is subjected to a tidal range of about 5ft. Here too the rainfall is seasonal, but the stratification lasts for only one month compared to the four or five months in the Swan River.

So far we have been considering normal positive estuaries, in which the river maintains sufficient flow to keep the mouth open throughout the year. In dry, tropical and sub-tropical areas the evaporation from an estuary may exceed the fresh-water inflow. Further, if the mouth of the estuary is small, so that limited exchange with the sea occurs, then the salinity of the estuarine water may rise above that of the sea. This results in a salinity gradient that is the reverse of that found in a positive estuary: an increase in salinity is found as one progresses inland from the mouth. In some areas the river flow may cease completely during the dry season, so that the estuary becomes a lagoon in which sea water becomes concentrated by evaporation. A striking example of this phenomenon has been described by Hedgepeth (1947) and Simmons (1957), who both studied the Laguna Madre in Texas. In the upper reaches of this lagoon the salinity reached nearly three times that of normal sea water.

Another example of a negative estuary is described by Day, Millard and Broekhuysen (1953) in their study of the St Lucia System in Natal. This system consists of three shallow lakes connected to the sea by a long narrow channel. During the dry season there appeared to be a slow movement of sea water inland through the narrow channel, and the salinity of the lakes increased above that of sea water because the fresh-water inflow was insufficient to counteract evaporation. In the summer of 1948 the salinity of the lake furthest from the sea had risen to 50‰, but in the following April there was heavy rainfall in the surrounding area, and the river began to flow strongly. The increased river flow washed out most of the saline water and a normal positive salinity gradient was established.

INTERSTITIAL SALINITY

The discussion so far has dealt with the salinity of open water, but there is also the problem of the salinity of the water in estuarine deposits. The animals burrowing in these deposits may gain a certain amount of protection from the rapid salinity changes which occur in open water. The first detailed studies on this subject were made by Reid (1930, 1932) who demonstrated that when a fresh-water stream flowed over intertidal sand the salinity of the interstitial water at a depth of 6 inches was well above that of the overflowing water, and remained so throughout the period of tidal exposure. The relationship between the overflowing and interstitial water is influenced by the slope of the shore. On steep shores the overflowing water travels at a greater pace, and the more saline interstitial water tends to drain away so that the fresh water penetrates to a greater depth in the sand. Fig. 11 compares a steeply sloping shore with a more gently sloping shore. The salinity of the interstitial water 6 in below the sand surface near HWM on the steep shore was below that on the other shore, but increased more consistently and rapidly as one proceeded down shore. In the middle section of the gently sloping shore there was a region where the interstitial salinity showed very little increase over a distance of 50 yards. This region was under sea water and fresh water alternately for approximately equal periods during the tidal cycle.

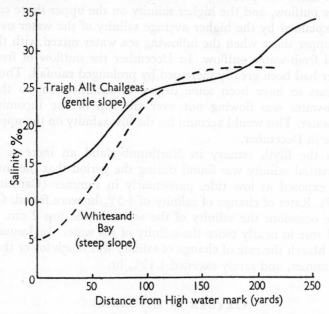

Figure 11. Salinity of interstitial water at depths of 6 inches in two beaches of differing slopes (after Reid)

The reduction in exchange between the overlying water and the interstitial water may also result in the salinity of the latter remaining low when a fresh-water seepage is covered by the sea. Smith (1955b) found that the water 6 in below the surface of the sand in some parts of Kames Bay in Scotland was of a lower salinity than the overlying sea water at high tide, and he was able to correlate the distribution of *Nereis diversicolor* with the belt of lowered salinity.

In a later paper Smith (1956) studied the interstitial salinities at various stations along the Tamar estuary in England. Large seasonal variations were found. In June at South Hooe the interstitial water in the upper part of the intertidal zone was about 26‰, while near LWM it was about 19‰. In December, along the same transect, the salinity of the interstitial water high on the shore was about 3‰, while lower on the shore it was about 7‰. In June the reduced salinity on the lower shore was probably caused by the relatively small fresh-water outflow, and the higher salinity on the upper shore can be explained by the higher average salinity of the water over the upper shore when the inflowing sea water mixed with the small fresh-water outflow. In December the outflow of fresh water had been greatly increased by prolonged rainfall. There appears to have been some degree of stratification, so that fresh-water was flowing out over the top of the incoming sea water. This would account for the low salinity on the upper shore in December.

In the Blyth estuary in Northumberland an increase in interstitial salinity was found during the period that the mud was exposed at low tide, particularly in summer (Capstick, 1957). Rates of change of salinity of 4·5‰/hr were found. On some occasions the salinity of the water in the top 2 cm. of mud rose to nearly twice the salinity of sea water. In January and March the rate of change of salinity was much lower than in summer, and rarely exceeded 1.15‰/hr.

OXYGEN IN ESTUARIES

The solubility of oxygen in water is influenced by salinity. Fig. 12 shows the relationship between salinity and oxygen in water saturated with air at two temperatures. This only covers the range from pure water up to the salinity of sea water. At higher salinities the solubility of oxygen falls more or less at the same rate, so that at a salinity of 195‰ the oxygen content at air saturation is only 2 ml/1 at 25°C.

Fig. 12 shows that temperature is a much more potent factor than salinity over the normal brackish range in determining the solubility of oxygen. Apart from the effect in lowering the solubility of oxygen, there will also be an increase in the metabolic rate of organisms when the temperature is raised, thus making greater demands on the available oxygen.

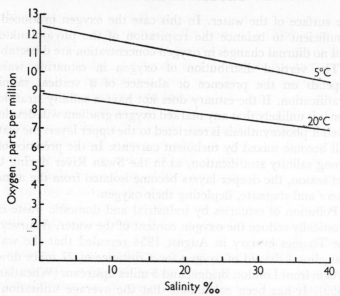

Figure 12. Salinity and the solubility of oxygen in water saturated with air
at two different temperatures

During the daytime the oxygen content of estuarine waters
may be increased by the photosynthetic activities of plants. At
night the plants cease photosynthesis, but continue respiring,
so that the oxygen content of the surrounding water might be
expected to be lower at night. An extreme example is given
by Broekhuysen (1935), who measured the concentration of
oxygen in the water of a *Zostera* bed. At night no oxygen was
detectable, but during the daytime the oxygen content rose
to 260 per cent saturation. Less extreme ranges have been
observed in salt marsh pools by Nicol (1935), but again the
daytime figures rose well above saturation.

In deeper waters the range is much less extreme, and the
oxygen concentration may not rise much above saturation, or
fall much below 30 or 40 per cent saturation. Where there are
large tidal movements the diurnal changes caused by photosynthesis may be masked by the influx of well-aerated sea
water.

The waters of many estuaries are so turbid that photosynthesis by planktonic algae is restricted to a narrow band near

the surface of the water. In this case the oxygen produced is insufficient to balance the respiration of the phytoplankton, and no diurnal changes in oxygen concentration are detectable.

The vertical distribution of oxygen in estuarine waters depends on the presence or absence of a vertical salinity stratification. If the estuary does not have a salinity stratification it is unlikely that any marked oxygen gradient will develop. Even if photosynthesis is restricted to the upper layers the water will become mixed by turbulent currents. In the presence of a strong salinity stratification, as in the Swan River during the wet season, the deeper layers become isolated from the atmosphere and stagnate, depleting their oxygen.

Pollution of estuaries by industrial and domestic waste can drastically reduce the oxygen content of the water. A survey of the Thames estuary in August 1954 revealed that the water was almost devoid of oxygen for a distance of 27 miles downstream from London Bridge, and 5 miles upstream (Wheatland, 1959). It has been calculated that the average utilisation of oxygen by the Thames Estuary is about 700 tons per day. This demand for oxygen is mainly the result of sewage discharge.

Not all the oxygen that is used in the oxidation of organic matter in the Thames Estuary is derived from the atmosphere. Gameson (1959) has shown that the reduction of nitrate and sulphate can be of importance in the oxygen balance of a polluted estuary.

About 110 tons of oxidisable nitrogen enters the Thames Estuary each day. If all this nitrogenous material was present in the form of ammonia the complete oxidation would follow the reaction

$$NH_3 + 2O_2 \rightarrow HNO_3 + H_2O$$

but if some nitrate is already present this may yield oxygen

$$2HNO_3 \rightarrow N_2 + H_2O + 5\ [O]$$

This reaction represents an oversimplification of some aspects of bacterial metabolism · the oxygen will not be liberated as molecular oxygen, but is used in the oxidation of organic matter.

Nevertheless, this utilisation of oxygen from nitrates reduces the demands made on the oxygen entering the estuarine waters from the atmosphere and may save the estuary from complete deoxygenation.

If there is no oxygen or oxidised nitrogen in the waters of an estuary any sulphate present is reduced to sulphide

$$H_2SO_4 \rightarrow H_2S + 4\,[O]$$

The oxygen which is freed is not available in molecular form but is used in the oxidation of organic matter. The fate of the hydrogen sulphide depends on the availability of iron salts (see p. 42). If the sulphide is removed from the estuary before it is reoxydised there will be an overall gain in available oxygen. Dredging of mud is estimated to remove from the Thames a quantity of sulphide equivalent to 5 tons of oxygen per day. Hydrogen sulphide also escapes to the atmosphere, and this may be as much as an oxygen equivalent of 8 tons per day for the whole Thames estuary. The total daily gain in oxygen by loss of sulphide may thus be about 13 tons per day. This is a relatively small fraction of the 700 tons normally used by the estuary. The reduction of nitrate is quantitatively more important, giving a total equivalent to 140 tons of oxygen per day for the Thames Estuary (Gameson, 1959).

All the chemical reactions discussed above will also take place in the interstices of mud and sand, but in this situation there will be the added complication of variation in permeability. A study by Brafield (1964) has shown that the oxygen content of the water in beaches is often very low when there is more than 10 per cent of sand grains with a diameter less than 0·25 mm. Table 3 compares three points on the shore at

TABLE 3

OXYGEN IN INTERSTITIAL WATER AT THREE STATIONS
ON THE SHORE AT WHITSTABLE
(Based on the data of Brafield, 1964)

Station	O_2 ml/l	Per cent fine sand smaller than 0·25 mm
A	0·26	95·5
B	0·62	23·75
C	3·99	4·00

Whitstable. At station A it was found that exposure to the air for 2½ hours at low tide doubled the oxygen content of the water at a depth of 2 cm below the sand surface, but at a depth of 5 cm the oxygen content of the water remained constantly low throughout the period of exposure.

OTHER CHEMICAL FACTORS

Sulphur is one of the most important elements in the biology of estuaries. The formation of ferrous sulphide in intertidal areas is responsible for the black coloration of sand and mud. The formation of a black layer beneath the surface formed the subject of a study by Bruce (1928) who reached the conclusion that the formation of ferrous sulphide was associated with diminished circulation of air and water in the mass of the sand. The presence of decaying organic matter was found to be essential to sulphide formation.

The depth of the black layer below the sand surface gives an indication of the depth to which aerated water can circulate. Fig. 13 shows the relationship found by Webb (1958) between

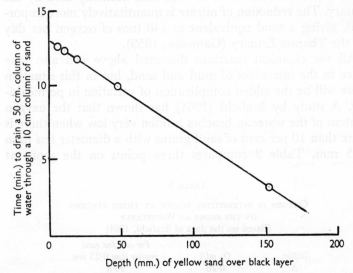

Figure 13. The relationship between the position of the black layer and the permeability of tropical lagoon sands (after Webb, 1958)

the depth at which the black layer commenced and the time for a standard column of water to drain through a 10-cm column of sand. It is clear that the more permeable the sand the greater the depth of sand above the black layer. A seasonal variation in the depth of the black layer at Whitstable has been found by Perkins (1957). In summer the black layer came much closer to the surface. This phenomenon can be explained as an effect of temperature on several processes. At increased temperatures oxygen is less soluble in water, and bacteria together with other organisms increase their activities, so that the overall effect will be a reduction in the amount of oxygen available to prevent the formation of ferrous sulphide.

The sulphur cycle in estuaries depends on the activities of a considerable range of micro-organisms (Baas Becking and Wood, 1955). Many of these organisms operate in anaerobic conditions. Sulphate reducing bacteria, such as *Desulphovibrio aestuarii*, are capable of using sulphate as a hydrogen acceptor, and liberate hydrogen sulphide. Iron salts in the surrounding sand or mud are converted to ferrous sulphide. If the amount of hydrogen sulphide is in excess of the amount of iron available, then free hydrogen sulphide permeates the interstitial water and may extend into the water above. In the presence of oxygen hydrogen sulphide is unstable, so that the presence of this malodorous substance can be taken as an indication of oxygen deficiency.

Other bacteria are capable of oxidising hydrogen sulphide. *Beggiatoa*, for instance, can produce elementary sulphur which is retained in the bacterial cells. Free sulphur is liberated by yet other bacteria, and can be metabolised in various ways. *Thiobacillus denitrificans* uses nitrate, which together with sulphur and calcium carbonate produces sulphate, calcium sulphate, carbon dioxide and molecular nitrogen:

$$6NO_3 + 5S + 2CaCO_3 \rightarrow 3SO_4 + 2CaSO_4 + 2CO_2 + N_2$$

At this point the sulphur cycle is intimately connected with the nitrogen cycle. There are also interconnections with the breakdown of cellulose. In anaerobic conditions the methane bacteria utilise fatty acids, alcohols and ketones which are produced by other bacteria in the breakdown of cellulose. The methane liberated by bacteria, such as *Methanococcus* and *Methanosarcina*, can be utilised by sulphate reducers in reactions of the following type:

$$CaSO_4 + CH_4 = CaCO_3 + H_2S + H_2O$$

Some of the sulphur bacteria are pigmented. For instance, the Thiorhodaceae are red or purple, while *Chlorobium* is green. These pigmented forms thrive best in light and can use hydrogen sulphide as a hydrogen donor in a photosynthesis of carbohydrate:

$$CO_2 + 2H_2S \rightarrow CH_2O + H_2O + S_2$$

Some varieties of pigmented sulphur bacteria also seem to be capable of chemosynthesis in reduced light conditions (Ruttner, 1963).

Most of the organisms involved in the sulphur cycle in estuaries can tolerate wide variations in salinity, pH and temperature. They are somewhat less tolerant in relation to oxygen. Some require low oxygen concentrations and others are strictly anaerobic, requiring low oxidation-reduction potentials. The oxidation reduction (redox) potential is a measure of oxidising or reducing power, usually designated by the symbol Eh or Rh. Oxidation need not involve oxygen. In the simplest form the loss of an electron constitutes oxidation, for instance the change from ferrous to ferric iron:

$$Fe^{++} \rightleftharpoons Fe^{+++} + e$$

The presence of free electrons hinders oxidation, and a measure of the redox potential can be made by drawing off the free electrons with a platinum electrode and measuring the electromotive force with a calomel electrode. The value obtained in this way is influenced by the relative proportions of oxidised and reduced compounds in solution, but not by their absolute concentrations. Thus it is possible to get very small changes in Eh over a wide range of oxygen concentrations, and yet get very large changes in Eh with small changes in oxygen concentration at critically low levels.

The redox potential is influenced strongly by pH, so that the value obtained with a calomel electrode has to be corrected to a standard pH. The zero point on the Eh scale is the potential of a gaseous hydrogen electrode with a pressure of one atmosphere. An important critical value is found at +200 millivolts, which is the critical point for the reduction or oxidation of iron. Well aerated water often has an Eh in the region of +500 millivolts, but in sediments the value often falls below zero. The green bacteria seem to have the most restricted range of tolerable Eh, from −100 to −300 millivolts, while some of the colourless sulphur bacteria can live in a much greater range of Eh, from −200 to +700 millivolts. The observed ranges of Eh and pH in estuarine muds are well within the tolerance of most micro-organisms concerned with the metabolism of sulphur (Baas Becking and Wood, 1955).

Phosphorus is important in ecological systems because organisms require a high proportion of this element in relation to its availability in Nature. This means that phosphates are likely to be limiting factors in the production of organic matter.

Phosphates are usually considered to be present in mud in three forms (Rochford, 1951):

1. Interstitial phosphate, which is present in the pore spaces between particles and is released if the mud is stirred.

2. Adsorbed phosphate, which is on the surfaces of silt particles; this can be released by digestive processes, and possibly by the action of some plant roots.

3. Insoluble phosphate, such as ferric or calcium phosphate.

The relationship between the phosphates in the mud and those in the overlying water is complex, and is summarised in Fig. 14.

The release of phosphate from mud is greatly enhanced by lack of oxygen. If the surface layer of the mud is well oxygenated it not only retains phosphate but prevents the diffusion of phosphates from the deeper layers. The iron cycle is intimately concerned in this process. In anaerobic conditions ferrous ions are liberated into the water, but when oxygen is present the ferrous ions are oxidised to ferric, and the combination of these with phosphate results in the formation of insoluble ferric phosphate.

In the Swan River Estuary it was found that there was a seasonal increase in phosphate which coincided with the seasonal discharge of fresh water (Spencer, 1956). The increase in phosphate could be related to the development of anaerobic conditions in the deeper parts of the estuarine basin which were cut off from communication with the sea by a sill and by over-flowing fresh water. In 1954 the fresh-water discharge was

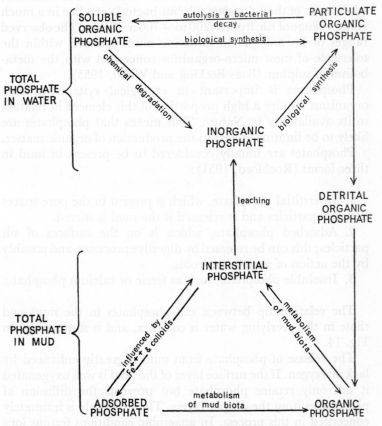

Figure 14. Relationship between phosphates in mud and overlying water (based on the data of Rochford, 1951)

less than usual and did not extend down to the top of the sill, so that the deeper parts of the basin were not isolated from the sea. As a result, the oxygen content of the deeper layers was not depleted, and the phosphate content of the water did not increase to the same extent as in years when the deep layers were isolated and became stagnant.

Because of the availability of organic debris for decay, and the presence of fine deposits in which anaerobic conditions can develop, the phosphorus content of estuarine water is often higher than that of the sea water outside the estuary. In contrast the nitrates and nitrites are often present in low concentrations (MacGinitie, 1935; Howes, 1939). It seems probable that in some unpolluted estuaries the production of phytoplankton is limited by a deficiency of nitrates. An increase in the nitrate content of the Swan River estuary has been correlated with the seasonal fresh-water discharge (Spencer, 1956). In a classification of estuarine zones, Rochford (1951) noted that the fresh-water zone was often richer in nitrates than the more seaward zones.

In polluted estuaries, nitrates may enter in high concentrations, but the oxidation of organic matter may result in reduction of nitrates to molecular nitrogen (see p. 39).

CHAPTER 3

ESTUARINE VEGETATION

TIDAL movement of water produces gradients of submersion and exposure. Even a superficial examination of an estuarine shore reveals that different levels are occupied by different plants. The zonation on a salt marsh is often striking, and has been studied in considerable detail by plant ecologists. Chapman (1960) has surveyed the extensive literature on this subject.

The effects of the tide are extended by the spring-neap cycle and by the increased tidal range at the equinoxes. In the upper regions of a salt marsh it is possible to find plants, such as the sea rush, *Juncus maritimus*, which may be submerged for only a few hours each month, while low down on the shore there are plants such as the glassworts, *Salicornia spp.*, which may be submerged for as many hours each day.

A prerequisite for colonisation of an intertidal area by rooted plants is that the surface should be stable. If the tidal flow is strong it can uproot seedlings and so prevent colonisation, even if the surface is fairly stable. Once a plant has taken root it may increase the stability of the surface, and the stem may help to slow the tidal current so that more material may settle from suspension. The limit of tolerance of tidal flow is often determined by the flow of the spring tides, which are more vigorous than the neaps. A plant such as *Salicornia* may take root and flourish under the flow conditions at neap tides, but be uprooted by the spring tides. Thus there are always areas of an estuary without rooted vegetation because the tidal flow is too powerful.

The vegetation of an estuarine shore does not form a continuous cover. In the early stages of colonisation small hummocks are formed around any plant that has succeeded in establishing itself. As the plants extend their range they promote the deposi-

47

tion of sediment in their immediate vicinity. Small islands may be formed, and the tide is forced to flow between them, establishing a diffuse system of wide channels. When the islands increase in size and coalesce the channels become narrower. The flow of the tide in these narrow creeks causes a certain amount of erosion so that the channels are cut deeper, often well below the original level colonised by the plants.

The coalescence of islands may also isolate bare areas, which fill with water when the spring tides cover the whole marsh. Such an isolated area is called a salt pan. The water in the pan may slowly evaporate until the next set of spring tides. Some pans dry out in hot weather, leaving crystals of salt on the mud surface. Rain dilutes the water in the pans, so that any organism living in such a pool may be subjected to a very wide range of salinities according to the climate.

Pans formed by the coalescence of islands are known as primary pans, but secondary pans can be formed by the blockage of creeks. A creek may be blocked when the vegetation meets across from one side to the other. This bridge of vegetation impedes the flow of water and encourages the deposition of silt. Later the bridge may collapse and block the creek. Pans formed in this way are usually elongate in form, whereas primary pans tend to be rounded.

The bottom of a salt pan is often devoid of vegetation. Variation in salinity, particularly the great increases in hot weather when the rate of evaporation is high, make the pans unsuitable for most plants. But there are some algae which can grow in such a situation, and at certain times of the year there may be conspicuous growths of *Vaucheria litorea*, *Enteromorpha torta*, *Lyngbya aestuarii* and *Cladophora fracta*. The diatoms *Diploneis didyma* and *Navicula pygmaea* are also often abundant in salt pans (Round, 1960).

The edge of an established salt marsh is often subjected to erosion by the tides so that a salting cliff varying in height from 1 to 3 ft is formed. Below the salting cliff there is bare mud, and above the cliff a dense sward of salt marsh plants. The salting cliff often marks the upper limit of ordinary tides, so that the marsh vegetation remains completely uncovered by neap tides.

Areas of bare mud are often colonised first by blue-green

algae, such as *Phormidium* and *Nostoc*, the filaments of which may increase the stability of the mud surface. The mud on the sides of creeks often has a covering of *Oscillatoria* and *Vaucheria*. Several species of the green alga *Enteromorpha* also act as primary colonisers of bare mud and sand. Once they have become established these algae can trap material on both the flood and ebb tides and so raise the general level of the mud surface. Some of the larger brown algae may also play a part in colonising bare mud. *Fucus vesiculosus* has several dwarf forms which occur in salt marshes, and at least one of these, the *volubilis* form, is found on mud banks below the rooted phanerogams. Other dwarf forms occur higher on the shore, sometimes mixed with the general salt marsh vegetation.

If the surface of a stabilised mud bank is watched after it has been uncovered by the tide, it may be seen to undergo a gradual change in colour. In bright sunlight the colour may become green or yellow. This colour change is caused by the migration to the mud surface of enormous numbers of *Euglena* and diatoms. The density of *Euglena obtusa* at the surface of the mud in the tidal reaches of the River Avon at Bristol has been found to exceed 100,000 per sq cm (Palmer and Round, 1965). These authors made a detailed study of the factors controlling the migration of *Euglena obtusa* and found that light was the dominant factor, but also that the organism had an inherent rhythm of activity.

During the daytime *Euglena* migrated to the mud surface when it was uncovered by the tide, and remained at the surface until a short time before the tide came in again. At night *Euglena* remained below the surface, and even in the daytime it could be made to burrow below the surface if placed in darkness. The vertical extent of this movement was 2 or 3 mm.

When mud was removed to the laboratory and placed in conditions of constant illumination a regular rhythm of migration to the surface was found. Once every 24 hrs *Euglena* would migrate to the mud surface, remain there for several hours, and then migrate down again. In continual darkness *Euglena* remained below the surface, and there was no rhythm of migration. The innate rhythm in continuous light provides the explanation of the apparent anticipation of the incoming tide in natural conditions. The time spent at the surface in continuous light was somewhat less than the interval between uncovering by the tide and the return of the tide.

The influence of the tide on the cycle of migration on the banks of the Avon is probably indirect in the sense that it exerts its influence by cutting out light. The water of the River Avon is very turbid, and once it has covered a mud bank it cuts out most of the light which formerly reached the mud surface. The resulting darkness inhibits *Euglena* from coming to the surface, and the inherent rhythm of surfacing is forced to wait until the banks are uncovered and illuminated again.

The movements of diatoms in the surface layers of the banks of the Avon have also been studied by Round and Palmer (1966). In their basic features the migrations of the diatoms are similar to those of *Euglena obtusa*, but some differences in detail between different species were found. When the banks were uncovered in the morning, *Navicula salinarum* and *Nitzschia tryblionella* reached their maximum abundance at the surface between 7 and 9 a.m., while *Cylindrotheca signata* and *Euglena obtusa* were somewhat later in reaching their peak at the surface. When removed to the laboratory the diatoms also showed persistent rhythms of migration in constant light, but some species exhibited less well-defined rhythms than others.

In the estuary of the River Eden in England, which has clear water, Perkins (1960) found that the diatoms *Pleurosigma aestuarii* and *Surirella gemma* (Fig. 15) did not burrow when covered by the incoming tide during the day. Instead these diatoms remained at the sand surface as long as the light intensity was above a certain level. This observation supports the view that the turbidity of the Avon water is important in inhibiting the upward migration of *Euglena* and diatoms during daylight hours.

Hopkins (1963) noticed that the diatoms in the mud of another English estuary, that of the Ouse, also remained at the surface while covered with water to a depth of 25 cm on a calm sunny day. On the banks of the Ouse very fine mud with particles ranging down to 0.5μ supported a denser population of diatoms than that supported by coarser mud (particles ranging from 2 to 20μ). The diatoms were also nearer to the

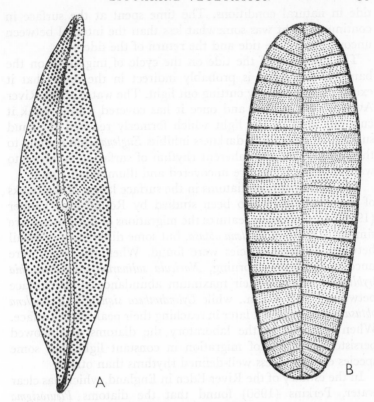

Figure 15. Two estuarine diatoms

A, *Pleurosigma aestuarii*, the length varies between 84 and 148μ. B, *Surirella gemma*, the length varies between 72 and 140μ

surface in the fine mud. This was associated with a more rapid extinction of light in the fine mud. Measurements made with a photoelectric cell showed that over 60 per cent of the incident light was absorbed by a layer of fine mud 0·2 mm thick, whereas a similar layer of coarser mud absorbed about 40 per cent of the incident light. Nearly all the incident light was absorbed in both muds at a depth of 4 mm. Very few diatoms were found deeper than this in the mud.

The distribution of different species of diatoms in estuaries is governed by a variety of factors. Salinity is important to many species, as, for instance, Kolbe (1927) has shown. The differing salinity ranges of a large number of species have been listed by Hustedt (1939). There are sufficient brackish water species to ensure that all parts of an estuary will have a variety of diatoms. Some of the estuarine species can tolerate and grow in a wide range of salinities. Fourteen species from the salt marshes of Georgia have been cultured under controlled conditions (Williams, 1964). All these species grew well at salinities between 10 and 30‰, and most of them continued to divide slowly in salinities as low as 1 or 2‰, and as high as 68‰.

Salinity is not the only important factor. Some species seem to be more tolerant than others to organic materials released in the black poorly oxygenated layer. Hopkins (1964a) found that *Pleurosigma baltica* was not much affected by a black layer only 5 mm below the surface, but *P. aestuarii* was greatly reduced in numbers in such a situation.

On the upper parts of a mud flat there is a danger that the surface may dry out, and some species of diatoms have developed the ability to secrete a mucilaginous envelope which protects them from this hazard. This ability has been found in several species of *Nitzschia* and *Pleurosigma*.

At the opposite end of a mud flat, near low water mark, the major problem faced by a diatom in an estuary with turbid water is lack of light. The mud may be exposed for only a couple of hours on each tide, and when the tide returns it cuts out the light. The species found near low water mark seem to have the ability to move rapidly, particularly at low temperatures. This enables them to reach the mud surface rapidly as soon as it is uncovered by the tide. *Nitzschia closterium* and *Stauroneis salina* have the ability to live under conditions of low illumination and are often abundant near low water mark. Both species are motile at 0°C and increase their rate of movement rapidly with increasing temperature. Species that are abundant higher on the shore, such as *Pleurosigma angulatum* and *Tropidoneis vitrea*, are not motile at low temperatures.

Not all the diatoms in estuaries are motile. A considerable number of species attach themselves to solid substrata such as stones, or to other plants, Some diatoms attach themselves to the surfaces of other diatoms. Solitary species of *Synedra* and *Licmophora* are frequently found attached to the colonial

Navicula grevillei, which itself may form a mat-like growth intermingling with *Enteromorpha* or with blue-green algae such as *Phormidium* and *Oscillatoria* (Hopkins, 1964a and b).

Diatoms in estuaries are important primary producers, and serve as food for a wide variety of animals. Perkins (1958), for instance, found diatoms forming part of the food of a range of Protozoa, Turbellaria, Nematoda, Ostracoda, *Nereis* and *Hydrobia ulvae*.

In the more marine areas of an estuary the primary colonists of bare mud are often species of *Zostera*, which is a grass-like perennial flowering plant. There are about eleven species of the genus found in the temperate regions. Some of these species, such as *Z. marina*, do not tolerate much exposure, and extend well below the low water of spring tides to form sub-marine meadows. Other species, such as *Z. hornemanniana* and *Z. nana*, are more tolerant of exposure and are found between half tide and low tide in the outer regions of estuaries. The creeping rhizomes of *Zostera* help to stabilise mud banks, and when *Zostera* is killed by disease, as during the 1930s around Britain, the mud banks may collapse and considerable erosion can occur.

In sheltered situations, where the tidal range is small and the salinity reduced, *Zostera* is often accompanied by *Ruppia*. In the brackish water along the south coast of Finland these two plants are found together, and are accompanied by *Zannichellia palustris* where the water is somewhat fresher. *Ruppia* and *Zannichellia* are slender plants with narrow, simple leaves; both genera may extend into fresh water where conditions are suitable. They frequently find conditions most suitable in ditches where there is open water that is only slightly brackish. An important point about these two genera is that they are not colonisers of bare mud, but live in permanent water, often overlapping with fresh-water species that penetrate into areas of stable low salinities.

The distribution of rooted vegetation in a Danish brackish waterway has been depicted in a diagrammatic manner by Mathiesen and Nielsen (1956). Randers Fjord is an inlet which extends about 16 km inland from the Kattegat. The channel is dredged to a fairly constant depth of 7 m, and since the tidal range in this area is small there is a fairly stable gradient of salinity along the length (Fig. 16A). The penetration of marine plants along the channel is shown in Fig. 16B, while the penetration of fresh-water plants is shown in Fig. 16c. *Zannichellia major* occupies an intermediate range of salinity, from 1·5 to 10‰.

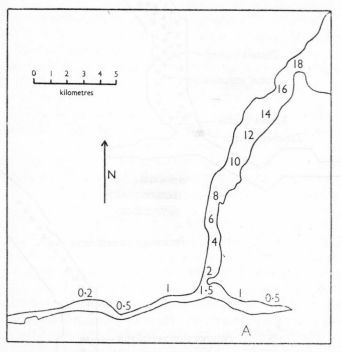

*Figure 16*A. Outline of Randers Fjord to show the salinity gradient

In the middle reaches of estuaries the primary coloniser of bare mud is often a species of *Salicornia*. The annual species of this genus, such as *S. stricta*, form open communities low on the shore, with considerable areas of bare mud between the plants. Higher on the shore the distance between the plants is much reduced. The stems of *S. stricta* are fleshy and succulent, projecting 10 to 30 cm above the surface of the mud. There are

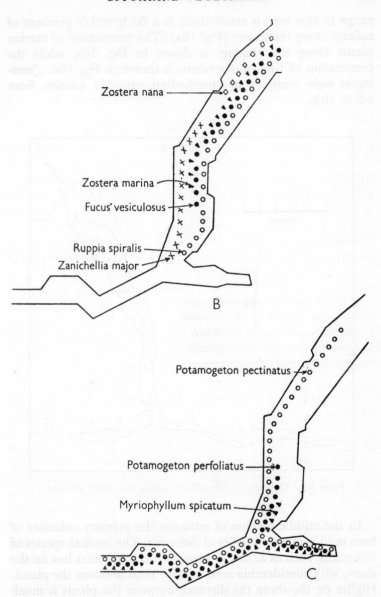

Figure 16 B and C. Diagrams to show the penetration of rooted plants into Randers Fjord: B, distribution of marine and brackish water plants; C, penetration of fresh-water plants into brackish water (after Mathiesen and Nielsen, 1956)

C

no free standing leaves, instead the bases are swollen and embrace the stem; the flowers are inconspicuous. The rooting system of *Salicornia* is shallow so that it cannot withstand strong currents, but in an area where the currents are slow a dense growth of *Salicornia* can reduce the current speed still further and so promote the deposition of mud from material suspended in the water.

In the upper parts of the *Salicornia* zone other plants take root. *Sueda maritima* is frequently found where the substratum is rather more sandy than muddy. The vegetational cover given by *Salicornia* and *Sueda* is often reinforced by algae, such as *Bostrychia scorpioides*, which may festoon the rooted plants and so increase their effectiveness in reducing the speed of tidal flow.

A striking contrast to *Salicornia* is found in another coloniser of bare mud, the cord grass, *Spartina*, about fifteen species of which are known from temperate regions. *Spartina* is a stout erect grass with a creeping rootstock that is effective in binding soft mud. Dense tussocks are formed which reach a height of over a metre. The effectiveness of *Spartina* in stabilising soft mud is a consequence of its rooting system, which consists of a series of stout roots descending vertically, and another series of more horizontally disposed roots which ramify in the surface layers of the mud. The rigid, erect stems are most effective in slowing water currents, and the rate of accumulation of new sediment is greater in an area of *Spartina* than in any other part of an estuary.

Where salinity is reduced by seepage of fresh water from the land, *Spartina* tends to be replaced by a tall marsh of *Phragmites communis* and *Scripus maritimus* (Ranwell, 1961). Both these plants are important members of the communities found near the head of an estuary.

Spartina may colonise a shore in the absence of *Salicornia*, or a zone of *Salicornia* may occupy the lower shore. One of the main factors governing the presence or absence of *Salicornia* appears to be the firmness of the substratum. Where the substratum is very soft *Salicornia* cannot take root, but *Spartina* often can.

The dominant plant on many of the sandier British salt marshes is the grass *Puccinellia maritima*. This species spreads by means of stolons which give rise to dense tufts of smooth

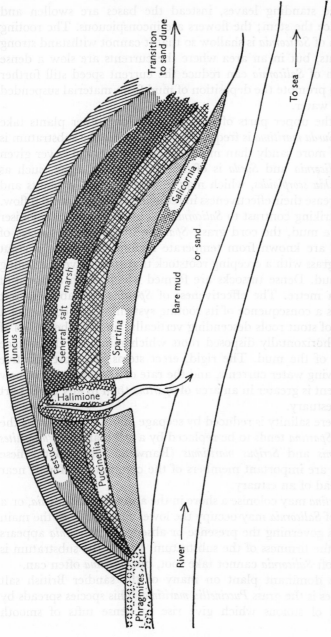

Figure 17. Simplified diagram of plant zonation in the intertidal region of a British estuarine shore. There are many variations of this zonation. Some of the zones shown may be absent and additional zones may be present. The diagram is based in part on the sequence found in the Gwendraeth Estuary in Wales

slightly inrolled leaves. In the upper parts of a salt marsh *Puccinellia* is often replaced by *Festuca rubra*, which is another perennial with a creeping rootstock and narrow inrolled leaves. *Festuca rubra* is not restricted to salt marshes, and grows in many inland areas away from open water.

Puccinellia is capable of acting as a primary coloniser of bare mud, but it usually lies above a zone of *Salicornia* or *Spartina*. In situations where *Spartina* is a recent coloniser, as on many parts of the British coast, the established zone of *Puccinellia* may be invaded from below. Chater and Jones (1957) estimate that *Spartina* can invade the *Puccinellia* zone at a rate of up to 15 cm per year.

The zone dominated by *Puccinellia* is often followed by a region containing a mixture of plants, known as a general salt marsh community. Thrift (*Armeria maritima*), sea spurry (*Spergularia salina*), sea lavender (*Limonium spp.*), scurvy grasses (*Cochlearia* spp.) and sea arrow grass (*Triglochin maritima*) are often common in this community. In some localities there may be a distinct zone where thrift is dominant, and in other localities the sea lavender may be so abundant as to dominate a zone above the *Puccinellia*. More usually there is a mixture of several co-dominant plants. The creeks running through such a community are often bordered by dense growths of *Halimione portulacoides*. This plant grows best in a well-drained situation, so that it might appear paradoxical for it to border the creeks. The paradox is resolved by a close examination of the creeks. The edges are usually raised slightly above the general level of the marsh. This physiographic feature is brought about by the material carried by the incoming tide being trapped as the tide flows slowly over the edges of the creeks. Material trapped in this way builds the creek borders up above the surrounding marsh. Once *Halimione* is established it produces tough woody stems which lie along the ground and give off erect branches at intervals. The whole structure is such that it promotes further deposition of material from the incoming tide, and so raises the edges of the creeks further above the surrounding marsh. If the whole marsh is well drained the areas between the creeks may be invaded by *Halimione*; when this happens few other plants survive.

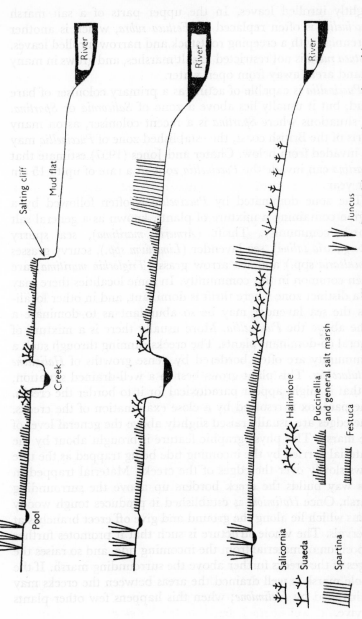

Figure 18. Three transects of the plants on an estuarine shore. The presence of a creek interrupts the zonation and interposes a band of *Halimione*. The three sequences are based on parts of the Gwendraeth Estuary, but similar sequences may be found elsewhere with local variations

The sea aster (*Aster tripolium*) is a yellow daisy which reaches a height of 60 cm. This plant sometimes invades the upper parts of the *Salicornia* zone, and may, as on parts of the East Coast of Britain, form a distinct zone in the lower part of the marsh. When such a zone is formed the sward of plants provides resistance to the tidal flow of water and promotes the deposition of silt.

As well as occurring low down on the marshes the sea aster may also occur in the upper regions, particularly where these are poorly drained. The yellow flowers attract many terrestrial insects to the marshes. On a warm day in August a patch of *Aster* may hum with the activities of bumble bees and hover flies.

At the landward edge of an estuarine salt marsh, at the limit of the spring tides, a zone dominated by rushes is often found. These rushes (e.g. *Juncus maritimus* and *J. gerardi*) are taller than the majority of plants inhabiting the middle zone of the marsh, and so form a conspicuous feature. The dense tufts shade the ground, reducing evaporation, so that the *Juncus* zone is often damper than the middle zones. This increase in dampness often leads to the reappearance of plants found in the lower zones. Some of the surface-dwelling algae become abundant between the tufts of rushes, and the sea aster may become quite abundant. Towards the head of an estuary, where the water is fresher, the *Juncus* zone may give way to a zone dominated by *Scirpus maritimus* and by *Phragmites communis*; the same two plants that replace *Spartina* in regions of fresh-water seepage.

ALGAE ASSOCIATED WITH SALT MARSHES

The part played by algae in colonising bare mud has already been mentioned (p. 48); there are also many algae associated with the higher plants in salt marshes. The red alga *Bostrychia scorpioides* often grows on plants in the lower parts of salt marshes, and prefers regions with a substratum of fine silt. The stem of *Halimione portulacoides* is a frequent substratum for this alga. *Catenella repens* is another red alga which often grows in areas similar to those in which *Bostrychia* is found; clefts between stones and well-shaded areas at the bases of tall plants are preferred substrata. Den Hartog (1958) found *Catenella* growing on the lower stalks in dense stands of *Scirpus maritimus*.

Green and blue-green algae are particularly widespread through salt marshes. Species of *Enteromorpha, Ulothrix, Phormidium* and *Phaeococcus* may be found at various levels from the *Salicornia* zone to the *Juncus* zone. The detailed distribution and seasonal succession of these algae have been studied by Carter (1932–3) and Chapman (1960). On the marshes of Canvey Island in the Thames Estuary a succession of algae occurs in the *Aster* zone during the course of a year. *Vaucheria* is the principal alga in this zone during the winter. *Ulothrix flacca* is dominant during the spring. Two species of *Enteromorpha* become abundant during the summer, and are joined in the autumn by blue-green algae and diatoms.

Several of the fucoid algae that are abundant on rocky shores also extend into estuaries wherever the substratum is suitable. In addition to colonising any rocky areas, some of the species also have dwarfed sterile forms which reproduce vegetatively in salt marshes. *Fucus spiralis, F. vesiculosus* and *F. serratus* are typical rocky shore forms which are also found among salt marsh vegetation. *Fucus ceranoides* normally inhabits rocky areas where fresh water enters the sea, and has several dwarf forms which occur in salt marshes. *Pelvetia canaliculata* and *Ascophyllum nodosum* also have dwarfed estuarine forms. Gibb (1957) has shown that dwarfed forms of the last species are frequently associated with conditions of low salinity. Sometimes these algae lie free on the surface of the marsh, or draped across the vegetation, but at other times the fronds become embedded in the mud. The normal attachment discs of fucoids growing on rocky shores are usually absent when they grow in salt marshes, and the fronds of all species tend to grow in a spiral or curled manner. This curling is thought to be caused by faster growth on the side of the frond in contact with the damp soil.

ESTUARINE FUNGI

The fungal flora of an estuary is richer in vegetated areas than in areas of bare mud. Pugh (1960) found 17 species in bare mud, but 27 species in soil colonised by *Salicornia stricta*. The presence of a primary coloniser would add to the diversity of the environment, and the roots would provide a substratum for fungal growth, so that an increase in the number of fungal species might be expected.

Many of the fungi in estuarine soils are widespread in other soils. In bare muds, species of *Mucor* and *Penicillium* are the most frequent, together with *Trichoderma viride* and *Aspergillus fumigatus*. In addition to the widespread species there are also some species characteristic of salt marshes. *Dendryphiella salina* has been found associated with the roots of *Salicornia* from a number of widely dispersed localities (Pugh, 1960,1962a), and is one of the dominant species colonising the seedling roots of *Halimione portulacoides*, although it tends to be infrequent on the roots of mature plants of this species (Dickinson and Pugh, 1965). The fungus does not penetrate the living roots, but once the *Salicornia* plant dies then the roots are invaded.

Ascochytula obiones is another fungus characteristic of salt marshes, and is closely associated with *Halimione portulacoides*. This fungus forms a component of the phylloplane microflora, which is an assemblage of plant micro-organisms growing on the surfaces of leaves. The microflora of the leaves of *Halimione* has been studied by Dickinson (1965). About twenty species were recorded, but of these only *Ascochytula obiones* and *Dendryphiella salina* were characteristic of salt marshes. Many of the other species were known to be widespread in non-saline habitats. *Ascochytula obiones* grows vegetatively on the surface of the leaves, and enters a reproductive phase when the leaves become moribund.

The typical salt marsh fungi are found more frequently in the upper regions of the intertidal zone. In contrast, many of the more widespread and transient salt marsh species are found more frequently low on the shore (Pugh, 1962b). This difference may be explained by the transient species being washed into the water from the land and finding good conditions for growth among the non-specific detritus low on the shore, while the upper shore is colonised by the fungi more adapted to live on specific salt marsh plants.

The *Spartina* zone yields fewer species of fungi than the

Salicornia zone (Pugh, 1961). This is probably associated with the low Eh and the toxic effect of accumulations of sulphides in the soft muds which are inhabited by *Spartina*. Some fungi appear to be adapted to such conditions, for instance *Nectiar inverta* and species of *Scopulariopsis* appear to be relatively more abundant under *Spartina* than under other salt marsh plants. Different salt marsh plants also differ in the development of a micorrhiza: *Puccinellia* is micorhizal, but *Salicornia* and *Sueda* are not (Pugh, 1962b).

MANGROVES

In tropical and sub-tropical regions the salt marsh is replaced by mangrove vegetation. Two of the most widespread genera in this type of vegetation are *Rhizophora*, the red mangrove, and *Avicennia*, the black mangrove. The members of the former genus are often pioneers, and grow into large trees occupying the outer zone of the mangrove. Some of the roots of this genus are borne on long, downward curving branches, or rhizophores, which are often referred to as prop roots. These rhizophores form tangles which reduce the tidal currents and promote the deposition of mud. The seeds of *Rhizophora* sprout while still on the tree, and the hypocotyl may reach a length of two feet before the seed drops. The young plant quickly develops roots and is capable of establishing itself much more rapidly than would be possible if it did not sprout before dropping.

Avicennia usually forms a zone to the shoreward side of the red mangroves. This genus does not have an extensive system of prop roots, but has aerial roots which emerge from the ground some short distance away from the tree. These contacts between the underground rooting system and the air are necessitated by the low permeability of the mud in which the trees grow.

It is possible to distinguish two major mangrove associations. The eastern mangrove extends from East Africa to Malaya and Australasia. In addition to species of red and black mangroves the association may include *Ceriops candolleana*, *Bruguiera gymnorhiza*, *Lumnizera racemosa*, and the dwarf palm, *Nipa fruticans*. The western mangrove is found in West Africa and the eastern coast of America from Florida to Brazil. In this association the red and black mangrove zones may be succeeded on the landward side by *Laguncularia, Hibiscus* and *Acrostichum.*

The mangroves of Florida have been described in detail by Davis (1940). *Rhizophora mangle* forms a pioneer community, with an association of marine plants and animals around its roots. Turtle grass, *Thalassia testudinum,* and manatee grass, *Cymodocea manatorum,* may be found with the pioneer community of red mangroves. Behind the *Rhizophora* zone the black mangrove *Avicennia nitida* forms a community together with certain salt marsh plants such as *Salicornia perennis* and *Spartina alterniflora.* The white mangrove, *Laguncularia racemosa* sometimes forms a zone to the landward side of *Avicennia,* but more often grows in a mixed community with *Avicennia,* or sometimes even with the pioneer *Rhizophora.* Even further towards the land the buttonwood, *Conocarpus erectus,* which is not a mangrove, often forms a transitional community between the mangroves and more strictly terrestrial vegetation. The cabbage palm, *Sabal palmetto,* and the strangler fig, *Ficus aurea,* are often associated with the buttonwood community. In the *Conocarpus* zone the water is often only slightly saline, although this plant does have the ability to grow in rather dry saline areas.

There is no doubt that mangroves are efficient at causing the accretion of material to form new land. The tangle of rhizophores slows down the currents to such an extent that fine particles can settle. A striking example is given by Davis (1940) of an open shoal that was transformed into thick swamp forest in the course of 30 years.

The algae associated with mangrove roots show a considerable resemblance to those found in salt marshes: *Bostrychia, Catenella, Caloglossa* and *Murrayella* are the most characteristic genera. These are usually attached to roots in partial shade, away from direct sunlight. The surface of the mud between the mangrove roots often supports a felt-like growth of *Vaucheria* and *Caulerpa.*

CHAPTER 4

INTRODUCTION TO ESTUARINE FAUNAS

COMPONENTS OF AN ESTUARINE FAUNA

If one considers the total assemblage of animals living in an estuary, it is clear that they have arrived at their present-day habitat from several different sources. The three major sources are from the sea (the marine component), from fresh water (the limnobiotic component) and from the land (the terrestrial component).

The marine component includes those animals which belong to groups of which the majority of members live in the sea. They have succeeded in penetrating estuaries to varying extents, and within this component three sub-components may be separated.

1. The stenohaline marine component includes those animals with limited powers of osmoregulation and restricted penetration of estuaries. This includes the large numbers of marine species which may be found near the mouths of estuaries in regions where the salinity does not fall below 30‰.

2. The euryhaline marine component includes animals which live both in the sea and in brackish water, with the limit of penetration lying below 30‰. A good example of this group is given by the common shore crab, *Carcinus maenas*, which is abundant on the rocky shores of the sea, but also penetrates to low salinities in estuaries.

On the basis of detailed studies over wide areas of stable low salinities in the Baltic, Remane (1958) has separated four grades of the euryhaline component, based on the lowest salinities to which they penetrate. It is doubtful if these groups can be applied to the more variable salinities in estuaries. For instance the lamellibranch *Scrobicularia plana* is placed in the second grade, with a limit of penetration at 10‰, but in many

65

estuaries this species is subjected to lower salinities for a part of the tidal cycle (see p. 160). In spite of such difficulties the limits of penetration are of interest in indicating the potentials of the various species as estuarine dwellers.

Remane's first grade, those which penetrate down to a salinity of 15‰, includes a large number of species: 100 polychaete species, 300 nematode species, and 200 copepod species, to mention only a few groups.

In the second grade are animals with limits of penetration between 15 and 8‰. The edible winkle, *Littorina littorea*, and the sea anemone, *Halcampa duodecimcirrata*, belong to this group.

The members of the third grade have limits of penetration between 8 and 3‰. Many of the tintinnid protozoans, and the jellyfish *Aurelia aurita*, are examples of this grade.

The fourth grade of euryhaline marine animals penetrate to salinities below 3‰. The planktonic cladoceran, *Evadne nordmanni*, for instance, penetrates to a salinity of 1·3‰, but also occurs in full strength sea water.

3. The brackish water component includes those animals of marine ancestry which live in salinities below 30‰ but do not normally occur in the sea. It is not always easy to differentiate this component from the euryhaline marine component, but an example will illustrate the criteria which may be used. In the genus *Sphaeroma* there are a number of species that live in salt marsh pools, and similar estuarine habitats. These are true estuarine forms. They are replaced at the seaward end of an estuary by other species of the same genus. In Britain *S. rugicauda* is found in salt marsh pools, but *S. serratum* replaces it in rocky areas at the mouths of estuaries and on the open coast.

The fresh-water component of an estuarine fauna includes those animals belonging to groups that are predominantly fresh water in habit, so that the estuarine forms may be presumed to have descended from closely related fresh-water forms. The size of the group may vary. For instance, *Asellus* is basically a fresh water genus, but *Asellus aquaticus* penetrates into brackish water. The larvae of the family Culicidae (mosquitoes) are also basically fresh-water dwellers, but a number of species have taken to breeding in salt marsh pools. The class Oligochaeta contains both terrestrial and fresh-water species, but some belonging to different families have penetrated downstream, and some members of the family Tubificidae have become important contributors to the biomass in estuarine muds.

This concept of the fresh-water component including animals which have penetrated well into estuaries differs from the definition given by Day (1951), who regarded this component as including only the animals which were restricted to low salinities.

It is possible to subdivide the fresh-water component in a manner similar to the subdivision of the marine component. Again these finer divisions probably apply only to areas of stable salinity, such as the Baltic.

Stenohaline limnobionts are restricted to regions where the salinity remains below 0·5‰, while euryhaline limnobionts enter brackish water. Remane (1958) has defined three grades of euryhaline limnobionts with limits of penetration of 3‰, 8‰ and above 8‰. The precise limits of penetration by members of the last group varies with the species. There are also some species which have become specialised for dwelling in brackish water, and rarely or never occur in fresh water. Some species of the beetle genus *Ochthebius* have become true brackish water dwellers (see p. 246). The animals of the fresh-water component are described in more detail in Chapter 8.

There remains an aquatic component which is difficult to classify as either fresh water or marine. This includes the migratory fish that pass through the estuary on the way from the sea to fresh water or *vice versa*. Obvious examples are the salmon (*Salmo salar*) and the eel (*Anguilla anguilla*). This migratory component needs to be clearly separated from the euryhaline marine forms which migrate into estuaries for breeding purposes, but do not pass into fresh water: the shrimp *Crangon vulgaris*, is an example.

Invasion of estuaries from the land has been effected mainly by arthropods. This aspect of estuarine biology has not received a great deal of study, so that subdivision of this component is probably premature. But it is possible to distinguish between those species which escape the effects of immersion by moving upwards when a spring tide floods the upper shore, and those species which remain on the shore and are capable of withstanding submersion for several hours. The terrestrial component of the estuarine fauna is described in Chapter 9.

TABLE 4

SUMMARY OF THE COMPONENTS OF AN ESTUARINE FAUNA

I THE MARINE COMPONENT
   (i) The stenohaline marine component, not penetrating below 30%
   (ii) The euryhaline marine component
      (a) first grade, penetrate to 15%
      (b) second grade, penetrate to 8%
      (c) third grade, penetrate to 3%
      (d) fourth grade, penetrate to below 3%
  (iii) brackish water component, lives in estuaries, but not in sea

II THE FRESH-WATER COMPONENT
   (i) The stenohaline fresh-water component, not penetrating above 0·5%
   (ii) The euryhaline fresh-water component
      (a) first grade, penetrate to 3%
      (b) second grade, penetrate to 8%
      (c) third grade, penetrate above 8%
  (iii) brackish water component, lives in estuaries, but not in fresh water

III THE MIGRATORY COMPONENT migrate through estuaries from sea to fresh water or vice versa
   (i) anadromous, ascending rivers to spawn
   (ii) catadromous, descending to the sea to spawn

IV THE TERRESTRIAL COMPONENT
   (i) tolerant of submersion
   (ii) intolerant of submersion

TABLE 5

NUMBERS OF SPECIES OF FISH AND LARGE CRUSTACEANS OF MARINE AND FRESH-WATER ORIGINS IN THREE ESTUARINE LAKES IN LOUISIANA
(After Gunter, 1956)

Name of Lake	Grand Lake	White Lake	Little Bay
Salinity range	0·08–2·70	0·11–2·70	0·16–4·05
No. of fresh-water species	10	6	5
No. of marine species	17	18	19
No. of fresh-water specimens caught	124	80	50
No. of marine specimens caught	1,771	1,647	2,457

When the relative numbers of species in each component of an estuarine fauna are considered it is found that the species of marine origin are predominant. Gunter (1956) has shown that

this is so for the estuarine fish of North America. Even in low
salinities he found that the marine species of fish and decapod
crustaceans outnumbered the species of fresh-water origin
(Tables 5 and 6).

TABLE 6

SALINITIES AT WHICH THE COMMON FISH AND LARGE CRUSTACEANS WERE CAUGHT
IN THREE ESTUARINE LAKES IN LOUISIANA
(After Gunter, 1956)

Common name	Species	Salinity range	Mean no. caught per haul
Menhaden	*Brevoortia* sp.	0·18–0·29	1·3
		0·41–0·93	69·2
		1·14–2·70	90·4
Bay Anchovy	*Anchoa mitchilli*	0·08	0·3
		0·10–0·19	23·6
		0·21–0·93	23·2
Croaker	*Micropogon undulatus*	0·08	0·7
		0·10–0·93	14·1
		1·84–4·05	35·8
Broad Sole	*Trinectes maculatus*	0·08	0·7
		0·10–0·93	6·2
		1·84–4·05	5·3
Blue Crab	*Callinectes sapidus*	0·10–0·73	1·3
		0·75–0·82	7·0
		1·14–4·05	4·1
White shrimp	*Penaeus setiferus*	0·44	0·3
		0·78–0·82	47·6
		1·84–4·05	27·2
Brown shrimp	*Penaeus aztecus*	0·78–0·82	4·5
		2·16–2·70	6·9

ESTUARINE HABITATS AND THEIR FAUNAS

In the last section the origins of the various animals inhabiting
estuaries were discussed, and in earlier chapters the physical
and chemical factors operating in estuaries were described. It
is now necessary to describe the various types of habitat
available in an estuary and to introduce the associations of
animals which will be discussed in greater detail in later
chapters.

The simplest habitat to define is the open water. By this
term is meant the water which occupies the estuarine basin and
is usually in contact, for part of the year at least, with the sea.

The animals in the open water are deemed to be pelagic if their powers of swimming are well developed, and are regarded as planktonic if their powers of locomotion are so feeble that they are easily transported by water currents. The herring (*Clupea harengus*) is a pelagic fish, but the copepods on which it feeds are planktonic.

TABLE 7

COMMON ZOOPLANKTERS IN BRITISH ESTUARIES AND BRACKISH LAGOONS

COELENTERATA	CRUSTACEA
Rathkea octopunctata	Acartia tonsa
Aurelia aurita	„ bifilosa
CTENOPHORA	„ discaudata
Pleurobrachia pileus	Centropages hamatus
	Cyclopina norvegica
ANNELIDA	Eurytemora velox
Pygospio elegans—larvae	„ affinis
	Podon polyphemoides
MOLLUSCA	Neomysis integer
Mytilus edulis—veliger larvae	Mesopodopsis slabberi
Littorina littorea—veliger larvae	Praunus flexuosus
UROCHORDATA	Gastrosaccus spinifer
Oikopleura dioica	Carcinus maenas—zoea larvae

The floor of an estuary provides several habitats for benthic organisms, which collectively constitute the benthos. Subdivisions of this group can be made in several ways. The infauna includes those animals which burrow and penetrate the substratum, while the epifauna moves over the surface, or else attaches to some firm part of the substratum. If an animal attaches to the surface of a stone it can be described as epilithic, or if it attaches to the surface of a plant the term epiphytic can be applied. Within the main groups of the benthos it is also possible to distinguish those forms which occur intertidally from those living in or on a substratum which is never exposed to the air. Not all animals fall neatly into such categories. For instance, many members of the Mysidacea spend a large part of the day at the bottom of the estuary, moving over the substratum like members of the epibenthos, but at night they ascend into the open water, and may then be regarded as members of the plankton. Some of the amphipods are also difficult to categorise precisely. Members of the genus *Gammarus* are often found under stones; they do not burrow into the substratum, but they are not found on the surface. In this book they will be treated as members of the epibenthos.

The intertidal zone, with its gradient of exposure, contains a wide range of habitats. Bare sand and mud are inhabited mainly by burrowing forms: *Nereis*, *Corophium*, *Mya* and *Scrobicularia* are examples. Sand and mud both provide shelter from predation and desiccation, but they also each present different problems to the animals dwelling in them. Sand tends to be unstable, so that permanent burrows are difficult to construct. Mud is generally more stable than sand, and allows the construction of permanent burrows. But mud is much less permeable than sand and is not so well aerated. This means that animals living at any depth in the mud must maintain connection with the surface; this they achieve by various means. Deep dwelling lamellibranches, such as *Scrobicularia* and *Mya*, use their siphons to transport aerated water to their mantle cavities, while *Corophium* constructs a U-shaped burrow which it irrigates, so maintaining a flow of aerated water over its body.

Along the length of an estuary there may be all gradations and mixtures of sands and muds, and many animals are distributed in relation to the nature of the substratum. For instance, among the amphipods, *Bathyporeia pilosa* lives in coarse sand, while *Corophium volutator* lives in fine sands and muds.

Solid structures such as stones and plant roots provide firm bases for the attachment of the sedentary epibenthos, such as mussels and barnacles. These animals live in one of the most variable habitats inhabited by any animal. At low tide the animals may be exposed to the air for five or six hours. In summer they will be subjected to heating and desiccation, while in winter they suffer the risk of being frozen. As the tide rises they will be subjected to the full range of any change in salinity that occurs as the tide interacts with the river. In this respect the epifauna is at a disadvantage compared to the infauna, which is often protected from the rapid changes in the open water by the slower changes in the salinity of the interstitial water.

The intertidal vegetation provides a variety of habitats. Different types of shelter are provided by the various plants on salt marshes. *Spartina*, with its tall stems, provides a different habitat from *Puccinellia*, which makes low dense tussocks of finer grass. Rhizophores of mangroves provide suitable substrata for the attachment of various barnacles, oysters and other sedentary invertebrates. The distribution of some of these members of the epifauna on *Rhizophora* in Kuramo Creek, near Lagos, has been studied by Sandison and Hill (1966). Fig. 19 shows how the epifauna in different parts of the creek was dominated by different species. Two species of barnacles were found: *Chthamalus aestuarii* was dominant near high water level at one end of the creek, while *Balanus pallidus stutsburi* became dominant lower in the water and further down the creek. About a quarter of the way along the creeks some oysters, *Gryphaea gasar*, were found near the lower ends of the rhizophores, and the importance of these oysters increased further along the creek. In turn the oysters were replaced by serpulid worms, with *Mercierella enigmatica* being more important than *Hydroides uncinata*. If an individual rhizophore was found with several of these sedentary animals they were dominant in the following order, from the water surface downwards: *Chthamalus aestuarii*, *Balanus pallidus stutsburi*, *Gryphaea gasar*, *Mercierella enigmatica*.

In some areas the salt marsh may end in a salting cliff from 6 in to 3 ft in height. This forms as a result of marginal erosion of an established salt marsh. This cliff provides a special habitat. The almost vertical face may be penetrated by burrowing forms. The isopod *Paragnathia formica* may be found together with some of the salt marsh Coleoptera in cavities excavated in the cliff, and the surface of the cliff is sometimes heavily colonised by the collembolan, *Anurida maritima*.

The surface of the salt marsh soil is inhabited by predatory beetles (e.g. *Dyschirius* and *Bembidion*), mites (e.g. *Punctoribates*), amphipods (e.g. *Orchestia*), isopods (e.g. *Sphaeroma*) and snails (e.g. *Phytia myosotis*).

Sphaeroma is also abundant in the pools in salt marshes, where it is accompanied by a variety of other Crustacea such as the prawn *Palaemonetes varians*, and copepods such as *Eurytemora* and *Onychocamptus*. The bottoms of these pools are

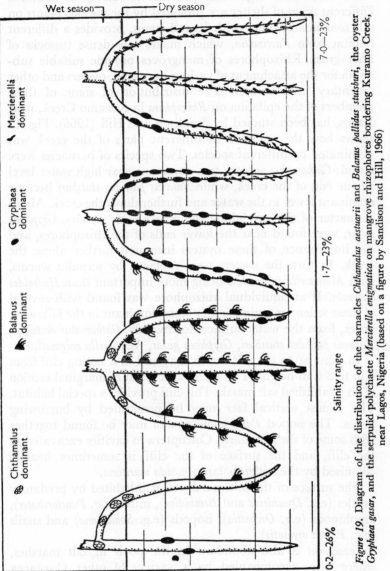

Figure 19. Diagram of the distribution of the barnacles *Chthamalus aestuarii* and *Balanus pallidus stutsburi*, the oyster *Gryphaea gasar*, and the serpulid polychaete *Mercierella enigmatica* on mangrove rhizophores bordering Kuramo Creek, near Lagos, Nigeria (based on a figure by Sandison and Hill, 1966)

often colonised by burrowing forms belonging to the bare mud fauna, and small fish, such as *Gobius microps*, may be abundant.

The bodies of animals living in estuaries serve as a habitat for parasites, which are just as abundant in estuaries as in other major habitats. Some estuarine parasites become exceptionally abundant because of the mass occurrence of their hosts in estuaries. *Hydrobia ulvae* and the cockle *Cardium edule* are but two of the species that occur in large numbers and serve as hosts to at least a dozen species of parasites. Some of the parasites of estuarine animals are discussed in Chapter 12.

The free-living animals that one might expect to find in a series of estuarine habitats in Britain are listed in Tables 7–12. These tables are not intended to be complete, but give an indication of the range of species that have been recorded. Many of the species found only near the mouths of estuaries, or in drowned valleys where the salinity is not significantly different from the sea, have been omitted.

TABLE 8

COMMON SPECIES OF THE MACRO-EPIFAUNA OF BRITISH ESTUARIES
(EXCLUDING THE SALT MARSH SURFACE)

(Based on data given by Bassindale, 1942; Crawford, 1937; Milne, 1939; Fraser, 1938; Spooner, 1947; Goodhart, 1941; Popham, 1966; Sexton and Spooner, 1940; and unpublished observation on the Gwendraeth Estuary)

COELENTERATA
Tubularia indivisa
Obelia flabellata
Gonothyrea loveni
Diadumene luciae
„ cincta
Cordylophora caspia
Sagartia elegans
POLYZOA
Membranipora crustulenta
ANNELIDA
Harmothoe spinifera
MOLLUSCA
Mytilus edulis
Littorina littorea
Littorina saxatilis
CRUSTACEA
CIRRIPEDIA
Balanus balanoides
„ improvisus
Elminius modestus

MYSIDACEA
Praunus flexuosus
Neomysis integer
ISOPODA
Sphaeroma rugicauda
„ monodi
„ hookeri
Jaera spp.
Ligia oceanica
Idotea viridis
TANAIDACEA
Heterotanais oerstedi
AMPHIPODA
Gammarus locusta
„ salinus
„ zaddachi
„ duebeni
„ chevreuxi
Marinogammarus marinus
„ stoerensis
„ pirloti

TABLE 8 (*continued*)

	DECAPODA
Melita pellucida	Carcinus maenas
„ palmata	Palaemonetes varians
Leptocheirus pilosus	Palaemon longirostris
Hyale nilssoni	Crangon vulgaris
Orchestia gammarella	INSECTA
Corophium insidiosum	Bembidion laterale

TABLE 9

COMMON SPECIES OF THE MACROFAUNA IN BRITISH ESTUARINE
SANDS AND MUDS

(Based on data given by Spooner and Moore, 1940; Bassindale, 1938; Beanland, 1940; Popham, 1966; and unpublished observations on the Gwendraeth Estuary)

NEMERTEA	MOLLUSCA
Lineus ruber	Cardium edule
Tetrastemma coronatum	„ lamarcki
ANNELIDA	Macoma balthica
Nereis diversicolor	Scrobicularia plana
„ virens	Mya arenaria
Nephthys hombergi	Hydrobia ulvae
Phyllodoce maculata	CRUSTACEA
Eteone longa	Cyathura carinata
Scoloplos armiger	Eurydice pulchra
Pygospio elegans	Corophium volutator
Heteromastus filiformis	„ arenarium
Arenicola marina	„ multisetosum
Ampharete grubei	Bathyporeia pilosa
Melinna palmata	INSECTA
Lanice conchilega	Bledius spectabilis
Clitellio arenarius	Heterocerus flexuosus
Tubifex costatus	Bembidion laterale
	Hydrophorus oceanus

TABLE 10

COMMON SPECIES IN THE FAUNA OF THE SALT MARSH SURFACE IN BRITAIN
(INCLUDING CRYPTOZOIC FORMS UNDER STONES, AND SPECIES CLOSELY ASSOCIATED
WITH SALT MARSH PLANTS)

(Based on data given by Nicol, 1935; Luxton, 1964; and unpublished observations on the Gwendraeth Estuary)

TURBELLARIA	CRUSTACEA
Uteriporus vulgaris	Orchestia gammarella
	Ligia oceanica
MOLLUSCA	Porcellio scaber
Hydrobia ulvae	Philoscia muscorum
Phytia myosotis	Spaeroma rugicauda
Limapontia depressa	Carcinus maenas

TABLE 10 (*continued*)

ARACHNIDA
Lycosa purbeckensis
Erigone arctica
" longipalpis
Rhombognathides spinipes
Agauopsis brevipalpus
Hermannia scabra
" pulchella
Platynothrus peltifer
Punctoribates quadrivertex
Hygroribates schneideri
Cheiroseius necorniger
Leioseius salinus
Macrocheles subbadius
Digamasselus halophilus

INSECTA
COLLEMBOLA
Archisotoma besselsi
Xenylla humicola

Isotomura palustris
HEMIPTERA
Salda littoralis
Saldula palustris
Halosalda lateralis
Orthotylus moncreaffi
" rubidus
Chaetaphis suaedae
Lipaphis cochleariae
Psammotettix halophilus
LEPIDOPTERA
Phthorimaea salicorniae
" suaedella
Coleophora salicorniae
" salinella
COLEOPTERA
Dicheirotrichus gustavii
Pogonus chalceus
Ochthebius auriculatus
Phaedon cochleariae

TABLE 11

COMMON SPECIES IN SALT MARSH POOLS IN BRITAIN (EXCLUDING PROTOZOA, TURBEL-
LARIA AND NEMATODES—MANY OF THE SPECIES IN TABLE 12 MAY BE FOUND IN THE
BOTTOM DEPOSITS OF THE POOLS)

(Based on data provided by Nicol, 1935; Sutcliffe, 1961; Galliford, 1956; Balfour-
Browne, 1940, 1950, 1958; Marshall, 1938; and unpublished observations on the
Gwendraeth Estuary)

COELENTERATA
Protohydra leuckarti
NEMERTEA
Lineus ruber
ROTIFERA
Brachionus plicatilis (= mulleri)
Notholca acuminata
ANNELIDA
Nereis diversicolor
Arenicola marina
Pygospio elegans
Manayunkia aestuarina
Ctenodrilus parvulus
Dinophilus taeniatus
Nais communis
MOLLUSCA
Littorina littorea
" saxatilis
Hydrobia ulvae
Potamopyrgus jenkinsi
Alderia modesta

Embletonia pallida
Limapontia depressa
Scrobicularia plana
Macoma balthica
Mytilus edulis
CRUSTACEA
DECAPODA
Carcinus maenas
Crangon vulgaris
Palaemonetes varians
MYSIDACEA
Neomysis integer
ISOPODA
Eurydice pulchra
Sphaeroma rugicauda
" hookeri
Jaera spp
AMPHIPODA
Gammarus zaddachi
" duebeni
Corophium volutator

TABLE 11 (*continued*)

Melita palmata
COPEPODA
Eurytemora velox
     ,,   affinis
Cyclopina gracilis
Halicyclops aequoreus
Ectinosoma curticorne
Enhydrosoma garienis
Nitocra typica
Nannopus palustris
Tachidius littoralis
Onychocamptus mohammed
Paronychocamptus nanus
Mesochra rapiens
Mesochra lilljeborgi
Stenhelia palustris
OSTRACODA
Heterocypris salina
Cyprideis litoralis
Potamocypris villosa
Leptocythere castanea
Loxoconcha gauthieri
     ,,   elliptica

INSECTA
HEMIPTERA
Sigara sahlbergi
     ,,   stagnalis
     ,,   selecta
Gerris thoracicus
ODONATA
Ischnura elegans
COLEOPTERA
Hydrobius fuscipes
Helophorus aequalis
     ,,   viridicollis
     ,,   minutus
     ,,   flavipes
     ,,   brevipalpus
Ochthebius marinus

Enochrus bicolor
Haliplus obliquus
     ,,   apicalis
Agabus conspersus
Colymbetes fuscus
Hygrotus parallelogrammus
Hygrotus inaequalis
Hydroporusangustatus
     ,,   planus

TRICHOPTERA
Limnephilus affinis

DIPTERA
Aedes detritus
     ,,   caspius
     ,,   dorsalis
Anopheles maculipennis
     ,,   claviger (bifurcatus)
Culicoides riethi
     ,,   salinarius
     ,,   maritimus
     ,,   halophilus
Cricotopus vitripennis
Chironomus aprilinus
Procladius choreus
Ephydra riparia

ACARINA
Copidognathus rhodostigma
Rhombognathides spinipes
Isobactrus uniscutatus
Agauopsis brevipalpus
Hyadesia fusca

FISH
Gobius microps
Clupea sprattus
Gasterosteus aculeatus
Platichthys flesus
Anguilla anguilla
Mugil sp. juveniles.

TABLE 12

COMMON MEMBERS OF THE MICROFAUNA OF ESTUARINE SANDS AND MUDS IN BRITAIN

(Based on data provided by Brady and Robertson, 1870; Capstick, 1959; Perkins, 1958; Webb, 1956; Wells, 1963; and unpublished observations on the Gwendraeth Estuary)

PROTOZOA
  CILIATA
  Aspidisca crenata

Blepharisma salinarum
Chaenea teres
Chilodon calkinsi

TABLE 12 (*continued*)

Chilodontopsis elongata
Chlamydodon mnemosyne
„   triquetrus
„   cyclops
Cohnilembus pusillus
Condylostoma patens
„   patulum
Cothurnia annulata
Cryptopharynx setigerus
Diophrys appendiculatus
„   scutum
Enchelys pectinata
Euplotes charon
„   harpa
„   aediculatus
Frontonia marina
Heminotus caudatus
Holosticha kessleri
Keronopsis rubra
Lacrymaria coronata
„   lagenula
„   olor var. marina
„   cohnii
Lionotus fasciola
„   folium
Loxophyllum fasciolatum
„   helus
„   meleagris
„   setigerum
„   undulatum
Mesodinum acarus
„   pulex
Micromitra retractilis
Nassula citraea
Opisthotricha parallelis
Pleuronema crassum
„   marinum
Placus socialis
Prorodon discolor
Spirostomum teres
„   loxodes
Strombidium elegans
„   styliferum
Stylonychia pustulata
Trachelocerca fusca
„   phoenicopterus
„   subviridis
Uronychia transfuga
„   marina
Vorticella microstoma

MASTIGOPHORA
Anisonema acinus
„   intermedium
Euglena obtusa
Oxyrrhis marina
Urceolus cyclostomus
Marsupiogaster picta

SARCODINA
Miliammina fusca
Trochammina inflata
Ammonia beccarii
Nonion depressulus
Elphidium striatopunctata
Lagena sulcata
Truncatulina lobatula

COELENTERATA
HYDROZOA
Protohydra leuckarti

PLATYHELMINTHES
TURBELLARIA
Macrostomum pusillum
„   appendiculatum
Gyratrix hermaphroditus
Thylacorhynchus conglobatus
Monocelis spp.

NEMATODA
Thalassoalaimus tardus
Oxystomatina elongata
„   unguiculata
Trefusia longicauda
Halalaimus gracilis
Enoplus brevis
Anaplostoma viviparum
Viscosia viscosia
Adoncholaimus thalassophygas
„   fuscus
Paracanthoncus elongatus
„   caecus
Halichoanolaimus robustus
Monoposthia costata
Metachromadora vivipara
Spirina parasitifera
Microlaimus tenuispiculum
Spilophorella paradoxa
Chromadorina microlaima
Hypodontolaimus inaequalis
„   butschlii
„   zosterae
Ascolaimus elongatus

TABLE 12 (*continued*)

Odontophora armata
Axonolaimus elongatus
   ,,   paraspinosus
Leptolaimus setiger
Eutelolaimus elegans
Bathylaimus filicaudatus
Tripyloides marinus
Paralinhomoeus lepturus
Theristus setosus
   ,,   tenuispiculum
Theristus acer
Sphaerolaimus gracilis
   ,,   hirsutis
   ,,   balticus
ANNELIDA
Manayunkia aestuarina
Ctenodrilus parvulus
Dinophilus sp.
Protodrilus sp.
Nais elinguis
Paranais litoralis
Tubifex nerthus
   ,,   costatus
Peloscolex benedeni
CRUSTACEA
OSTRACODA
Leptocythere castanea
Cyprideis torosa
Loxoconcha pusilla
Hirschmannia impressa
   ,,   tamarindus

Paradoxostoma variabile
   ,,   abbreviatum
   ,,   ensiforme
Cytherois fischeri
Heterocypris salina
Cypridopsis aculeata
COPEPODA
Ectinosoma melaniceps
   ,,   curticorne
   ,,   gothiceps
Horsiella brevicornis
Tachidius discipes
   ,,   incisipes
Microarthridion littorale
   ,,   fallax
Stenhelia palustris
Amphiascoides limicola
Nitocra typica
   ,,   lacustris
Mesochra lilljeborgi
   ,,   rapiens
   ,,   aestuarii
Onychocamptus mohammed
Paronychocamptus nanus
Nannopus palustris
Enhydrosoma longifurcatum
   ,,   garienis
Platychelipus littoralis

FACTORS LIMITING THE DISTRIBUTION OF ANIMAL SPECIES IN ESTUARIES

In the previous section a number of animals were listed under the habitats in which they are normally found. This implies that each species has certain environmental requirements, some of which it shares with other species. Two major factors in estuaries are salinity and the substratum. Both must be suitable for an animal to inhabit an area. To take an extreme example, it does not matter how suitable the salinity may be if there is no hard substratum to which a barnacle can attach itself.

For shore-dwelling animals the duration of exposure is also important, so that animals often have a characteristic position on the shore. Looked at from the opposite point of view this factor can be regarded as the duration of submersion. This can be important in two respects: in the duration of time available for feeding, by suspension feeders for instance, and secondly in the duration of stress by submersion, as affecting terrestrial animals.

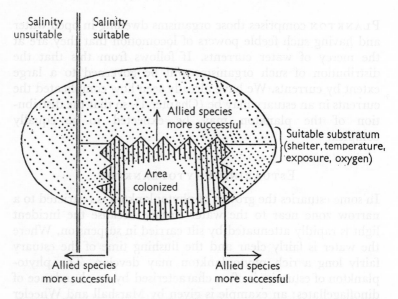

Figure 20. Diagram to show how the distribution of an estuarine animal is limited by environmental factors. The zig-zag lines at the borders of the distributions of allied species indicate that the two species may overlap parts of their ranges

If we use the term substratum in its widest sense to include all the physical and chemical attributes of a substratum, then even if the substratum is suitable it does not necessarily follow that a given species will inhabit all the areas in which it could survive. Biotic factors intervene. As we review the various animals in estuaries it is a frequent phenomenon to find a series of allied species occupying different parts of an estuary. This general situation is shown diagrammatically in Fig. 20. The area actually occupied by the particular species in this example is limited by the occurrence of three allied species.

CHAPTER 5

ESTUARINE PLANKTON

PLANKTON comprises those organisms dwelling in open water and having such feeble powers of locomotion that they are at the mercy of water currents. If follows from this that the distribution of such organisms will be governed to a large extent by currents. We have already seen how complicated the currents in an estuary may be (Chapter 1), so that the distribution of the plankton may be expected to be equally complex.

ESTUARINE PHYTOPLANKTON

In some estuaries the growth of phytoplankton is restricted to a narrow zone near to the water surface because the incident light is rapidly attenuated by silt carried in suspension. Where the water is fairly clear and the flushing time of the estuary fairly long a rich phytoplankton may develop. The phytoplankton of estuaries is often characterised by an abundance of dinoflagellates: an example is given by Marshall and Wheeler (1965) who found that these organisms dominated the plankton of the Niantic Estuary.

At the mouth of the Navesink River in New Jersey a complex succession of phytoplankton was found by Kawamura (1966). Starting at the fresh end of the estuary it was possible to distinguish a zone dominated by euglenoids at salinities below 20‰. This was followed by a zone in which *Rhizosolenia* was dominant and in which the salinity lay between 20 and 22‰. At higher salinities (22–25‰) *Cerataulina bergonii* was dominant. In the outer region of the estuary the phytoplankton was dominated by an assemblage of dinoflagellates, including *Peridinium conicoides*, *P. trochoides* and *Glenodinium danicum*. Finally, in the open water beyond the mouth of the estuary the

81

diatom *Skeletonema costatum* was dominant, although *Cerataulina* was still present together with some of the dinoflagellates.

In discussing this work in relation to that done in other areas Kawamura (1966) gave the following scheme for regions with a fairly stable salinity gradient.

Salinity	Dominant forms
2–5‰	*Anabaenopsis* sp., *Microcystis* sp., *Synedra ulna, Melosira varians.*
9–10‰	*Anabaena flos-aquae, Melosira varians, Chaetoceros* sp., *Biddulphia* spp. *Coscinodiscus* sp.
16‰	Euglenoids
20‰	*Melosira varians, Chaetoceros debilis Ditylum brightwelli,* Peridinians.
24–31‰	*Skeletonema costatum, Rhizosolenia longiseta, Biddulphia aurita, Ditylum brightwelli,* Dinophyceans.

The first category includes a basically fresh water assemblage, and the middle categories show considerable variation from one locality to another according to the local conditions.

Wood (1964) has made a detailed study of the phytoplankton in a number of Australian estuaries, and reached the conclusion that each estuary had its own communities of phytoplankton which differed from those found in other estuaries. But there were some dinoflagellate species which tended to dominate the phytoplankton in several estuaries. Among these dominant species *Goniaulax spinifera, G. monacantha, Ceratium buceros, Peridinium subinerme* and *P. ovatum* were the most widespread.

In estuaries which become stratified for part of the year the breakdown of stratification may lead to a renewal of nutrients in the upper layers, and this may be followed by an outburst of phytoplankton. But outbursts of phytoplankton are not always associated with a breakdown of stratification. Sometimes it is possible to find blooms of different organisms at different depths. In the plankton of Lake Macquarie after heavy rain in June, Wood (1964) found that the upper layers of the lake had a salinity of 25‰ and contained a bloom of

Ceratium furca, while the middle depths had a salinity of 29‰ and a bloom of *Cyclotella menenghiniana* with *Thalassiothrix nitzschioides*. The bottom layers of the lake had a salinity of 30‰ and an association including *Melosira sulcata*, *Pleurosigma* spp. and *Navicula subcarinata*, all of which are species usually found in sediments. Such species are not usually found in the true phytoplankton, but the occurrence in the plankton of species which grow on a solid substratum is a fairly common phenomenon. On some occasions these diatoms may dominate the estuarine phytoplankton. For instance, at Port Hacking in Australia a plankton sample was taken which consisted entirely of *Striatella unipunctata*, *Licmophora flabellata*, *Achnanthes longipes* and *Climacosphenia moliligera*. These represented the shedding of cells from growths attached to stones and other solid objects.

Distribution of the Zooplankton

In a small shallow estuary from which the salt water drains on each tide there can be no permanent plankton. The plankters brought in by the tide are carried out again on the ebb. There is thus only a temporary plankton whose stay in the estuary is brief, being limited to the duration of a single tide. Planktonic organisms of fresh-water origin may be brought into the estuary by the river, but they are usually carried directly out to sea and so perish. A good example of a small estuary without permanent plankton is given by Rogers (1940), who took samples from the harbour bridge over the Margaree River, which flows into the Gulf of St Lawrence. At low tide the samples contained a typical fresh-water assemblage including *Daphnia*, *Bosmina*, *Holopedium*, *Cyclops* and *Diaptomus*. At high tide the plankton was marine in character, containing *Podon*, *Evadne*, *Calanus*, polychaete larvae, ctenophores and medusae.

In estuaries with a long flushing time and a fairly stable salinity gradient there is the possibility of a permanent plankton. An example of the zooplankton found in a large American estuary is given in Table 13. The distribution of this plankton will vary with the state of the tides, particularly during the spring-neap cycle where this is strongly developed. Weather conditions and day length may also influence the distribution

TABLE 13

PRINCIPAL ZOOPLANKTON IN THE DELAWARE ESTUARY
(After Cronin, Daiber, and Hulbert, 1962)

$+ = < 0.1 \text{ M}^3$　　　　$xxx = 10.0-99.9 \text{ M}^3$

$x = 0.1-0.9 \text{ M}^3$　　　　$xxxx = 100.0-999.9 \text{ M}^3$

$xx = 1.0-9.9 \text{ M}^3$　　　　$xxxxx = 1,000.0-9,999.9 \text{ M}^3$

Numbers given are average counts at station of highest abundance

	Fall	Winter	Spring	Summer
COELENTERATA				
Aglantha digitale	+	x	+	+
Bougainvillia sp.	+	+	+	x
Nemopsis bachei	x	+	xx	x
Blackfordia virginica	+	+	+	xx
Dactylomera quinquecirrha	+	+	+	+
CTENOPHORA				
Beroe ovata	+	+	+	+
Mnemiopsis leidyi	+	+	+	+
CHAETOGNATHA				
Sagitta elegans	x	xx	+	x
„　enflata	xx	x	+	+
ANNELIDA				
Autolytus cornutus	+	+	x	+
Tomopteris sp.	+	x	+	+
CRUSTACEA				
CLADOCERA				
Evadne nordmanni	+	+	xxxx	+
Podon leuckarti	xx	+	+	+
Penilia avirostris	+	+	+	xxxxx
Fresh-water Cladocera	xx	+	xxxx	xxxx
COPEPODA				
Acartia tonsa	xxxxx	xxxxx	xxxxx	xxxxx
Centropages spp.*	xxxx	xxxx	xxxx	xx
Eurytemora spp.**	xxxxx	xxxxx	xxxxx	xxxx
Labidocera aestiva	xxx	+	xxx	xxx
Paracalanus parvus	xxxx·	+	+	xxxx
Pseudocalanus minutus	+	xxxx	xx	+
Pseudodiaptomus coronatus	xxxx	xxxx	xxxx	xxxx
Temora longicornis	xxx	xxx	xxxx	xx
Corycaeus americanus	xx	+	+	+
Cyclops viridis	xxx	xxxx	xxxxx	xxxx
Euterpina acutifrons	+	+	xx	+
CIRRIPEDIA				
Barnacle larvae	+	+	xxxx	xxx
MYSIDACEA				
Neomysis americana	xxx	xxx	xxx	xxx
AMPHIPODA				
Gammarus fasciatus	xxx	xx	xxxx	xxxx
ISOPODA				
Aega sp.	x	x	+	x

	Fall	Winter	Spring	Summer
DECAPODA				
Crangon septemspinosa	+	x	x	+
Zoea larvae	+	+	xxx	xxx
TUNICATA				
Oikopleura dioica	+	+	+	x
FISH				
Anchoa mitchilli	x	+	+	+
Micropogon undulatus (young)	x	x	+	+
larval fish (unidentified)	+	x	x	x

* Includes C. typicus and C. hamatus
** Includes E. affinis and E. hirundoides

of the zooplankton. The fact that some zooplankters undergo a diurnal cycle of vertical migration leads to the possibility of very complex distributions. Many zooplankters ascend when the light intensity decreases and descend when the light intensity is high. Young smelt (*Osmerus mordax*) in the Miramich River show a diurnal migration of this type. Rogers (1940) found that a higher proportion of the population was in the upper layers on a dull cloudy day than on a bright sunny day.

On a very bright day the zooplankton will descend lower, so that if the estuary is one where a tongue of sea water penetrates under outflowing fresh water the extent to which the plankton is carried towards the head of the estuary will be influenced by the light intensity. Clearly the more the zooplankton descends into the penetrating tongue the further it will be carried upstream.

The vertical migration cycle may also be influenced by vertical salinity stratification in estuaries (Lance, 1962; Grindley, 1964). The marine forms penetrating in the lower more saline layers may be inhibited from ascending into the upper layers of lower salinity. Lance (1962) found that a salinity discontinuity layer inhibited the upward migration of six species of copepods. When there was a large difference between the upper and lower layers the copepods were unable to enter the upper layer. Their swimming activity was depressed by lowered salinities so that they sank downwards into the more saline water.

Similarly Lyster (1965) found that when larvae of the polychaete *Phyllodoce maculata* were placed in a vertical salinity gradient they tended to congregate at a salinity of about 12‰. They were unable to ascend further because lower salinities reduced their ability to swim. This mechanism tends to keep the larvae within an estuary. By remaining in the lower, more saline layers, they will avoid the outflowing low salinity water near the surface. This is of particular importance when the river is in flood.

Seasonal variation in the horizontal distribution of the zooplankton reflects the varying tolerance of dilution of sea water with variation in temperature. In temperate regions many marine plankters are able to tolerate lower salinities in summer than in winter, so that they penetrate further into estuaries during the summer.

PLANKTONIC COPEPODS IN ESTUARIES

In common with the zooplankton of coastal waters the zooplankton of estuaries may be considered under two categories. The permanent plankton includes those animals that are planktonic for the whole of their lives, while the temporary plankton includes forms which become planktonic at certain seasons, or for parts of their life cycles. The most numerous of the permanent plankters are the copepods of the order Calanoida.

In different parts of the world one can trace sequences of calanoid species which replace one another along the length of an estuary. In the genus *Acartia* there are several species which are essentially marine, but some of them penetrate considerable distances into estuaries. Jeffries (1962a) found that in the estuaries of New England *Acartia clausi* was dominant in winter and spring while *A. tonsa* was dominant in summer and autumn. Replacement of *A. clausi* started at the head of an embayment and spread seawards, and it appeared that *A. tonsa* was more resistant to dilution of sea water than its congener. The resistance to dilution of sea water shown by three species of the genus *Acartia* has been studied experimentally by Lance (1963). Using specimens collected from Southampton Water she found that *A. tonsa* was the most resistant; next came *A. bifilosa*, while *A. discaudata* showed the least resistance to dilution. There was some interaction between temperature and salinity tolerance, and this interaction was modified by temperature acclimation. Resistance to dilution at a particular temperature varied with the degree to which the copepods had been acclimated to that temperature. An example is given in Table 14 where the copepods were collected and tested at a temperature close to that of the sea at the time of capture, and at various other temperatures. It is clear that they were most resistant to dilution at the temperature nearest that at which they were living in Southampton Water.

TABLE 14

EFFECT OF TEMPERATURE ACCLIMATION ON THE SALINITY TOLERANCE OF THREE SPECIES OF 'ACARTIA'
(After Lance, 1963)

Species	$T°C$ at time of capture	Order of tolerance to dilution of sea water
A. tonsa	19·7	20°C>17°C>24°C>10°C>4·5°C
A. discaudata	15·3	16°C>10°C>4·5°C
A. bifilosa	8·1	10°C>16°C>4·5°C>20°C

In Australasia the family Centropagidae has produced a complex of species inhabiting brackish water, fresh water and inland saline waters. These forms have been studied in detail by Bayly (1961, 1962, 1963, 1964, 1965). The genus *Gladioferens* contains five species, one of which, *G. spinosus*, is confined to fresh water and lives at the fresh water end of estuaries, or in closed fresh lakes near to the sea. The most widespread estuarine species of the genus is *G. pectinatus*. This is an open water euryhaline species which dominates the zooplankton of long stretches in the fresher parts of estuaries, but also extends into nearly full strength sea water. The seaward extension of this species was found to be seasonal, reaching its maximum in the coolest months. In the Brisbane River estuary *G. pectinatus* was accompanied by *Sulcanus conflictus* over the fresher part of its range, and overlapped *Isias uncipes*, an *Acartia* species, and three species of *Pseudodiaptomus* at the seaward end. The extent of this overlapping was found to be influenced by temperature. *Gladioferens pectinatus* only extended into the more saline water at low temperatures and appeared to be incapable of withstanding the combined effects of high temperature and high

D

salinity. On the other hand the *Pseudodiaptomus* species appeared to be incapable of tolerating the combined effects of low temperature and low salinity, so that they penetrated furthest into the estuary during the warm season.

The genera *Boeckella*, *Hemiboeckella* and *Calamoecia* also belong to the Centropagidae, but they differ from *Gladioferens* in that they are basically fresh-water genera. Nevertheless the various species occupy a wide range of salinities in inland habitats. *Calamoecia salina* and *Boeckella triarticulata* live in inland saline lakes, and *Calamoecia tasmanica* is found in acid swamp waters which are remarkably deficient in ions. This wide range of habitats can be regarded as an indication of the estuarine origins of these genera. Bayly (1964) has proposed a series of centropagid genera which represent a sequence of penetration of inland habitats in Australia. Starting with the marine *Centropages* he traces one line to the estuarine genera *Isias* and *Gladioferens*, with *G. spinosus* in fresh water, and another line through *Boeckella* and *Hemiboeckella* to *Calamoecia*. It is noteworthy that the last two genera show structural reductions, such as fewer spines and setae on the limbs, and can to some extent be regarded as neotenic derivatives of *Boeckella*.

The complex of centropagids in Australian estuaries and coastal habitats does not have a parallel in any other part of the world. But there are indications of a relationship between *Centropages* and *Limnocalanus* in the Northern Hemisphere. The former genus contains about 23 species, mostly marine, but some species penetrate into brackish water. In British waters *C. hamatus* enters estuaries, but is never found in fresh water, and rarely in salinities below 14‰. *Limnocalanus* has several features in common with *Centropages*, and a detailed comparison by Burckhardt (1913) revealed that some details of its structure were reduced from the condition in *Centropages*. *Limnocalanus macrurus* lives both in the Baltic and in fresh-water lakes, though some authorities regard the brackish water and fresh-water forms as separate species, and give the name *L. grimaldi* to the brackish water form. The fresh-water localities for *Limnocalanus* are often the same as those inhabited by *Mysis relicta*. Like the latter, *Limnocalanus* is also a winter breeder. Egg laying begins in October and ceases in the spring. The eggs are laid freely and sink to the bottom of the lake. Nauplii emerge in March and April and the adult form is assumed in about two months, but breeding is delayed until the autumn when the water temperature falls to 7°C.

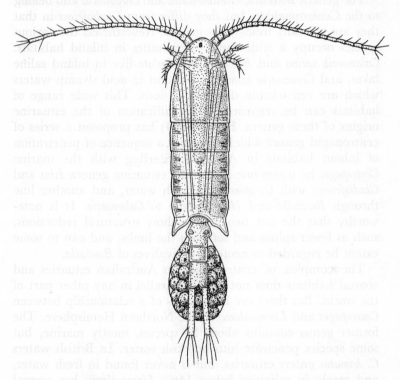

Figure 21. Adult female of *Eurytemora hirundoides* (Copepoda: Calanoida). Actual length 1·7 mm (after Sars)

The family Temoridae also has a number of brackish water species, and a few which have penetrated into fresh water. The best known brackish water genus is *Eurytemora*, with about 16 species. Some of these species, such as *E. velox*, can live in both fresh and brackish water. Others live only in brackish water. *Eurytemora affinis* and *E. hirundoides*, which is regarded as a mere variety of *E. affinis* by some authorities (e.g. Gurney, 1931), are both widespread in European and North American estuaries. The euryhalinity of *E. affinis* is paralleled by its tolerance of oxygen deficiency. At the mouth of the River Tyne *E. affinis* is often abundant in water which appears to be completely lacking in oxygen (Bull, 1931).

The life cycle of a calanoid copepod involves six naupliar stages and five copepodid stages. The nauplius which emerges from the egg has only three pairs of appendages: the antennules, antennae and mandibles, and these remain as the only functional appendages for the next five moults. The copepodids resemble the adults in general body form, but the limbs are not fully developed. Each successive copepodid adds to the number and completeness of the limbs until the adult condition is reached. The duration of development from the first naupliar stage to the adult varies with temperature. In cold regions

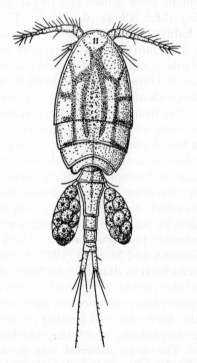

Figure 22. Adult female of *Cyclopina norvegica* (Copepoda: Cyclopoida). Actual length 1·5 mm (after Sars)

there may be a single generation in a year, but in the tropics it is possible for a generation to be passed in a month.

The calanoids are not the only copepods living in the plankton of estuaries. Cyclopoids (e.g. Fig. 22) are also found, sometimes in abundance. *Cyclopina norvegica* is a species which penetrates from the sea into many British estuaries. A much smaller number of species of harpacticoid copepods are also planktonic. Most members of this group are found in the micro-benthos, but a few, like *Euterpina acutifrons*, may be found in the plankton near the mouths of estuaries.

MYSIDACEA IN ESTUARIES

The Mysidacea form an important element in estuarine plankton, although some authorities might have reservations about regarding them all as planktonic. These reservations arise from the habit of several species of remaining close to the surface of the substratum for a large part of the day, and entering the plankton only under certain conditions. In superficial appearance the mysids resemble small shrimps. They have elongated bodies, stalked eyes, a well-developed carapace and a tail fan. They differ from the adult decapod shrimps in having well-developed exopods on their thoracic limbs, and the females have a brood pouch formed by oostegites on the bases of some of the thoracic limbs. Most of the estuarine mysids have large statocysts in the bases of the inner branches of the uropods. This is an easy character to use in distinguishing them from larval decapods which do not have statocysts in this position.

The food taken by estuarine mysids varies over a wide range of size. An elaborate filter mechanism, which has been fully described by Cannon and Manton (1927) is capable of collecting small particles such as diatoms and finely divided detritus. The mysids are also capable of individual acts of capture, and may pounce upon smaller crustaceans, such as copepods. Some species have also been observed feeding on larger dead crustaceans, such as amphipods, which they carried about in their thoracic limbs. The dead material was pressed against the mandibles, which bit pieces off and pushed them up into the oesophagus. This wide range of feeding habits, coupled with the medium size of mysids, makes them important links in the estuarine food web, utilising various grades of material and providing food for a wide range of fish.

The distribution and seasonal movements of mysids in the Brisbane River illustrate some of the differences which may be found between species. Hodge (1963) found two species, *Rhopalophthalmus brisbanensis* and *Gastrosaccus dakini*. The latter penetrated further into the estuary, and most of the juveniles were found in salinities between 2 and 6‰. The largest numbers of *Rhopalophthalmus* juveniles occurred over a wider range of higher salinities, from 7 to 17‰. Both species showed a seasonal movement upstream, reaching its maximum in October; after which the counter movement downstream began. Both species also produced three broods of young in a year, but *Rhopalophthalmus* produced its brood in the summer, while *Gastrosaccus* produced its young in the autumn and spring.

The abundance of estuarine mysids is sometimes impressive. *Neomysis integer* (=*N. vulgaris*) often congregates near the edge of the tide as it rises slowly over estuarine mud banks. A dense, easily visible band, sometimes a yard wide, is formed and the mysids are so crowded that it is difficult to see what advantage is gained by this behaviour, unless some food material is floated off the surface of the mud as the tide rises. Another species, *Mesopodopsis slabberi*, has been observed in enormous numbers over the oyster beds at Ostend, and is one of the most abundant species in the brackish lakes in the delta of the Danube.

In estuaries on the Atlantic coast of Europe *Neomysis integer* penetrates further towards fresh water than other mysids. It can survive for several days in fresh water, and there are records of it living in pools and lakes close to the sea, often in the company of typical fresh-water animals such as frogs and pond snails. It does not, however, penetrate far inland, and the apparently fresh localities in which it is found are usually very slightly brackish.

Other estuarine mysids in Europe include *Praunus flexuosus*, *Mesopodopsis slabberi*, *Gastrosaccus spinifer*, and *Leptomysis gracilis*. The order in which they are listed gives an approximate indication of their ability to penetrate from the sea into

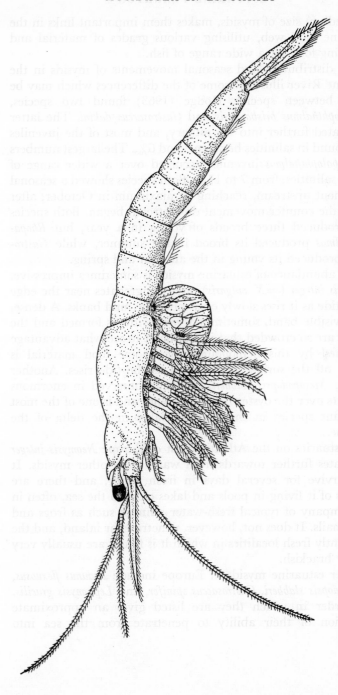

Figure 23. Lateral view of adult female of *Neomysis integer* (Peracarida: Mysidacea). Actual length 17 mm. Note the brood pouch bulging behind the thoracic limbs

estuaries. *Praunus flexuosus* extends into water with a salinity as low as 0·3‰ (Percival, 1929), while *Leptomysis gracilis* lives near the mouths of estuaries, usually in fairly deep water, ranging from 12 to 530 metres. All these species may be found in the open sea, so that they are basically marine species, whereas *Neomysis integer* is a more typically estuarine species, although there are a few records of it being captured in the open sea.

Mysids in general seem to be sensitive to any substantial reduction in the oxygen content of their medium. The work of Jørgensen (1929) in the Tyne area shows that *Neomysis integer* is rare in estuaries where the oxygen content of the water is depleted by pollution, particularly in warm weather. A parallel is found in *Mysis relicta*, which lives in various fresh-water lakes in Europe and North America, as well as in the brackish waters of the Baltic. This species requires well-oxygenated water, and has been recorded rarely in fresh water with an oxygen content less than 4 ml/l, although in some parts of the Baltic it has been recorded in brackish water with only 1·6 ml/O_2/l. It seems that to survive in fresh water *M. relicta* needs more oxygen than in brackish water.

The breeding cycle of *Neomysis integer* has been worked out in detail by Kinne (1955). In the Kiel Canal the breeding season lasted from April to September, and there were three generations in a year. A summer generation, born in May, lived until September or October, and produced a generation which hatched in July or August and lived until the following June. Further south the breeding season may be longer: Tattersall and Tattersall (1951) record breeding extending from February to November at Plymouth.

The summer generation at Kiel became mature about a month and a half after leaving the parental brood pouch. But the winter generation, born in August, did not become mature until the following April, so that they were by that time eight months old. A few females born in July became mature in September, but the third generation produced by these females was so small in number that it played a negligible part in the population cycle.

The duration of development of *Neomysis* from egg to neonate varies with temperature. At 11°C the developmental period was 20 days, and at 18·5°C the time taken was 14 days. Each female of the overwintering generation produced 3 or 4 broods before dying, while some of the summer generation produced up to 6 broods. The number of young per brood varied with the size of the female; some of the larger females produced 50 young at a time.

Praunus flexuosus breeds throughout the year at Roscoff, but further north, on the Danish coast, it stops breeding in winter. The duration of embryonic development is similar to that of *Neomysis*, and varies from 2 to 3 weeks according to the temperature.

In contrast to *Neomysis* and *Praunus* the main breeding period of *Mysis relicta* is during the winter. This species does not live in areas where the summer temperature is higher than 14°C, and it lays its eggs at temperatures below 7°C. The limits of the breeding season lie between October and the beginning of May.

There are several records of mysids living in coastal localities which have become isolated from the sea. Pesta (1935) found *Mesopodopsis slabberi* in a lake on the island of Corfu where the salinity was only 1·3‰. More recently Hodge (1964) has rediscovered *Tenagomysis chiltoni* in a coastal dune lake in the North Island of New Zealand. The total concentration of salts in this lake was about 0·2‰, so that it can be regarded as a true fresh-water habitat. The only previous record of *T. chiltoni* was from a tidal inlet at Parakai, New Zealand. Other members of the genus *Tenagomysis* live in the open sea around New Zealand and Japan, but another brackish water species, *T. nigerianus*, has been found in Lagos Harbour, where considerable seasonal salinity fluctuations are found (Tattersall, 1957).

TEMPORARY PLANKTON

In estuaries it is possible to distinguish two types of temporary plankton. The first category includes those animals which themselves are permanent members of the zooplankton, but which can only enter estuaries at certain seasons when the combinations of temperature and salinity are suitable. In many temperate regions the estuarine plankton is augmented by the invasion of marine forms during the summer, when the salinity is generally higher and the water warmer. Hartley (1940) has described how the jellyfish *Aurelia aurita* invades the Tamar Estuary in the summer, and sometimes becomes so abundant as to hinder the activities of fishermen by filling their nets.

The second category of temporary estuarine plankton includes the larval stages of the estuarine benthos. For example, the veliger larvae of molluscs such as *Cardium*, *Mytilus* and *Macoma* become part of the plankton for a short period during the spring and summer. The seasonal occurrence of these larval forms is governed by the breeding season of the adults. Once the larval forms have entered the plankton their fate depends upon the salinities and temperatures that they encounter, and whether they are carried by water currents to a place suitable for settlement.

In tropical regions where the temperature range is small and there is markedly seasonal rainfall the occurrence of certain larval forms in the plankton may be governed almost entirely by the salinity regime. A striking example of seasonal limitation by salinity change has been described by Webb (1958). Larvae of the lancelet, *Branchiostoma nigeriense*, enter Lagos Lagoon only when the salinity is above 13‰. In a normal year the salinity of the lagoon is below 13‰ between the end of May and the beginning of December. During the latter month the salinity begins to rise as the effects of the wet season die away. Larvae derived from the spawning of lancelets in the sea invade the lagoon and may remain in the plankton for a month or two. The larvae that metamorphose and settle in the lagoon in January become mature and spawn in March and April. In May the wet season begins to exert its effects, the salinity in the lagoon falls, and the lagoon population of *Branchiostoma* is wiped out. There is thus no permanent population of lancelets in Lagos Lagoon, and the cycle of invasion by planktonic larvae is dependent on the spawning of marine populations outside the lagoon.

ZOOPLANKTON IN STABLE BRACKISH WATERS

An example of the plankton found in a region of low and fairly stable salinity is given by Lafon, Durchon and Saudray (1955), who studied a canal leading from Caen to the sea. They made collections from two stations in the canal throughout the year, so that they could distinguish between the permanent and the temporary plankton. Taking the two stations together the salinity ranged from 2·6 to 8·1‰ during the course of a year. The zooplankton was dominated in summer by *Acartia tonsa*, which was accompanied in late spring by small numbers of *Eurytemora hirundoides*. Among the Protozoa there were two species of *Tintinnopsis* which became abundant in the canal, but in the sea outside they were replaced by two other tintinnids: *Amphorella subulata* and *Parafavella media*. The rotifer *Synchaeta littoralis* was present in the canal throughout the year, and became abundant in September and October. This rotifer probably entered the canal from the sea, since it is one of the

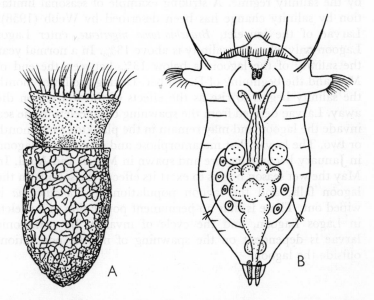

Figure 24. A, Tintinnid (Protozoa: Ciliata), total length about 100μ. B, *Synchaeta* sp. (Rotifera), length about 240μ

species which have frequently been recorded as marine in habit. Four other rotifers: *Keratella cochlearis*, *K. aculeata*, *Brachionus urceolaris* and a species of *Colurella*, were found in small numbers in summer and autumn. All four are typically fresh-water species, but have some ability to penetrate into waters with a stable low salinity.

Two fresh-water Cladocera, *Scapholeberis mucronata* and *Chydorus sphaericus*, and a brackish water form, *Bosmina coregoni* var. *maritima*, were also found in the canal plankton for short periods. A marine cladoceran, *Podon maritimus*, was found in the late summer. *Bosmina* and *Podon* are important members of the plankton in the stable low salinities of the Baltic Sea.

The temporary plankton of the canal included larvae of the zebra mussel, *Dreissena*, and larvae of the fan worm *Mercierella enigmatica*. The adults of both species are members of the epibenthos. An interesting feature was noted concerning the larva of *Nereis succinea* in the canal. Many estuarine nereids do not have truly planktonic larvae, and this has been considered as a means of avoiding the danger of being carried out to sea by the ebb from a tidal estuary. In the canal, where the tidal flow was restricted by lock gates at the seaward end, there was no such danger, and the larvae of *N. succinea* were found in the plankton samples taken at night, but not during the day. One other major component of the canal plankton showed a similar nocturnal occurrence. *Neomysis integer* was abundant in the samples taken at night, but during the day it retreated away from the light and lived near the bottom of the canal.

PLANKTON OF THE BRACKISH SEAS

a. *The Baltic Sea*

In the Gulfs of Bothnia and Finland there are large areas with stable surface salinities ranging from 7‰ at the mouths to less than 2‰ at the extreme ends of the gulfs. The transition from 7‰ to 2‰ extends over a distance of 400 miles in the Gulf of Bothnia. Along the shores of the gulfs there are local gradients where rivers enter. In the mouths of the rivers the salinity is somewhat more variable than in the open sea, and a considerable reduction in salinity is found when the ice melts in the spring.

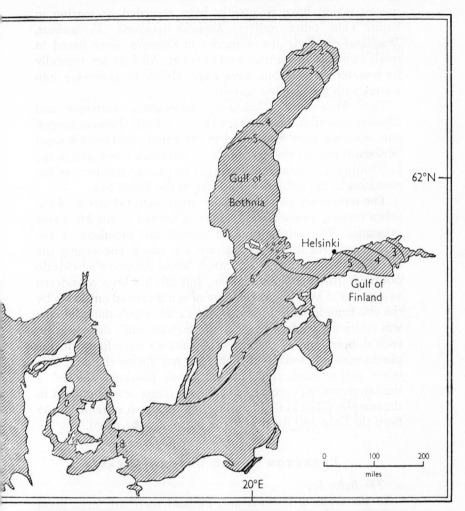

Figure 25. Map of the Baltic Sea showing average surface salinities in ‰ (after Segerstrale)

The large areas of stable low salinity provide ideal conditions for the gradual adaptation to almost fresh conditions by brackish water dwellers. Also, the mixing of fresh and low salinity water at the mouths of rivers provides a gentle salinity gradient which enables fresh-water animals to penetrate into the open waters of the gulfs. The composition of the zooplankton provides a good illustration of the admixture of fresh and brackish water forms.

TABLE 15

COMPOSITION OF THE ZOOPLANKTON IN THE HARBOUR OF HELSINKI

(Data from Valikangas, 1926)

	Total species	Fresh-water species	Percent fresh-water species
Ciliata	31	20	65
Rotifera	37	23	62
Copepoda	7	4	57
Cladocera	7	4	57
Cirripedia	1 (larvae)	–	–
Bryozoa	1 (larvae)	–	–
TOTAL	84	51	61

A thorough study of the plankton in the harbour at Helsinki was made by Valikangas (1926). The salinity of the water in the harbour varied from 0·25 to 4·96‰ in different parts, and the list of species found in the zooplankton included a considerable proportion of fresh-water forms, which were found mainly in the lowest salinities (Table 15). The species which regularly occurred in salinities over 2‰ were as follows:

CILIATA

Tintinnidium fluviatilis*
Tintinnopsis tubulosa
      ,,   brandti
      ,,   relicta
      ,,   bottnica

Cothurnia maritima
Euplotes harpa*
Coleps hirtus*

ROTIFERA

Floscularia pelagica*
Asplanchna brightwelli*
Synchaeta baltica
      ,,   littoralis
      ,,   fennica
      ,,   monopus
Filinia longiseta*

Brachionus angularis*
      ,,   calyciflorus*
Keratella quadrata*
      ,,   cochlearis*
      ,,   eichwaldi
Hexarthra oxyura

*The species marked * are widespread in fresh water*

COPEPODA
Limnocalanus grimaldi   Eurytemora affinis

CLADOCERA
Bosmina coregoni v. maritima   Podon polyphemoides
                                 Evadne nordmanni

CIRRIPEDIA
Balanus improvisus (nauplii and cyprids)

BRYOZOA
Membranipora crustulenta (cyphonautes larvae).

In the vicinity of Tvärminne, which lies 60 miles to the west of Helsinki, the plankton is somewhat richer in species. In the open water the salinity is about 6‰. The dominant zooplankters at Tvärminne are four Cladocera (*Evadne nordmanni, Podon polyphemoides, P. intermedius, Bosmina coregoni* v. *maritima*), six calanoid copepods (*Acartia bifilosa, A. longiremis, Eurytemora hirundoides, Temora logicornis, Pseudocalanus elongatus, Limnocalanus grimaldi*), and rotifers of the genera *Synchaeta* and *Keratella*. The jelly fish *Aurelia aurita* is often common in the plankton in the summer, and at times there are incursions, from deeper more saline water, of the ctenophore *Pleurobrachia pileus*, and the appendicularian *Fritillaria borealis*. During the summer there are also many larvae in the plankton. Nauplii and cyprids of *Balanus improvisus* and the cyphonautes larvae of *Membranipora crustulenta* are common, as well as veligers of the bivalves *Macoma balthica, Mytilus edulis, Cardium edule* and *Mya arenaria*. The trochophores of the polychaete *Harmothoe sarsi* also contribute to the zooplankton of the Tvärminne area.

Five species of Mysidacea are known from the Tvärminne area: *Neomysis integer, Mysis relicta, M. mixta, Praunus inermis* and *P. flexuosus*, but these are not all strictly planktonic. The last species for instance is found in wave beaten *Fucus* beds.

In the Gulf of Bothnia the zooplankton has been studied in detail by Lindquist (1959), who gives many references to earlier work on the zooplankton of the Baltic. The list of species given by Lindquist is almost identical with that at Tvärminne, but there is an addition to the calanoid copepods in the form of *Centropages hamatus*. This species is rare in the Gulf of Bothnia, and seems to require a salinity above 7‰ to survive permanently.

Further south, outside the two large gulfs, there is an increase in salinity and some additional species are found. For instance, *Acartia tonsa* is found in the Central Baltic, and there is a brackish water variety of the marine arrow worm, *Sagitta elegans*, particularly in the western Baltic.

b. *The Caspian Sea*

There is great variation in the salinity of different parts of the Caspian Sea (Zenkevitch, 1957, 1963). In the central and southern regions the surface salinity is about 12 or 13‰, but in the north the inflow from the Volga dilutes the salinity down to 10 or 5‰, or even lower near the delta. On the eastern shore there are no rivers of any importance, and the rainfall is slight. Evaporation exceeds precipitation and the sea water tends to become concentrated, especially in enclosed bays with limited access to the main body of the sea. In the Gulf of Kara Bugaz, which lies three metres below the mean sea level of the Caspian, evaporation is rapid and the inflowing water becomes so saturated that salts are precipitated.

The salts in Caspian water occur in different proportions from those in other seas. The Caspian is relatively richer in magnesium, calcium and sulphate. In the Gulf of Kara Bugaz the ratio changes in favour of an increase in sodium sulphate, which is precipitated in large amounts when the temperature falls in the winter.

The zooplankton shows variations which correspond with the variations in salinity. In the northern region, particularly on the western side near the mouth of the Volga, there is a predominance of fresh-water species. Rotifera are numerous, including *Brachionus plicatilis*, *Asplanchna priodonta* and *Synchaeta vorax*, while the 20 species of Cladocera which have been recorded include a high proportion of fresh-water forms.

The most remarkable feature of the Caspian zooplankton is the presence of a number of endemic genera and species of

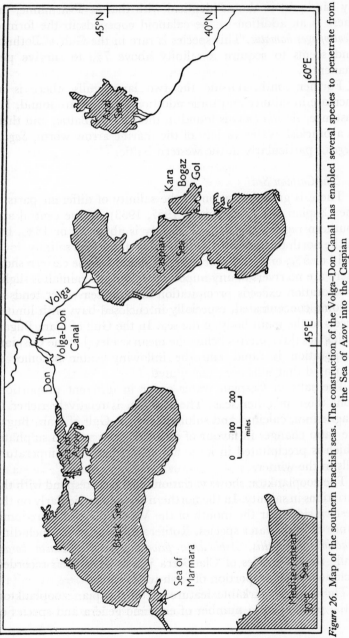

Figure 26. Map of the southern brackish seas. The construction of the Volga–Don Canal has enabled several species to penetrate from the Sea of Azov into the Caspian

the cladoceran family Polyphemidae. In a classical paper, Sars (1897) described a number of these species, and several have since been added (Sars, 1902; Mordukhai-Boltovskoi, 1964). Among these endemic species *Polyphemus exiguus* differs only slightly from the common European fresh-water species *P. pediculus*, but the species of *Cercopagis* and *Apagis* (Fig. 27) appear to be bizarre relatives of the fresh-water genus *Bythotrephes*.

Among the copepods the presence of *Limnocalanus grimaldi* is notable, and there are a number of other calanoids, such as *Calanipeda aquae-dulcis*, *Heterocope caspia*, *Popella guerni*, and several species of *Eurytemora*, of which *E. grimmi* is the most important in the open sea.

In the northern Caspian one of the dominant copepods is *Calanipeda aquae-dulcis*, which forms over 90 per cent of the zooplankton biomass in winter. In summer *Halicyclops sarsi* becomes abundant.

In the central and southern Caspian *Limnocalanus grimaldi* is the dominant zooplankter, at least in the superficial waters. The deeper waters tend to be dominated by mysids. There are about 20 species of the family Mysidae in the Caspian, and some of these undergo extensive vertical migrations. *Mysis microphthalma*, *M. amblyops* and *Austromysis loxolepis* are the most abundant of these vertical migrants. They remain at a depth of 300 metres during the day, and migrate to the surface at night.

In the spring and summer the zooplankton is augmented by larval forms, especially by lamellibranch larvae. Fish fry also join the zooplankton in the summer and participate in the vertical migrations.

Planktivorous fish are abundant in the Caspian. The genera include *Clupeonella*, *Sprattus*, *Atherina* and *Caspialosa*, all of which utilise the zooplankton as their main source of food.

c. The Black Sea

The surface layers of the Black Sea have a salinity between 17 and 18‰, except near the mouth of rivers where the salinity varies seasonally. The zooplankton of this sea is basically an impoverished version of that found in the Mediterranean Sea.

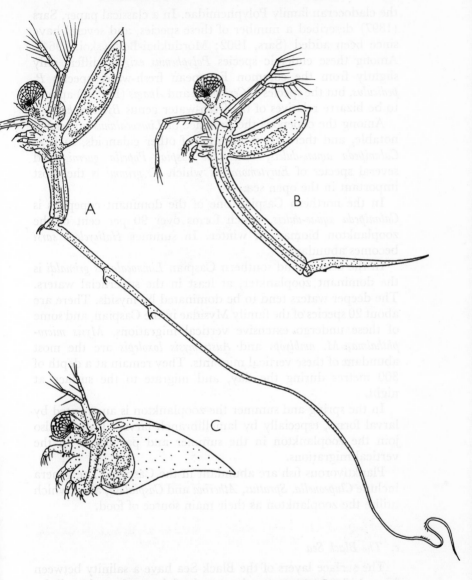

Figure 27. Planktonic Cladocera from the Caspian Sea
A, *Cercopagis tenera*, length about 1·5 mm not including the caudal process. B,
Apagis cylindrata, length about 2·3 mm not including the caudal process. C, *Evadne
producta*, total length about 1·4 mm (after Sars)

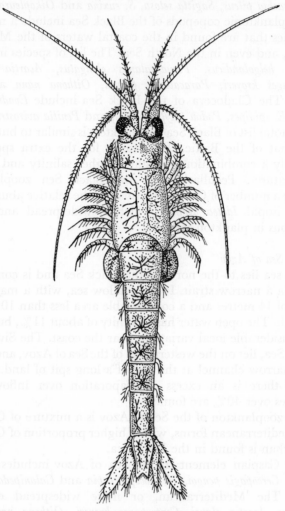

Figure 28. Dorsal view of adult female of a *Mesomysis* species from the Caspian Sea. Actual length 18 mm (after Sars)

In the open waters there are several tintinnids, including three species of *Tintinnopsis* and two of *Tintinnus*. The jelly fish *Aurelia aurita* is common, and may be accompanied by the larger species *Rhizostoma pulmo*. Other common forms include

Pleurobrachia pileus, Sagitta setosa, S. euxina and *Oikopleura dioica.*

The planktonic copepods of the Black Sea include a number of species that are found in the coastal waters of the Mediterranean, and even in the North Sea. The list of species includes *Calanus helgolandicus, Pseudocalanus elongatus, Acartia clausi, Centropages kroyeri, Paracalanus parvus, Oithona nana* and *O. similis.* The Cladocera of the Black Sea include *Evadne nordmanni, E. spinipes, Podon polyphemoides,* and *Penilia avirostris.*

The total list of Black Sea zooplankters is similar to but richer than that of the Baltic. The reason for the extra species is probably a combination of slightly higher salinity and higher temperatures. Peculiarities of the Black Sea zooplankton include a number of Caspian mysids and the relative abundance of the isopid *Idotea algirica* which is widespread and often numerous in plankton samples.

d. The Sea of Azov

This sea lies to the north of the Black Sea and is connected to it via a narrow strait. It is a shallow sea, with a maximum depth of 14 metres and a considerable area less than 10 metres in depth. The open water has a salinity of about 11‰, but there are considerable local variations near the coast. The Sivash, or Putrid Sea, lies on the western side of the Sea of Azov, and opens via a narrow channel at the end of a long spit of land. In the Sivash there is an excess of evaporation over inflow, and salinities over 40‰ are found.

The zooplankton of the Sea of Azov is a mixture of Caspian and Mediterranean forms, with a higher proportion of Caspian forms than is found in the Black Sea.

The Caspian element in the Sea of Azov includes *Evadne trigona, Cercopagis pengoi, Heterocope caspia* and *Calanipeda aquaedulcis.* The Mediterranean, or more widespread element includes *Acartia clausi, Centropages kroyeri, Oithona nana* and *Podon polyphemoides.*

On the north-eastern side, in the Gulf of Taganrog, where the salinity is lower than in the rest of the sea, there is a considerable admixture of fresh water forms in the zooplankton. In the eastern parts of the gulf the rotifers *Asplanchna priodonta, Brachionus plicatilis* and *Synchaeta vorax* are abundant, together with *Keratella cochlearis* and *K. quadrata* where the salinity falls below 4‰.

There are some striking resemblances between the zooplankton of the Gulf of Taganrog and the Gulf of Finland, but the former includes species of the Caspian element which are absent

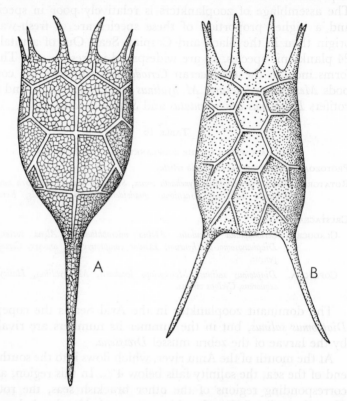

Figure 29. A, *Keratella cochlearis* and B, *Keratella quadrata.* Both these rotifers are found in the brackish seas in regions where the salinities fall below 4‰. Both are a little over 200μ in length

from the Baltic. The mysids in the Sea of Azov include a Caspian element in the form of *Mesomysis kowalewskyi*, but *Paramysis helleri* is also present, and together with *Mesopodopsis slabberi* represents an invasion from the Mediterranean.

e. The Aral Sea

The salinity of the Aral Sea is about 10‰, but the proportions of the various salts differ from normal sea water. The percentages of magnesium, calcium, sulphate and carbonate are all significantly higher than in oceanic water. In this respect the Aral Sea shows some resemblances to fresh waters. The assemblage of zooplankters is relatively poor in species, and a higher proportion of these species are of fresh-water origin than in the Black and Caspian Seas. Out of a total of 24 planktonic species, 10 are widespread in fresh water. These forms include the cladoceran *Ceriodaphnia reticulata*, the copepods *Mesocyclops leuckarti*, *M. hyalinus* and *Cyclops viridis*, and the rotifers *Brachionus quadridentatus* and *Synchaeta vorax*.

TABLE 16

THE MAIN SPECIES OF THE ZOOPLANKTON IN THE ARAL SEA

PROTOZOA—TINTINNOIDEA. *Codonella relicta.*

ROTATORIA. *Floscularia mutabilis, Synchaeta vorax, S. neopolitana, Trichocerca marina, Colurella adriatica, Brachionus quadridentatus, B. plicatilis, Keratella quadrata.*

CRUSTACEA

CLADOCERA. *Ceriodaphnia reticulata, Moina microphthalma, Alona rectangula, Diaphanosoma brachyurum, Evadne camptonyx, E. anonyx, Cercopagis pengoi.*

COPEPODA. *Diaptomus salinus, Mesocyclops leuckarti, M. hyalinus, Halicyclops aequoreus, Cyclops viridis.*

The dominant zooplankter in the Aral Sea is the copepod *Diaptomus salinus*, but in the summer its numbers are rivalled by the larvae of the zebra mussel *Dreissena*.

At the mouth of the Amu river, which flows into the southern end of the sea, the salinity falls below 4‰. In this region, as in corresponding regions of the other brackish seas, the rotifer *Keratella quadrata* is found and is accompanied by the cladoceran *Diaphanosoma brachyurum*.

The Caspian element in the zooplankton of the Aral Sea is represented by *Cercopagis pengoi*. In recent years the Russians have made several introductions of fish into the Aral Sea, and this has resulted in the introduction of three species of Caspian Mysidacea: *Mesomysis kowalewskyi, M. intermedia* and *Paramysis baeri.*

The only indigenous planktivorous fish in the Aral Sea is the stickleback *Pungitius platygaster*, but in recent years the Baltic Herring, *Clupea harengus membras*, has been introduced. The herring has been successful in its new environment, and now grows faster and reaches a larger size than in the Baltic.

CHAPTER 6

ESTUARINE MACROBENTHOS

MANY of the surveys of estuarine faunas have been restricted to the large animals, easily visible to the naked eye and retained by a sieve with a mesh of 1 mm. This somewhat arbitrary size limit is discussed further at the beginning of the chapter on estuarine micro-fauna. Retention of the division into macrobenthos and microbenthos is a matter of convenience, but it is also possible to place whole groups of estuarine animals into one or other of the categories, so that a systematic account does not become too disjointed.

The macrobenthos may live on or in the substratum. Some groups, such as the barnacles, are exclusively epibenthic, but other groups, such as the amphipods, contain forms which remain on the surface, and forms which burrow. Two approaches may be followed in discussing such animals. All the epibenthos may be discussed together, and the burrowing forms discussed separately, or a systematic scheme can be adopted. The two approaches are complementary, but to use both would involve repetition. In this chapter the systematic approach has been adopted because in this way the ecological differences between allied species can more clearly be demonstrated. A synthesis in terms of habitat can be made using Tables 7–12 in Chapter 4, and the food webs given in Chapter 13.

ESTUARINE COELENTERATES

The coelenterates are usually divided into three classes: the Hydrozoa, Scyphozoa and Anthozoa. The only fresh-water coelenterates are found in the Hydrozoa, and it is to this group that most of the brackish water species belong. The species which has been studied in the greatest detail is the colonial *Cordylophora caspia* (Roch, 1924; Kinne, 1956, 1957). This species flourishes best at a salinity about 15‰, but it can also live in fresh water, where its proportions are somewhat different from those in brackish water, and it was described as a separate species, *C. lacustris*.

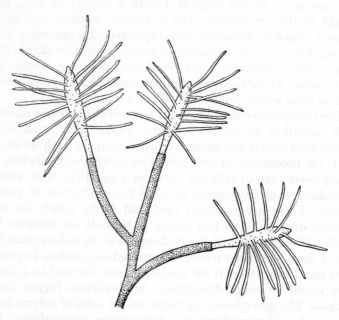

Figure 30. Part of a colony of *Cordylophora caspia* (Coelenterata: Hydrozoa). Each polyp protrudes about 2 mm from the chitinous sheath

Cordylophora is a gymnoblastic hydroid; the stolons and stalks of the colony are covered by a chitinous sheath, but this does not extend around the individual polyps. In adverse conditions the polyps may be reduced and the colony retreats within the sheath. Each polyp has a number of tentacles emerging from its surface (Fig. 30). The precise number and the length of these tentacles vary with salinity and temperature. The size and shape of the polyps also vary with environmental conditions. In general a high temperature reduces the length of the body, and at low temperatures a greater body size is reached. In fresh water the body tends to be shorter and broader than at the same temperature in higher salinities. Any increase in salinity above 20‰ causes a reduction in body length, and the polyps are more slender. These gross variations in form are reflected in cellular detail. In fresh water the ectodermal cells are tall and columnar. In an optimum salinity of 15‰ these cells are more or less cubical, but in a salinity of 30‰ the height of the ectodermal cells is greatly reduced and their general shape is flattened and expanded. In assuming this flattened shape each cell covers an increased area, so that in the more saline conditions fewer ectodermal cells are present in each polyp. At 20°C each polyp has about 22,000 ectodermal cells in fresh water, but only 4,500 in a salinity of 30‰.

The reproductive capacity of *Cordylophora* also shows variation with variation in salinity (Kinne, 1956). This variability affects both sexual and asexual reproduction. In fresh water at 16°C the production of new polyps by asexual reproduction is much lower than at salinities between 5 and 16‰. The sexual reproduction of *Cordylophora* involves the production of gono-phores. These are modified medusoid forms which do not become detached and free living, as they do for instance in *Obelia*. This retention of the medusoids on the colony may be one of the adaptations which has enabled *Cordylophora* to pene-trate into fresh water. If the medusae were liberated in a river they would be swept downstream into salinities higher than optimal. The gonophores are borne on the sides of polyps, and the number of gonophores per polyp varies with salinity. In fresh water, and in a salinity of 30‰ only one gonophore is produced at a time on a polyp. Under these conditions each gonophore produces only two or three eggs. If the salinity is about 5‰ the reproductive capacity of the gonophores is increased; a polyp may produce up to four large gonophores, each producing about nine eggs. It is clear from these results that the extremes of salinity, both sea water and fresh water, adversely affect the reproduction of *Cordylophora*.

Another indication that *Cordylophora* is not in ideal conditions in fresh water comes from the work of Roch (1924), who found that this hydroid would only survive in fresh water if well aerated. In brackish water *Cordylophora* can survive in poorly aerated water. This can be used as an example to illustrate the general hypothesis that maximum acclimatisation to one environmental variable is only possible when other conditions are optimal. Thus, when the salinity is optimal, *Cordylophora* can tolerate poorly aerated water, but in fresh water the oxygen content of the water must be optimal to enable survival in the low salinity.

Cordylophora is not the only brackish water hydroid. In European waters *Laomedea loveni* is a euryhaline marine form extending into low salinities in the Kiel Canal, and *Perigonimus megas* may be found growing with *Cordylophora* in salinities between 1 and 10‰. In Lake Pontchartrain, Louisiana, the hydroid *Bimeria franciscana* has been found in great abundance (Crowell and Darnell, 1955). This species will tolerate salinities from 1 to 35‰ in the laboratory, and can also survive temporary, but not permanent, exposure to fresh water.

In the warmer parts of the world various members of the family Moerisiidae have penetrated into brackish water. The members of this family release their medusae, so that they are more likely to be successful in brackish lakes or lagoons than in rivers. The family has been monographed by Valkanov (1938, 1953) and examples of the group are *Moerisia lyonsi* from Lake Quarun in Egypt, and *Halmomises lacustris* from Trinidad.

The Scyphozoa are mostly pelagic or planktonic, but most do have an attached stage, the scyphistoma, which may penetrate into estuaries. The most successful in this respect seems to be *Aurelia aurita*, the scyphistoma of which can tolerate low stable salinities down to 6‰ in the Baltic (Segerstrale, 1957).

Few members of the Anthozoa have penetrated into low salinities. Several marine species such as *Actinia equina* can tolerate salinities down to about 8‰ provided that the salinity is stable, but cannot dwell permanently in water of variable salinity. There are a few species which can dwell permanently in quite low salinities, although none have become adapted to fresh water. One of the most interesting of these species, *Metridium schillerianum*, has been described in detail by Annandale (1907). This form was found in brackish ponds near Port Canning in Lower Bengal. The adult anemones were found buried in mud at the bottom of ponds where the salinity sometimes fell as low as 2‰. The column of the fully grown

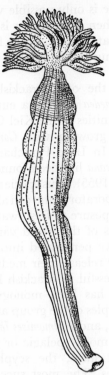

Figure 31. Metridium schillerianum (Coelenterata: Anthozoa), actual length 12 cm
(after Annandale)

individual was found to be greatly elongated, resembling the
burrowing anemone *Cerianthus*. The largest individuals were
found buried in the mud as far as the bases of their tentacles.
The young specimens had much shorter columns and attached
themselves to solid objects, but were also found in large numbers inside the water channels of the sponge *Spongilla cerebellata*,
which is a member of a fresh-water genus. *Metridium schillerianum* appears to be the only member of the Anthozoa to
have penetrated so far into brackish water that it overlaps the
seaward penetration of a fresh-water sponge.

Another anthozoan has been found in brackish pools on the
Isle of Wight (Stephenson, 1935) and in some brackish lagoons
near the mouth of the River Ore (Robson, 1957). The range

of salinity from which this small anemone, *Nematostella vectensis*, has been recorded extends from 8·7‰ to 37·7‰. *Nematostella* is peculiar in the possession of small bodies, or nematosomes, which lie free in the body cavity and move about under the influence of the cilia on the mesenteries of the anemone. These small bodies are densely packed with nematocysts, but their function is unknown.

ESTUARINE TURBELLARIANS

The majority of estuarine turbellarians are small enough to be included in the microfauna, but there are a few species in Britain which are large enough to be regarded as members of the macrofauna, and one of these, *Procerodes* (= *Gunda*) *ulvae*, has been the subject of some classical studies (Pantin, 1931a and b; Beadle, 1934). *Procerodes* lives in small streams which flow across the intertidal zone. This means that at low tide the worms are subjected to fresh water and at high tide to sea water. Survival of the worms in fresh water depends on the presence of calcium salts. These presumably act by reducing permeability. If the water is too soft the osmotic uptake by the worms becomes excessive and they swell until they disintegrate. The normal sequence of events when *Procerodes* is immersed in a hard fresh water is that it swells rapidly in the first 20 minutes, so that the volume may increase by 70 per cent. After this initial swelling the volume is slowly reduced.

In its normal habitat *Procerodes* does not have to cope with the problem of living permanently in fresh water. When immersed in fresh water the worms lose salts until the concentration of the tissue fluids is only about one-tenth that of sea water. When the tide returns and the worm is surrounded by sea water the loss of salts is rapidly made good.

By losing salts when in fresh water *Procerodes* reduces the concentration difference between the tissues and the external medium, so that the osmotic inflow of water is reduced to a level within the capacities of the excretory system. The gut cells play a part in dealing with the inflow of water; large vacuoles appear in these cells when the worm is in fresh water and disappear when it is transferred to sea water.

Transference of *Procerodes ulvae* from sea water to fresh water is accompanied by an increase in oxygen consumption (Beadle, 1931), and this has been interpreted as a reflection of the increased osmotic work performed by the animal.

Another flatworm found in British estuaries is the red *Uteriporus vulgaris*. This species has not been recorded very often, probably because it lives on the surface of damp soil at the base of salt marsh plants, and has been overlooked. The osmotic problems faced by this species differ from those faced by *Procerodes*. The worms in the salt marsh are not subjected to the complete range from fresh to sea water, but have to face the problem of living in damp air for several hours each day, and may be subjected to osmotic stress in heavy rain. The physiology of this species has not been studied, but would probably make an interesting comparison with *Procerodes*.

The flatworms are hermaphrodites and lay their eggs in protective capsules which may be stuck on the undersides of stones. The young of *Procerodes* emerge as miniatures of the adult, so that there is no larval stage. Some of the marine flatworms have planktonic larvae, but these are not advantageous in estuaries.

ESTUARINE ANNELIDS

If faced with the choice of a single polychaete to represent the estuarine fauna, most zoologists would undoubtedly choose a nereid, and if they were European the choice would fall on *Nereis diversicolor*. This species reaches its greatest abundance in estuarine muds and thrives in conditions of varying salinity. In recent years its ecological distribution has been studied particularly by Smith (1955a, 1955b, 1956, 1964), but there have also been many earlier studies on the distribution of this species, some of which have been summarised by Spooner and Moore (1940).

An apparent paradox in distribution is found in the Baltic Sea. In the Gulf of Finland, Smith (1955a) found that *N. diversicolor* was restricted to regions where the salinity remained above 4‰, and yet in many estuaries in Britain this species is found where the salinity falls well below 1‰. One might have thought that in an area with fairly stable salinity this species would have been able to penetrate further towards fresh water. The anomalous distribution in the Gulf of Finland is probably caused by a combination of low temperature and low salinity at the time when *N. diversicolor* breeds.

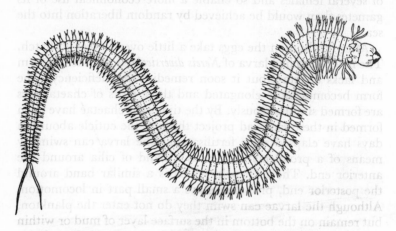

Figure 32. Nereis diversicolor (Annelida: Polychaeta), actual length 10 cm

In the South of England *Nereis diversicolor* spawns in February, as the water temperature rises above 5°C (Dales, 1950). The spawning period on the south coast of Finland has not been determined, but from such data as are available it seems that the gametes are released early in May, a week or two after the break-up of the ice.

Females of *N. diversicolor* release their eggs by rupture of the body wall and die after spawning. Males appear to release their spermatozoa through the nephridia. In both males and females there is considerable histolysis of the body wall before spawning, so that the worms become abnormally fragile and liable to burst when handled.

Males are generally less abundant than females. Several authors have noted this low proportion of males, and at Chalkwell in Essex, Dales (1950) found that the percentage of males varied between 1 and 10. Higher ratios, up to 33 per cent males, have been found in other areas. The reason for this abnormal sex ratio is unknown, but it may be connected with reports of a form of sexual congress. Several authors have recorded finding tangled masses of several females around a single male. If this phenomenon occurs regularly it would enable a single male to release its spermatozoa in the proximity of several females and so enable a more economical use of its gametes than would be achieved by random liberation into the sea.

After fertilisation the eggs take a little over a week to hatch. The newly hatched larva of *Nereis diversicolor* is rounded in form and lacks chaetae, but it soon remedies this deficiency. The form becomes more elongated and three pairs of chaetal sacs are formed simultaneously. By the time that chaetae have been formed in these sacs and project through the cuticle about ten days have elapsed since fertilisation. The larva can swim by means of a prototroch, which is a band of cilia around the anterior end. The telotroch, which is a similar band around the posterior end, probably plays a small part in locomotion. Although the larvae can swim they do not enter the plankton, but remain on the bottom in the surface layer of mud or within the parental burrow. After the first three pairs of chaetal sacs have been developed the internal anatomy of the larva becomes more highly differentiated. The gut is formed and a muscular pharynx with jaws is developed in the region of the first chaetigerous segment. At this stage tentacles begin to develop on the head. For the first seven weeks of life the larva does not feed, but relies upon its yolk.

After the non-feeding period the larva becomes increasingly active. It is probably at this stage that a certain amount of active dispersal takes place. These active larvae have about ten chaetigerous segments and are 2 mm long. When they have doubled this size they begin to behave like adults. Specimens with length of 4 mm and a body with 20 segments will burrow in mud and construct U-shaped tubes which they irrigate by undulating their bodies.

The whole larval life of *N. diversicolor* can be spent in the surface layers of mud. Avoidance of a planktonic stage can be regarded as an adaptation to estuarine life. To enter the plankton would involve the risk of being swept out to sea,

E

away from the region where *N. diversicolor* can thrive in the absence of its marine competitors which do not have the same powers of osmotic regulation.

The superior osmoregulatory powers of *N. diversicolor* are demonstrated most readily by comparing it with allied species. The simplest method of comparison is to place the worms in diluted sea water and then weigh them at intervals. If *N. diversicolor* is placed in 20 per cent sea water its weight increases because the osmotic intake of water exceeds the rate of excretion. After several hours the weight begins to fall and eventually is brought to a level just a little above that at the start of the experiment. If *Perinereis cultrifera* is treated in the same way it is found to be incapable of reducing its weight. It has been found that *N. diversicolor* can tolerate fresh water for several days, while *Nereis virens* dies within an hour when similarly treated.

More elaborate experiments, using radioactive sodium (^{24}Na) as a tracer, have shown that *Perinereis cultrifera* is about three times as permeable to sodium as *Nereis diversicolor* (Fretter, 1955). Similar experiments, using radioactive chloride (^{36}Cl) showed that *Nereis virens* and *N. pelagica* were also roughly three times as permeable as *N. diversicolor* (Jørgensen and Dales, 1957). More recent experiments by Oglesby (1965a and b) show that *N. limnicola*, which can live in fresh water, has a much lower rate of chloride exchange than *N. succinea* and *N. vexillosa*. At very low salinities *N. limnicola* showed no exchange over a period of 4 hrs. It seems that a reduction in permeability is an important part of the adaptation of estuarine nereids to the fluctuations in salinity which they encounter.

The importance of permeability is also indicated by the experiments of Beadle (1937) and Ellis (1937) who showed that if calcium is removed from the experimental medium there is a reduction in the ability of *Nereis diversicolor* to regulate its weight in dilute media. Calcium ions are well known for their effect in reducing the permeability of biological membranes (see also p. 116 on *Procerodes*). But it is also clear from the work of Beadle (1937) that some active process is also involved. When *N. diversicolor* is treated with cyanide, to inactivate its respiratory enzymes, worms lose the ability to regulate their weight in dilute media.

The precise nature of the active regulation of water and ions in nereids is still undetermined. At present the major differences between species are related to permeability. Oglesby (1965b) has suggested that food might be a major source of salts, and that these salts are conserved when the worm is in dilute media by the production of a urine which is hyposmotic to the blood.

Quite apart from differences in ability to regulate their internal environments the different species of *Nereis* may also vary in the tolerance which their tissues show towards changes in the fluids surrounding them. Isolated muscles of *N. diversicolor* will function in lower salinities than those of *N. pelagica* (Wells and Ledingham, 1940). Although *N. diversicolor* regulates its water content with some precision over a very wide range of salinities, it is not so precise in its regulation of the concentration of its blood. When the external sea water medium has a chloride content over 200 m equiv. the ionic concentration of the blood varies directly with that of the external medium, at least up to 400 m equiv. Cl. At low salinities the concentration of blood is kept above that of the external medium, but the concentration of the blood may fall as low as 100 m equiv. Cl. A normally functioning *N. diversicolor* may thus have body fluid which varies over a fourfold range of total ionic concentration, so that the tissues have need of the tolerance to change which was shown by the isolated muscles in the experiments of Wells and Ledingham.

Although *Nereis diversicolor* can tolerate very low salinities it cannot live and breed in fresh water. The larvae cannot tolerate such low salinities as the adults. The closely allied *N. limnicola*, which lives in the fresh-water lake Merced in California, as well as in estuaries, can breed in fresh water. This ability is achieved by the assumption of viviparity. Eggs are retained in the coelom, and the young are bathed by coelomic fluid, the concentration of which does not fall below 150 mM/Cl/(= 27 per cent sea water), so that they are not subjected to the osmotic stress of the dilute external medium. By the time the young are released from the parent coelom they have developed osmoregulatory abilities comparable to those of the adult.

Rees (1940) noted that there was an inverse relation between the numbers of *Nereis diversicolor* and the numbers of certain members of the microfauna of a mud flat in South Wales. An examination of the gut contents of small *N. diversicolor* by Perkins (1958) revealed that small copepods, ostracods, nematodes and Foraminifera were included in the diet, along with diatoms and organic debris. The inverse relationship found by Rees was probably due to the feeding activities of *Nereis* reducing the numbers of the microfauna.

Large specimens of *N. diversicolor* have a wide range of foods. In a small, drying salt marsh pool I have seen large specimens seizing the dying young of the grey mullet and dragging them head first down into the mud. At the other extreme, Harley (1950) has described a mechanism by means of which *N. diversicolor* can filter small particles from suspension. The mechanism involves the secretion of a mucous net near the entrance to the burrow. The burrow is then irrigated by undulations of the body so that a current of water passes through the net. Small particles carried by the current are trapped on the net, and when a sufficient quantity has accumulated the net and the particles are eaten. This cycle of net secretion, irrigation and ingestion has been seen to be repeated at fairly regular intervals when *N. diversicolor* was kept in a glass tube in the laboratory, but it is not known how much this mechanism is used in nature.

Many points of contrast to *Nereis* are found in the lugworms of the genus *Arenicola* which are often abundant in muddy sand. These worms feed by ingesting sand and digesting any organic matter that may be present. A clean sand will not provide sufficient food, so that coarse clean sands are often without *Arenicola*.

The anatomy and mode of life of *Arenicola* has been the subject of special study by Wells (1937, 1945, 1949a, 1949b, 1950, 1952, 1957, 1959, 1962), and the species which has received the greatest attention is *A. marina*. The body of this species is formed of three regions: head, trunk and tail. The head is roughly conical extending forwards from the first chaetigerous segment. The mouth opens anteriorly on the head, and an eversible proboscis is present. The middle region or trunk consists of 19 segments (rarely 20), all of which bear chaetae, but the first 6 lack gills. Segments 7 to 19 bear hollow contractile gills in a dorso-lateral position above the parapodia. The tail is narrower than the trunk and consists of up to 60 or 70 segments lacking chaetae and gills. Many specimens of *A. marina* have fewer than 60 tail segments because they lose parts of their tails to predatory birds which attack them as they move tail first to the sand surface to defaecate.

Figure 33. Arenicola marina (Annelida: Polychaeta), actual length 16 cm

Arenicola marina lives in a burrow in sand, and may remain in the same burrow for many months. There is some variation in the form of the burrow, but in general it is L-shaped with a tail shaft which continues downwards into a gallery with walls that are consolidated by secretion from the surface of the worm. The lower end of the gallery turns more or less horizontally and ends against the head shaft which is not open but filled with sand. As the worm ingests sand from the lower end of the head shaft the sand above slowly sinks downwards, leaving a depression at the surface. At intervals of about 45 minutes the worm moves backwards up to the tail shaft and defaecates on to the surface of the sand, depositing cylindrical worm casts that are common on sandy beaches with a slight admixture of silt. After defaecation the worm descends again into the gallery.

The normal activity of *A. marina* in the gallery includes intermittent bursts of irrigation. During this process the worm lies with the ventral and ventro-lateral surfaces in contact with the walls of the gallery, but with a gap between the dorsal surface and the gallery wall. Waves of swelling pass along the body, closing the dorsal gap and pushing water through the burrow. The normal direction of travel of this swelling is from the tail towards the head. This means that water is driven from the gallery up through the head shaft. This forwardly directed current helps to keep the sand in the headshaft soft, but it may also serve to increase the supply of available food. The water drawn in down the tail shaft will contain some organic particles in suspension, including algae. Now the sand at the base of the head shaft will form an efficient filter, retaining these small particles, so that when the sand is ingested by the worm it will be enriched with food brought in by the irrigation current (Krüger, 1959).

The sand is ingested by repeated extrusion and retraction of the proboscis, which does not bear jaws. This process is rhythmic and is controlled by a pacemaker mechanism in the oesophagus. If the proboscis is isolated with a short length of oesophagus and kept in sea water it will remain active for over 24 hours and show a rhythmic pattern of contraction alternating with periods of rest. This feeding rhythm interacts with the irrigation and defaecation cycles. Wells (1949b) found that when the proboscis was active there was inhibition of the body wall contractions in the trunk segments responsible for creeping and irrigation.

When the tide retreats the possibility of irrigating the burrow is removed. This does not have any serious effect on *A. marina*, for several reasons. First, there is the possibility of the worm creeping tail first up the tail shaft and trapping bubbles of air which are passed forwards by irrigation type movements. Secondly, the haemoglobin of the blood is capable of taking up oxygen from the surrounding water at very low oxygen concentrations. Thirdly, *Arenicola marina* has a considerable resistance to anaerobic conditions, being able to survive for about 9 days in the absence of oxygen (Hecht, 1932). This ability enables it to penetrate upshore to such an extent that it may occur above the high water of neap tides and may not be covered by the tide for several days. Dales (1958) found that when subjected to anaerobic conditions *A. marina* utilised glycogen as an energy source, but did not accumulate any lactic or pyruvic acid, which are often end products of anaerobic metabolism. This lack of accumulation of lactic acid after a period of anaerobiosis accounts for the finding by Borden (1931) that *A. marina* does not show an increased rate of oxygen consumption when returned to aerobic conditions after suffering a period of oxygen lack.

If *Arenicola marina* is removed from its burrow and then placed on the surface of wet sand it will burrow back into the sand. The first movements are relatively gentle probings of the sand with the proboscis directed downwards. These movements bury the first few chaetigerous segments, and then the worm begins a series of powerful contractions. First the circular muscles contract, and this is followed by a contraction of the longitudinal muscles of the trunk. This sequence is repeated at intervals of 5 to 7 seconds. These contractions bring a high hydrostatic pressure to bear on the substratum. Truman (1966) has recorded hydrostatic pressure peaks up to 110 cm of water during burrowing. The normal pressure in the coelom of a resting *Arenicola* is about 2 cm of water. During the peaks of pressure the anterior end of the worm is forced forwards, and the pressure also serves to anchor the anterior end in the sand. Subsequent longitudinal contractions draw the worm into the sand. *Arenicola* is structurally adapted to produce high hydrostatic pressures by lacking septa through most of the trunk. This allows the pressure produced by the posterior trunk muscles to be transmitted to the anterior end, and it spreads the work of producing a high pressure over all the body wall muscles of the trunk.

The spawning of *Arenicola marina* in Britain has been surveyed by Duncan (1960). Most populations spawn in autumn, with maximum spawning on neap tides in October, November or December. Some populations spawn epidemically, with all the worms spawning in a few days, but in other populations the spawning period may extend over several weeks. At Whitstable in the mouth of the Thames Estuary, the spawning period was sharply defined, lasting 14 days between the new moon and full moon spring tides in the second half of October, with a maximum at the intervening neap tides (Newell, 1948). This autumn spawning may be triggered off by the first sharp fall in temperature, and the similarity in the dates of spawning in different years may be due to the fact that the first sharp decrease in temperature tends to take place in the same two or three weeks each year.

There are some exceptions to the general rule of autumn spawning. Howie (1959) found that a population at Fairlie Sands, near Millport, spawned in the spring, and gave evidence which indicated that there may be other populations of spring spawners.

The larvae hatch from the eggs about 4 or 5 days after fertilisation. In shape the newly hatched larva resembles a pear, broader anteriorly than posteriorly. There is a small apical tuft of cilia, a large prototroch encircling the anterior part of the body, and a telotroch near the end of the body. Ventrally there is a longitudinal cilated band between the prototroch and telotroch. Near the anterior end of the larva there are two dark brown eyes. At this stage there are no chaetae. When the larva reaches an age of about 14 days it has developed two chaetigerous segments.

Although the larva of *Arenicola marina* can swim it rarely does so, and never enters the plankton but spends its time in silty deposits. The young worms are carried up the shore by the tide, and it is a common feature to find a distinct zone of young *A. marina* higher on a sandy beach than the region where the mature individuals are found. The juvenile worms and the adults are capable of leaving their burrows and swimming for some distance, then forming new burrows. On beaches with an upper zone of small worms, the population on the lower shore must be maintained by the migration of small worms from the upper shore. Two years normally pass before *A. marina* spawns for the first time and, after spawning, about 40 per cent of the adults die (Newell, 1948).

Arenicola marina does not penetrate as far into brackish water as *Nereis diversicolor*, but it can tolerate reduced salinities down to about 8‰. The tolerance of *A. marina* to reduced salinity is not related to any capacity for regulating the concentration of its blood. Schlieper (1929) found that the blood was isosmotic with the external medium over a wide range of concentrations. More recently Duchateau-Bosson, Jeuniaux and Florkin (1961) have shown that there is adjustment of the amino-acid concentrations in the muscle cells of *A. marina* when kept in 50 per cent sea water. The major adjustments are made in the concentrations of alanine and glycine. By reducing the concentrations of these amino acids the osmotic concentration within the muscle cells is brought down to a level similar to that of the blood when the worm is in brackish water. This reduces the

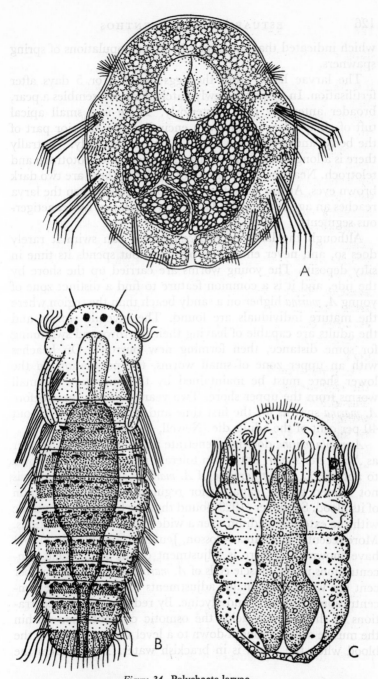

Figure 34. Polychaete larvae

A, *Nereis diversicolor,* larva with three chaetigerous segments (after Dales). B, *Pygospio elegans,* late larva with 11 chaetigerous segments (after Thorson). C, *Eteone longa,* metatrochophore with the beginnings of three segments (after Rasmussen) danger of osmotic swelling of the muscle cells. Clearly there must be a limit to this method of tolerance, and in *A. marina* this limit is reached when the salinity of the external medium falls to 8‰. The active method of regulation of *Nereis diversicolor* is clearly superior, and by having the capacity to regulate the concentration of the blood in dilute media *N. diversicolor* has overcome a major problem in penetrating from the sea into fresh water. *Nereis limnicola* finally solves the complete problem by overcoming viviparous, so protecting its young from the osmotic stress which appears to prevent *Nereis diversicolor* from breeding in fresh water.

Other brackish water annelids have not been studied in the same detail as *Nereis* and *Arenicola*, but there are numerous species which penetrate at least a short distance into estuaries. One of the better known species is the catworm, *Nephthys hombergi*. Several species of the genus *Nephthys* live in marine sands, but around Britain *N. hombergi* penetrates further into estuaries than the other species. This species does not construct a permanent burrow, but moves through the sand in search of prey.

In general form *Nephthys* resembles *Nereis*, but if one handles the two worms, *Nephthys* appears to be much more muscular. This difference may be connected with the difficulty that the worm may encounter in moving through sand, which tends to be unstable and hard to penetrate. If *Nephthys* is removed from the sand and placed on the surface it will rapidly bury itself again. At first the worm executes vigorous undulatory swimming movements which drive the head into the sand. Once the first few segments have penetrated the sand the undulatory movements stop. The worm then shows a series of longitudinal contractions of the body which force the coelomic fluid anteriorly and evert the proboscis.

Nephthys is remarkable in lacking circular body wall muscles. When the longitudinal muscles contract the body wall is prevented from swelling by the combined action of dorso-ventral muscles and a series of ligaments which brace the body wall (Clark and Clark, 1960). The first 34 segments, which house the proboscis when it is retracted, are without septa. This allows the transmission of pressure from the most posterior of these segments to the anterior end of the worm. The violent extrusion of the proboscis forces a hole in the sand, enabling the worm to penetrate.

Near the mouths of estuaries two worms, *Scoloplos armiger* and *Heteromastus filiformis* eat mud in much the same manner as *Arenicola*, by extrusion and retraction of the proboscis. The former species belongs to the family Ariciidae, and produces balloon-shaped egg capsules which are anchored in the sand by a stalk. Each capsule contains about a thousand eggs. The larvae remain in the capsule and hatch in a crawling stage with a number of well developed segments. There is no planktonic stage.

Heteromastus filiformis is a representative of the Maldanidae or bamboo worms. The latter name is given because the worms have a relatively small number of elongated segments with slight swellings resembling those of a cane. The development of *Heteromastus* is not properly known, but two other members of the same family have been studied; *Axiothella mucosa* has been studied by Bookhout and Horn (1949), and *Clymenella torquata* has been studied by Newell (1951). The former species lays its eggs in jelly cocoons during a lengthy breeding season extending from April to October. The larvae remain within the cocoon for about two weeks until they have acquired twelve chaetigerous segments. When they leave the cocoon the young worms make tubes in the sand and adopt the adult mode of life. *Clymenella torquata* does not lay its eggs in cocoons, but has a short breeding season of only two or three days in May when the gametes are shed near the mouths of the tubes. The young worms spend the whole of their developmental period in the benthos, again omitting any planktonic stage. The first larval chaetae appear on the fifth day, and by the time that the worms are four or five weeks old they have developed the adult number of 22 segments and construct tubes like those of the adults.

Another burrowing polychaete, *Pygospio elegans*, collects its food with two long grooved tentacles. These tentacles are ciliated and are waved about in the water or moved across the surface of the sand in which the worm burrows. Small particles, including diatoms, are collected in the grooves and transported towards the head. The bases of the tentacles are wiped across the underside of the proatomium at intervals to transfer the collected material to the mouth. This species sometimes occurs in enormous numbers in estuarine sands; Thamdrup (1935) found densities up to 20,300 per sq m on the Danish coast.

In stable brackish waters *P. elegans* penetrates down to a salinity of 8‰ (Remane, 1958), and from observations on the population in the Gwendraeth Estuary it appears to be able to tolerate salinities as low as 2‰ for short periods.

A female *Pygospio* may produce up to 16 egg capsules in the spring breeding season. Each capsule is anchored to a sand grain by a thin thread. When first laid each capsule contains about 50 eggs, but only 2 to 9 of these develop, the others serve as nurse cells which disintegrate and act as nourishment for the embryos. There seems to be considerable variation in the stage of development of the larva when it leaves the egg capsule. If 9 larvae develop in a capsule they emerge at an earlier stage of development than when only two larvae develop in a capsule. Thorson (1946) suggests that sometimes a single larva may develop in a capsule, and that when this happens the larva does not become planktonic but passes through its early stages in the benthos. The smaller larvae normally become planktonic and are often a prominent component of the temporary plankton of estuaries. Quite early in development the larva develops the two tentacles characteristic of the adult.

ESTUARINE NEMERTEA

The nemertines have a superficial resemblance to very elongated flatworms. They have a ciliated ectoderm, and there is usually no body cavity around the viscera. There is, however, a blood system, and in some species the blood contains haemoglobin. The outstanding characteristic of the nemertines is the elongated eversible proboscis, which may reach a third the length of the body, and is used in the capture of prey. All the nemertines are carnivores. Although they are common and widespread in the sea there are few species found in estuaries, and these have not been studied much from the point of view of the ways in which they cope with the estuarine environment.

Lineus ruber is the best known and most widespread of the

estuarine nemertines (Coe, 1943). The geographical range of this species extends from Alaska to South Africa, and it is common on the shores of Europe. The lower limit of salinity tolerance of *L. ruber* appears to be about 8‰ (Remane, 1958), but there does not appear to have been any detailed experimental work on this topic.

When mature *L. ruber* is about 10 cm long, sometimes reaching a length of 20 cm. The sexes are separate. During the breeding season the worms come together in pairs and secrete a layer of mucus into which the female deposits her eggs, and the male fertilises them as they are laid. The embryos pass though a 'larval stage' within the gelatinous mass and emerge in a crawling stage. By this means the planktonic larval stage found in some of the more strictly marine nemertines is omitted.

Apart from *L. ruber* the only nemertine found well inside the Gwendraeth Estuary is a species of *Tetrastemma*. This genus is sometimes called *Prostoma*, but some authors restrict this name to the few fresh-water species. Some species of the genus *Tetrastemma* are hermaphrodites, and one species *T. obscurum* is known to be viviparous (Thorson, 1946), so that this makes it a likely candidate for entry into an estuary. The fresh-water species *Prostoma rubrum* produces eggs in a gelatinous sheath, and from this larvae like small ciliated flatworms emerge and swim for a short while before adopting a creeping mode of life. This may represent a secondary development in a species which has established itself in static fresh waters and is no longer faced with the dangers of estuarine life, where a free swimming larval stage is likely to be disadvantageous.

ESTUARINE ECTOPROCTA

The ectoprocts bear a superficial resemblance to colonial hydroid coelenterates, but they are coelomate animals, with a U-shaped gut having an anus opening outside the circle of tentacles surrounding the mouth. Most species are enclosed in an exoskeleton which may be horny or partly calcified. In some of the encrusting forms the individual exoskeletons fit closely together and form regular geometric patterns. The tentacles can be retracted within the shelter of the exoskeleton when the colony is uncovered by the tide.

The most widespread of the estuarine ectoprocts is *Victorella pavida*, which was first described from the Victoria Docks in London, where it was found entangled with *Cordylophora*. *Victorella* has since been found in many different parts of the

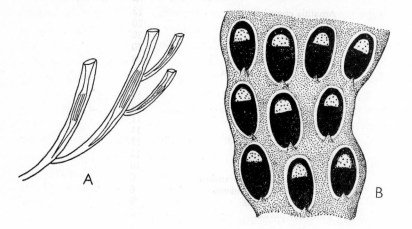

Figure 35. Colony form in estuarine ectoprocts
A, *Victorella pavida*, showing resemblance to a small hydroid colony. B, *Membranipora crustulenta*, an encrusting form, each polyp is housed in a separate case or zooecium

world, from Egypt to Japan, and from Australia to Brazil (Brattstrom, 1954). A closely related, or possibly identical, species, *V. bengalensis*, is found at the mouth of the Ganges entwined with the hydroid *Irene ceylonensis*. Another form, known as *Tanganella symbiotica*, which may be merely a variety of *Victorella pavida*, has been found in the fresh waters of Lake Tanganyika growing among fresh-water sponges.

The salinity tolerance of *Victorella pavida* appears to extend from fresh water up to about 27‰. In Chesapeake Bay it was found at salinities between 3 and 27‰, and was most abundant at salinities between 10 and 12‰ (Osburn, 1944). The salinity tolerances of other Ectoprocta occuring in the same area are given in table 17.

TABLE 17

LOWER LIMITS OF SALINITY TOLERANCE OF ECTOPROCTS
IN CHESAPEAKE BAY

(Based on data given by Osburn, 1944)

Species	Lowest salinity at which found in Chesapeake Bay ‰
Amathia convoluta	22
Microporella ciliata	20
Alcyonidium polyoum	20
Bugula turrita	20
Schizoporella unicornis	18
Alcyonidium parasiticum	15
„ verrilli	13
Anguinella palmata	13
Aeverrillia armata	12
Amathia vidovici	11
Conopeum truitti	11
Electra pilosa	11
Hippothoa hyalina	11
Bowerbankia gracilis	10
Membranipora membranacea	6
„ crustulenta	6
Acanthodesia tenuis	6
Victorella pavida	3

Victorella pavida broods its eggs in small groups in the coelom. The larvae that develop from these eggs are rounded and ciliated, but they do not swim for more than an hour or so before they metamorphose into the adult structure. The colonies of *V. pavida* in temperate regions die back in the winter, leaving contracted dormant bodies which give rise to new colonies in the spring.

Another widespread estuarine species, *Membranipora crustulenta*, has different reproductive habits. This species produces larger numbers of small eggs which are not brooded and develop into cyphonautes larvae (Fig. 36). These larvae swim and feed in the plankton for up to two months before settling.

The salinity tolerance of *M. crustulenta* seems to vary with its geographical distribution. In Chesapeake Bay it was found in salinities down to 6‰, but in Tunis it has been found in two small fresh rivers (Remane, 1958). The polyps survive through the winter in the Baltic, but in the spring the old polyps degenerate and new ones develop. During the period from April to June there is active sexual reproduction. After this period there is a period of asexual reproduction in which there is repeated budding of the zooids to form large colonies. In good conditions the asexual reproduction of *M. crustulenta* can proceed at a very high rate. Osburn (1944) records that a colony 50 mm across was formed from a single larva in four weeks.

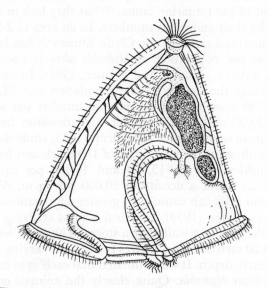

Figure 36. Cyphonautes larva of *Membranipora* (after Atkins)

When abundant, *M. crustulenta* can become a pest in oyster beds. The encrusting colonies render the surfaces of stones and other solid objects unsuitable for the settlement of oyster spat.

Competition for space sometimes occurs between *M. crustulenta* and another ectoproct, *Acanthodesia tenuis*. This species can also tolerate salinities down to 6‰, and grows so rapidly that it can cover over growing colonies of *M. crustulenta*. A small colony of *Acanthodesia* may produce two buds a day in summer. The settled cyphonautes larva produces an ectoproct polyp which serves as the founder, or ancestrula, of a new colony. *Acanthodesia* is remarkable in that the ancestrula is twinned, so that two polyps are formed simultaneously as founder members. It may be that this twinning of the ancestrula is related in some way to the rapid growth rate of this species, which gives it an advantage over its competitors.

ESTUARINE GASTROPODS

The small snails of the family Hydrobiidae form an important component of the estuarine fauna. What they lack in size they make up by their enormous numbers. In an area of 2·5 square miles of intertidal sands in the Clyde Estuary it has been estimated that the population of *Hydrobia ulvae* is about thirty thousand millions (Hunter and Hunter, 1962). In one region of the estuary the density of the population was 42,012 per sq m, but in most regions the normal number per sq m lay between 5,000 and 9,000. Similar high densities have been recorded from other areas: Thamdrup (1935) studied the mud flats at Skalling on the west coast of Denmark and frequently found densities between 15,000 and 46,300 per sq m, and exceptionally found a density of 60,000 per sq m. Abnormal weather and tides can cause even greater aggregations of these small snails. Linke (1939) described a mass of living *H. ulvae* found at the base of a wall after a storm. The snails were piled in a band 20 metres long, 2 metres broad, and varying between 2 and 20 cm in depth. It was estimated that each sq m contained over a million *Hydrobia*. Quite clearly the animals could not live long at this density and would have to disperse again as soon as the weather permitted.

The dispersal of *Hydrobia ulvae* is aided by the habit of floating on the underside of the surface film of the water. Newell (1962) has described how *Hydrobia* at Whitstable creep up the sides of sand ripples and float in the water contained between them. When the incoming tide flows over the ripples the snails are carried upshore. While floating at the surface *Hydrobia* secretes a mucus raft, which not only serves as a flotation device, but also traps small particles such as diatoms which are then eaten by the snail. When the tide ebbs the snails detach from the surface film, withdraw into their shells and sink to the bottom. Here some time is spent crawling and browsing on detritus and diatoms. This phase is following by a period of burrowing under the mud surface, but before the next incoming tide *Hydrobia* emerges and crawls up the sides of the ripples to float in the shallow water. By ascending to the surface on the incoming tide *Hydrobia* can take advantage of the food available in the plankton and by descending early on the ebb it can avoid being carried away from its preferred level on the shore.

Several species of the Hydrobiidae are found in estuaries and show differences in their ecological preferences. In Denmark it was found by Muus (1963) that the approximate salinity ranges of four species were as follows:

Potamopyrgus jenkinsi	0–15‰
Hydrobia ventrosa	6–20‰
Hydrobia neglecta	10–24‰
Hydrobia ulvae	10–33‰

Apart from these differences in salinity range there were differences in the type of habitat in which each was dominant. *Hydrobia neglecta* was most abundant in rich vegetation, while *H. ulvae* was more abundant on bare exposed areas. Sheltered, vegetation-covered areas were also preferred by *H. ventrosa*, and when the salinity was about 15‰ the numbers of this species were about equal to those of *H. neglecta*. At lower salinities *H. ventrosa* became the more abundant of the two.

The salinity ranges found in one area do not necessarily apply to other areas. There is some evidence that there are physiological races of *Hydrobia ulvae* which differ in their salinity tolerance. Specimens collected from Barton Marsh in Cheshire could tolerate a salinity of 1·7‰ while specimens from Ireland died in a salinity of 7·7‰ (McMillan 1948).

The upstream distribution of *Hydrobia ulvae* has recently been studied in the River Crouch by Newell (1964). In this estuary there is only a slight vertical salinity gradient (less than 2 per cent sea water). In the upper reaches of the estuary *H. ulvae* is confined to the higher parts of the intertidal zone. This restriction is to be expected if the distribution is influenced by salinity. In an estuary without a strong vertical salinity stratification the upper parts of the shore will be subjected to higher and less variable salinities (see Chapter 2). An interesting feature of Newell's work is that it shows how the distribution can be governed by the behaviour of the animal and not by the direct effect of a lethal factor. The habit of floating is restricted at low salinities. A floating snail closes its operculum and sinks if the medium is diluted. Snails which are subjected regularly to low salinities will not adopt the floating habit and they maintain their position on the shore by crawling and burrowing. By not floating they avoid subjecting themselves to the risk of being transported to an area where they may encounter lethal dilutions of the medium.

The upshore distribution of *Hydrobia ulvae* is governed by its tolerance of desiccation. Stopford (1951) found that this species began to die after four days in dryness, although some individuals could survive for about two weeks. This indicates that the upper limit will be about the level of the high water of spring tides which cover the upper shore once a fortnight, but in general the snails will find better conditions lower on the shore.

The food of *Hydrobia ulvae* varies from one locality to another. The specific name *ulvae* implies an association with the green alga *Ulva* and indeed the snail will feed on this alga when the opportunity arises, but the association is not specific, and various species of *Enteromorpha* as well as blue-green algae and diatoms are eaten. On bare mud flats the food seems to consist mainly of bacteria attached to organic debris (Newell, 1965).

The breeding season of *Hydrobia ulvae* extends over several months. In the North Sea most eggs are laid in May and June, but in the Plymouth area there are records of breeding in February and March as well as in the autumn. The eggs are enclosed in a gelatinous capsule which is usually fixed to the shell of another *Hydrobia*. Each capsule contains between 3 and 25 eggs, and a female will produce a number of such capsules during the course of a season, so that her total output is of the order of 300 eggs (Linke, 1939). Veligers emerge from the eggs after two or three weeks, and may spend a month in the plankton, but this is not certain. Smidt (1944) for instance thinks that the larval stage may be suppressed in some areas. This would be advantageous to an estuarine animal because entering the plankton involves the risk of being carried out to sea on the ebb tide. In *Hydrobia ventrosa* the larval stage is suppressed and each capsule contains a single egg. These capsules are attached to stones and the emerging animal is a miniature of the adult.

The mode of reproduction of *Potamopyrgus jenkinsi* is even more advantageous to an estuarine animal. The species is viviparous and parthenogenetic. Its success is emphasised by its sudden appearance and spread in Britain during the latter half of the nineteenth century. The first records were from brackish water (Smith, 1889), but these were soon followed by records from fresh water, and it has now been recorded from most parts of England and Wales, as well as several parts of Scotland. This species has successfully made the transition from living in brackish water to living in fresh water. The details of the physiological adjustments made by this species have not been worked out, but Todd (1964) has shown that the urine produced in fresh water is hyposmotic to the blood. In higher salinities the osmotic pressure of the urine increases with that of the external medium. The kidney of *P. jenkinsi* is similar to that of *Hydrobia ulvae*, and differs from that of fresh-water

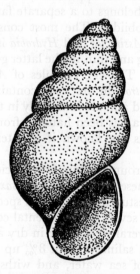

Figure 37. Potamopyrgus jenkinsi (Gastropoda, Prosobranchia), actual height of shell
5 mm

gastropods such as the pulmonate *Limnaea* which has a tubular kidney capable of producing urine with a concentration about 70 per cent of the blood.

There are other hydroblids and allied snails in Britain which live in salt marshes and brackish regions. *Truncatella subcylindrica* is a small hydrobiid with a restricted distribution on the South Coast of England. It lives high on the shore near the mouths of estuaries, often among *Sueda* and *Halimione*. The young resemble a slender *Hydrobia*, but the shell of a mature specimen lacks the point of the spire which breaks off after the internal cavity has been sealed. This leaves a short subcylindrical shell quite different in appearance from that of *Hydrobia*. The living animal also differs from *Hydrobia* in having short tentacles and in moving about like a geometrid or looper caterpillar. The tip of the snout can be extended and then attached, like the anterior end of a leech, the small rounded foot then detaches and is drawn up to a position immediately behind the snout, which can then release its hold and extend to take another step forwards. The reproductive biology of this species is not well known, but there does not appear to be a larval stage.

Assiminea grayana belongs to a separate family, but is cloely allied to the Hydrobiidae. The most conspicuous difference between the living *Assiminea* and *Hydrobia* is found in the tentacles which are long and thin in the latter genus, but short and broad in *Assiminea*. The egg capsules of *A. grayana* resemble those of *Hydrobia ventrosa* in that each contains a single egg, but they are not attached and are laid freely in the surface layers of mud. A veliger larva has been hatched from these eggs in the laboratory, but it is not known if it enters the plankton in natural conditions.

On rocky shores around Britain four species of periwinkle are common. Two of these species, *Littorina saxatilis* and *L. littorea* penetrate well into estuaries. The former species has a somewhat greater tolerance to severe environmental conditions. It can go without food for several weeks, live in dry air for several days, live permanently in salinities from 8‰ up to more than twice the concentration of sea water, and withstand immersion in fresh water for almost a week. Desiccation and osmotic stress are resisted by quiescence; the body is withdrawn into the shell and the mouth of the shell closed with the operculum. *Littorina littorea* is not so tolerant, but can live in salinities down to 10‰, and is often found in abundance in salt marsh creeks. The two species feed on detritus, diatoms and other small algae which they scrape up from the surfaces over which they move.

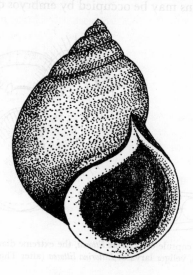

Figure 38. Littorina littorea (Gastropoda, Prosobranchia), actual height of shell 26 mm

In some populations of *L. littorea*, such as that at Whitstable, studied by Smith and Newell (1955), the average length of life is about 2 years but it is possible for individual winkles to live much longer. Comfort (1957) gives a record of 20 years for the lifetime of an individual kept in captivity. The smaller species *L. saxatilis* probably does not live as long.

There are some interesting differences in the breeding biology of *L. saxatilis* and *L. littorea*. The latter species lays eggs in small capsules shaped like a British infantryman's tin hat (Fig. 39A). Each capsule contains 2 to 4 (sometimes more) eggs, and the whole structure floats freely in the plankton. Veligers (Fig. 39B) emerge from the capsules about 6 days after laying. The duration of the planktonic life of the veliger is variable, extending from a fortnight to over a month (Thorson, 1946). In contrast *L. saxatilis* is viviparous. The genital system is modified by having one of the ducts enlarged to form a brood pouch in which the eggs are retained. The brood pouch is subdivided by transverse folds of the wall, and the different regions may be occupied by embryos of different ages,

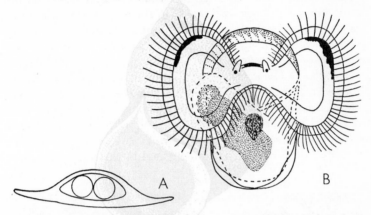

Figure 39. A, egg capsule of *Littorina littorea*, the extreme diameter is about 1 mm. B, veliger larva of *Littorina littorea* (after Thorson)

with the youngest near the upper end. A large specimen may carry up to 900 young in the brood pouch. At the end of the brooding period the young emerge as miniature winkles. The advantages of viviparity to an estuarine animal have already been mentioned (pp. 121 and 128) so that it is not surprising that of the four British species of *Littorina* the one that penetrates the furthest into estuaries is *L. saxatilis*.

Two small species of nudibranch gastropods are common in the salt marshes of Europe, and one of these, *Alderia modesta*, has been found in similar situations on the Pacific Coast of North America from California to the Canadian border (Hand and Steinberg, 1955). The other common species, *Limapontia depressa*, is found associated with mats of *Vaucheria*, where this alga grows on mud in shaded places near the bases of salt marsh grasses and rushes. On the Gwendraeth estuary, *L. depressa* is particularly abundant in the *Juncus* zone. This nudibranch feeds by slitting the wall of the alga and sucking out the green contents. The radula is reduced to a single row of teeth, and the leading tooth alone is used in slitting the alga (Gascoigne, 1956).

A closely allied species, *Limapontia capitata*, lives in rock pools on the sea shore. The two species show an interesting difference in the position of the anus. In *L. capitata* the anus opens on the dorsal surface at some distance from the hind end of the body. The faeces are liquid and as the animal is nearly always in water they are easily disposed of. *Limapontia depressa* spends much of its time out of water in moist air, and the disposal of its faeces has to be arranged in a different manner. The anus of *L. depressa* is subterminal, so that the faeces can be ejected beyond the end of the body without coming into contact with the dorsal surface. It is noteworthy that the anus of *Alderia modesta* opens in the same position as in *L. depressa*, and can be protruded on a papilla beyond the end of the body.

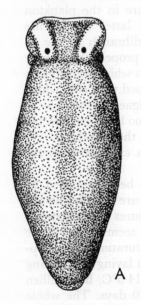
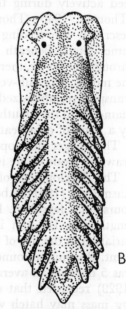

Figure 40. Small estuarine nudibranchs
A, *Limapontia depressa*, length about 6 mm. B, *Alderia modesta*, length about 10 mm

There are also interesting differences in the mating habits of *Limapontia depressa* and *L. capitata*. Both species are hermaphrodite, and in each there is a male duct which leads to a penis bearing a hollow spine or style which functions like a hypodermic needle. The oviduct opens separately behind the penis. *Limapontia depressa* has a third genital opening, into the bursa copulatrix, which receives the spermatozoa from another individual during mating. In *L. capitata* the outer wall of the bursa is closed. The penis of the latter species bears a smooth sharp style which pierces the outer wall of the bursa copulatrix. In *L. depressa* the penial style bears three or four small hooks which serve to retain the style in the bursa while the spermatozoa are transferred.

The two species of *Limapontia* lay their eggs in globular or sausage-shaped masses up to 5 mm in length. Each mass contains between 10 and 450 eggs. The veliger larvae which hatch from these eggs about 7 days after fertilisation have characteristic black patches around the mouth. These veligers feed actively during the time that they are in the plankton (Thorson, 1946; Thompson, 1959). The latter author has described the feeding mechanism of nudibranch larvae in some detail. Beneath the velum, which propels the veliger through the water, there are sub-velar ridges which converge on the mouth. The sub-velar ridges are equipped with cilia which transport suitable-sized particles, such as algae less than 15μ in diameter, to the mouth. Particles that are too large are rejected by a band of cilia beating backwards from the mouth.

The veliger develops a shell, but this is cast off when the crawling mode of life is adopted.

The development of *Alderia modesta* has been described by Rasmussen (1951). About four egg masses are laid during the course of a few days. Each egg mass measures up to 5 mm in length and 1·5 mm in diameter. There seems to be some variation in the size of the eggs and in the duration of development. Rasmussen found that the time from laying to hatching was 5 days at an average temperature of 14·5°C, but Gallien (1929) recorded that development took 20 days. The whole egg mass may hatch within the same hour, releasing typical veligers which swim in the plankton. The duration of planktonic life is not known. The larvae metamorphose into small snails, about 0·5 mm long, and immediately after settling they may still carry shells, but these are soon discarded and the cerata (Fig. 40B) are developed on the dorsal surface.

ESTUARINE LAMELLIBRANCHS

Wherever there is a solid substratum, and the salinity does not fall too low the common mussel, *Mytilus edulis*, is likely to be abundant. Near the mouths of many estuaries there are extensive areas dominated by *Mytilus*, which at first sight appears to be attached to the surface of the mud. Closer inspection generally shows that the tough byssus threads, which are secreted by a gland in the foot, penetrate through the surface layers and are attached to a stone. New, longer treads can be made if the mud threatens to cover the mussels completely. The mass of shells and byssus threads encourages the settlement of further mud, so that the level of the mussel bed tends to rise above that of the neighbouring mud flats. On dense beds the newly settled spat attach to the shells of older individuals, and the original attachment to a stone may be several shells down.

Mytilus edulis forms a dominant component of the fauna of buoys in estuaries (Frazer, 1938; Milne, 1940). In this situation there are frequently rapid changes in salinity, because the buoys often mark channels where the rate of tidal flow is rapid. The mussel can of course close its shell, and so avoid the most severe effects of the rapid salinity change. This ability gives the mussel an advantage in relation to other sessile animals which lack closing devices. But quite apart from the effects of lowered salinity, the sponges, compound ascidians and hydrozoans which require clear water are put at a disadvantage by the conditions created by dense settlements of mussels. Where the estuarine water has a high silt content the accumulation of shells and byssus threads encourages the collection of silt on the buoys and renders them unattractive to other sessile animals.

The concentration of the blood of *Mytilus edulis* follows that of the external medium down to a salinity of 10‰, below this salinity there appears to be some active osmotic regulation

(Krogh, 1939), but the mussel cannot survive permanently in salinities below 4‰.

The feeding mechanism of *Mytilus* resembles that of many other lamellibranchs in that it relies upon the separation of small particles from a current of water that flows through the gills (Atkins, 1936–8). In *Mytilus* each gill consists of two equal branches, and each branch is folded on itself so that the whole

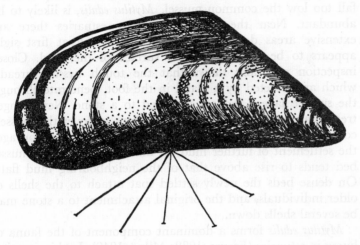

Figure 41. Mytilus edulis, length of shell about 6 cm; note the byssus threads which anchor the mussel to a solid substratum

gill is W-shaped in cross section. One gill lies on each side of the foot in a space between the foot and the mantle, which secretes the shell valves. Each gill is formed by a series of filaments loosely connected at intervals. The water current is propelled by lateral cilia, which lie on the sides of the filaments adjacent to neighbouring filaments. The water current flows in through a restricted opening between the mantle lobes, passes through the gills and then out above the inflowing current. In spite of their complicated shape the gills can be regarded as a screen separating the ventral, or incurrent part of the mantle cavity from the dorsal or excurrent part. The incurrent aperture, which lies posteroventrally, is larger than the exhalent aperture, so that the outflow is more rapid. This serves to project the outflow to a region where it is less likely to be taken in again by the inhalent current. Water flowing out from the mantle cavity will be partly deoxygenated by the gills, and may contain waste products, so that the advantages of not recirculating it are obvious.

The outer faces of the gill filaments also bear cilia, the frontal cilia, which beat down towards the free edges of the gill branches. At the two free edges and at the three bases of the W-shaped gill there are ciliary tracts running forwards towards the mouth. Particles carried in by the inhalent current are trapped on the gills and transported by the frontal cilia to the apical and basal tracts where they are moved towards the mouth. On either side of the mouth there are flaps, or labial palps, which are also ciliated and transport suitable food particles from the gills to the mouth.

This mechanism could operate without the aid of mucus, but if a layer of mucus is secreted over the outer face of the gill, then its efficiency of trapping very small particles is increased. There are indications from the work of MacGinitie (1941) and Jørgensen (1949) that *M. edulis* can vary the secretion of mucus according to the nature of the particles it encounters in the inhalent current. Jørgensen found that when he fed *Mytilus* on graphite particles with a diameter of 4 or 5μ only a small percentage of the particles was retained by the gills, but when he supplied flagellates of the same size a very high proportion was retained. These results show that when the particles carried in by the inhalent current are suitable as food the gills are capable of increasing their retentive capacity. The method of increasing retention appears to involve the secretion of a mucous sheet over the outer side of the gill. The presence of this mucous sheet in lamellibranchs has been observed directly by MacGinitie, who made glass windows in the shells, and so could observe the gills functioning without undue disturbance of the animals.

The breeding cycle of *Mytilus edulis* in British waters has been studied in detail by Chipperfield (1953). Spawning usually coincided with a period of fine weather in late spring and lasted two to four weeks, although on some occasions the spawning period was more extended. The rate of spawning seemed to be dependent to some extent on the rate of increase of temperature at spawning time. For instance, in a comparison of a pool and a creek at Brancaster Staithe, Norfolk, it was found that the temperature in the pool rose more rapidly than that of the creek during the first week of May. By the 14th of May 75 per cent of the mussels in the pool had spawned, while none in the creek had spawned. It was not until the beginning of June that the population in the creek had spawned to the same extent as those in the pool.

The spring spawning period of *M. edulis* in the Bay of Arcachon extends over twelve weeks, beginning at the end of March and ending early in June, but spawning is not continuous. Lubet (1956) found six separate peaks which coincided with the new and full moons. After each peak of gamete liberation there was a period of gametogenesis so that the mussels would be ready for spawning at the next spring tide. A cycle of neurosecretion in the cerebral ganglia coincided with the spawning cycle, indicating that the maturation and liberation of gametes was controlled in part at least by hormones. A second period of spawning was found in December and January. This spawning period appeared to be stopped by falling temperatures in February. At temperatures below $7.5°C$ gametogenesis was completely inhibited.

At spawning time the eggs and sperm are shed freely into the surrounding water, and may be so numerous as to give a milky appearance to wide areas. The fertilised eggs give rise to veliger larvae which become planktonic and spend several weeks drifting in coastal waters. The veliger possess a large ciliated lobe or velum which is used in swimming. As the veliger grows older the velum gradually diminishes in size as the foot increases. When the larva is capable of both swimming and crawling it is known as a pediveliger. This is the stage which settles.

The growth of the larva to the pediveliger stage is influenced by temperature, food and salinity. From the viewpoint of estuarine biology the last factor is of the greatest interest. Bayne (1965) has shown that the influence of salinity varies with the locality in which the larvae are found. Thus larvae from the sound between Denmark and Sweden, where the surface waters have a lower salinity than the North Sea, showed a diminution in growth rate when the salinity fell below 18‰, but also showed a diminution when the salinity was raised above 26‰. Larvae from the Menai Straits, where the salinity is somewhat higher, did not show any diminution in growth rate at salinities between 26 and 30‰, but the growth rate fell when the salinity was lowered below 20‰.

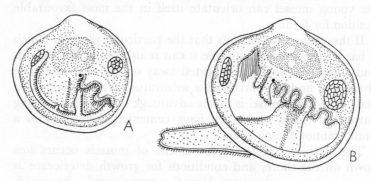

Figure 42. Pediveligers of *Mytilus edulis*
In A, the foot is small and the velum large, while in B, the foot has increased greatly in size while the velum has shrunk. Length of the shell in B is about 0·4 mm

There is considerable variation in the size of larvae settling on the shore. This variation is caused by the ability of the larvae to delay settling until a suitable substratum has been found. The type of substratum to which the larvae most readily attach, both in the field (Blok and Geelen, 1958) and in the laboratory (Bayne, 1965), is one which includes filaments of algae. There appears to be some tactile stimulus involved in this selection. When nylon and silk threads are offered together as substrata in the laboratory the larvae prefer to settle on the silk, which has a rough surface with small threads projecting, while the nylon is smooth.

A comparison of pediveligers from the Menai straits with those from Helsingor revealed that in the latter locality the mean size was greater. Bayne (1965) attributed this greater size to a delay in metamorphosis which was caused in part at least by the relative scarcity of suitable filamentous substrata for mussel settlement in the vicinity of Helsingor.

The primary attachment to a solid object is made by the foot, but attachment is soon made permanent by the secretion of byssus threads. In the early stages of attachment the young mussel can move about by extending its foot, secreting new threads at the new position then detaching the old threads and pulling itself up to the new position by contraction of the foot. In this way a suitable position on the shore can be found and the young mussel can orientate itself in the most favourable position for feeding.

If the young mussel finds that the particular place in which it has settled is not favourable it can resorb the byssus threads and allow itself to be transported away on the next tide. The chances of finding a favourable substratum are thus increased and the young mussel is at an advantage compared to young barnacles which, once they have cemented themselves to a stone, cannot detach again.

The heaviest intertidal settlement of mussels occurs low down on the shore, and conditions for growth deteriorate as one progresses up the shore. Very few mussels settle above mid-tide level.

If the settlement of spat is heavy the young mussels may literally smother their parents. The most spectacular example of this phenomenon has been described by Savage (1956). In the spring of 1940 there was an immense spatfall of mussels at the mouth of the River Conway. Many of the older mussels were completely covered with young mussels—over 700 young were counted on some of the larger adults. These young effectively deprived their parents of food, so that the adults lost condition and died. The young mussels were so abundant that there was not sufficient solid substratum for their attachment, so that they attached to each other. Clumps of up to a thousand were found lying loosely on the surface of the sand. Some of the young mussels attached themselves by their byssus threads to areas of stable sand, particularly near the mouths of worm burrows. But in such a situation the attachment was very feeble, and in rough weather the mussels were washed off the sand.

The growth of young mussels once they have settled is very variable. Low salinity reduces both the growth rate and the final size. In the Baltic Sea the maximum size of specimens growing in a salinity of 5‰ is only a quarter of that of North Sea specimens growing in a salinity of 35‰. Havinga (1929) also found local variations in growth rate in the Zuider Zee which were related to variations in salinity.

The growth of the very dense settlements at the mouth of the river Conway was found to be relatively slow. By December the average length was 11 mm, and a year later most of the mussels were 29 mm long. In favourable conditions on the Dutch coast a length of 34 mm has been observed at the end of their first December and 57 mm by their second December. This indicates that the crowded conditions at the mouth of the Conway were reducing the growth rate to under half that possible in favourable conditions.

Mussels attached to buoys may grow at rates faster than those found on the shore. Ritchie (1927) found rapid growth on an experimental buoy anchored in the Firth of Forth. The buoy was anchored in May, and by October the spat which had settled on it had reached a length of 24 mm.

Mussels which are permanently submerged are obviously at an advantage from the feeding point of view. The subtidal mussel has the potentiality of continuous feeding and so increases its chances of reaching a large size. Loosanoff (1942) found that at temperatures between 5 and 18°C the valves of *Mytilus edulis* were kept open about 98 per cent of the time, and that food was always present in the stomach. This may indicate that if given the opportunity *Mytilus* will feed continuously. The large size and good condition of subtidal mussels is well known to commercial collectors. Many of the commercial fisheries operate only on subtidal populations, and either transfer young mussels from the shore to permanent channels or provide suitable substrata, such as thick ropes for the settlement of spat.

Oysters (family Ostreidae) differ from mussels in a variety of ways. The left shell valve is normally cemented to a solid substratum at the end of larval life. The pediveliger has a byssus gland in its foot, but this gland is capable of discharging once only. When a suitable substratum is found the pediveliger pours out secretion from the byssus gland and then turns so that the left shell valve is pressed against the secretion, which hardens rapidly. If the newly settled spat is dislodged after this process has been completed it is incapable of reattaching itself. The byssus gland degenerates once settlement has been effected.

Within the family Ostreidae the two main estuarine genera are *Ostrea* and *Crassostrea*. Both are basically marine forms with abilities to tolerate brackish water. *Crassostrea* is capable of penetrating further into estuaries than *Ostrea*. In some American estuaries *Crassostrea* lives permanently in salinities down to 12‰, and can tolerate lower salinities for short periods. In general *Ostrea edulis* does not occur where the salinity falls below 20‰ for any length of time.

Crassostrea also has some morphological adaptations which enable it to survive in silty conditions that would smother *Ostrea*. The left shell valve of *Crassostrea* forms a deep bowl which raises the edge of the shell above the surface of the substratum so that sediment does not enter the mantle cavity as easily as in the flatter *Ostrea*. The adductor muscle of *Crassostrea* is also capable of more rapid and more powerful contractions than that of *Ostrea*. This means that the shell valves can be brought together smartly, and the resulting outflow of water will dislodge any sediment that has accumulated in the mantle cavity.

The breeding biology of oysters is remarkable. Most individuals start life as males, but change to females after liberating their sperm. The females may then change back to males, after liberating eggs. This change can occur several times during the lifetime of an individual. The change from male to female is generally slower than the change from female to male. The latter process may be complete in a few days after the eggs have been shed. At high temperatures an individual may change its functional sex several times in a season, but at lower temperatures the change may occur only once a year.

Liberation of the sperm is a straightforward process; they are shed into the cavity above the gills and carried out with the exhalent water current.

The eggs are released in a more complex manner. The genital ducts discharge into the cavity above the gills, but instead of passing out with the exhalent current the eggs are forced through the interstices of gills into the inhalent part of the mantle cavity. In the genus *Ostrea* the eggs are retained in

F

the inhalent cavity and are fertilised by sperm brought in by the inhalent current. But in members of the genus *Crassostrea* the eggs are ejected from the inhalent cavity by vigorous rhythmic adductions of the shell valves.

The effect of temperature on the breeding of *Ostrea edulis* has been reviewed in detail by Korringa (1957). In the main Dutch oyster beds at the mouth of the River Scheldt, spawning never occurs before the water has reached a temperature of 15°C, and the main peak of spawning occurs between the 20th June and the 10th August. In the region of Naples, *O. edulis* starts to breed in April, but again the water temperature is over 15°C, and a similar relationship is found in the Bay of Arcachon.

On the Atlantic coast of Spain, in the Bay of Vigo, oyster larvae may be found from March until December. Surprisingly for a locality so far south the larvae seem to develop quite normally at temperatures of 13 to 14°C.

Yet a further paradox is found in a fjord near Bergen in Norway. Here the larval oysters appear in masses when the water temperature rises above 25°C. This remarkably high temperature for such a northern locality is made possible by a superficial layer of fresh water which acts like the glass of a greenhouse. The saline layers underneath reach a much higher temperature than the superficial layer. The density of a salt solution is much more strongly related to salinity than to temperature, so that conventional mixing is prevented by salinity stratification. Heat can thus accumulate in the lower saline layers which are insulated against losing heat to the air by the overflowing fresh water.

The population of oysters near Bergen seems to be particularly demanding in terms of the temperature required for the liberation of larvae. Over a period of ten years larvae were consistently liberated in July, and never at a water temperature under 25°C.

The relationship between temperature and spawning in *Crassostrea* appears to follow a more logical pattern than that just discussed. In general *C. virginica* spawns when the temperature reaches 20°C, but in a few northern areas spawning may occur at 17°C. In Florida spawning does not occur at temperatures below 25°C. There seems to be a fairly direct relationship between latitude and the minimum temperature required for spawning, but in reality the situation is not so simple. In some localities, such as near the mouth of the Delaware River, it is possible to distinguish between bay oysters, which spawn at 25°C, and open coast oysters, which spawn at 20°C.

It is evident from the facts presented above that both *Ostrea* and *Crassostrea* can occur as distinct physiological varieties, each with their own optimum temperatures for spawning. Knowledge of these differences is important from a commercial point of view. Clearly it would be inadvisable to transplant stock requiring a high temperature for spawning to a region where this temperature is never, or rarely, reached. Simple consideration of latitude is not sufficient. For instance, it is unlikely that the oysters from Bergen would ever be warm enough to spawn if they were transferred to the Dutch coast.

Mussels and oysters are members of the epibenthos, but the next example belongs to the infauna, and the succeeding examples penetrate deeper and deeper into the substratum.

The common cockle, *Cardium edule*, has a large active foot, but it does not burrow deeply. The most frequent habitat of the cockle is in slightly muddy sand, but it is not confined to such a substratum, and can inhabit coarse gravels and stiff muds. When uncovered by the tide the cockle may retreat a couple of centimetres below the surface, but when the tide returns the shell is pushed up to the surface so that the short siphons may be extruded into the water. The food taken in by the inhalent siphon appears to be mainly small plankton organisms in suspension, but there is little precise information on the actual food ingested. The results obtained by Wernstedt (1942) indicate that cockles may thrive better when the food includes bottom deposits and bottom-dwelling diatoms. He found that growth was better on such food material than when the cockles were fed on pure cultures of planktonic algae such as *Nitzschia* and *Chlorella*.

The growth rate of cockles varies greatly from one locality to another. Kristensen (1957) has made a detailed study of this variation in the Dutch Wadden Sea. He was fortunate in being able to study an extensive new population which fell as spat

on the Dutch tidal flats in the summer of 1947 after the intertidal population had been destroyed by the severe winter of
1946–7. This enabled a comparison to be made between
populations of the same age in different parts of the area. One
important factor was found to be the period of submersion. A
cockle needs five hours in which to gather enough food to
maintain a rapid rate of growth. If the period of submersion is
reduced below five hours the rate of growth is reduced. A
large amount of silt in suspension was also found to be deleterious to growth. The feeding efficiency of the gills was reduced
by having to cope with the rejection of this material.

Certain areas of the flats were found to be very favourable
for the settlement of spat. In these areas there were dense
populations of slow-growing cockles with a high mortality rate.
Other areas were not so favourable for spat settlement, but the
sparser populations which developed were made up of fast-
growing individuals with a low mortality. One of the major
differences between the two types of locality was the degree
of exposure. Where the current were faster the spat had
difficulty in settling, and once settled they were in danger of
being washed away. In sheltered areas the spat could settle
successfully, but then had to contend with competition from
their numerous neighbours. In spite of the crowded conditions
and reduced growth rate in sheltered localities the production
of eggs per unit area was about twice as high as in the exposed
localities with their sparser populations.

The growth rate of the cockle was also found to be reduced
by the presence of *Mytilus edulis*. Kristensen observed the
development of a new mussel bank and found that the growth
rate of cockles close to the banks was reduced to about half
that of more distant cockles.

In sparse populations, with a rapid growth rate, individual
cockles may live for at least 8 years, but in dense populations
the duration of life is reduced. Mortality can be also greatly
increased by unfavourable weather, particularly by severe
winters. Low temperatures decrease the digging speed and so
increase the risk of being washed away. More severe conditions
kill by freezing or by covering the mud flats with a layer of
ice beneath which the oxygen becomes depleted.

Spawning by cockles begins in March and extends through the summer. The fertilised eggs give rise to veligers which may spend two or three weeks in the plankton before they settle. There is considerable variation in the size at which they change into settled spat. Jørgensen (1946), working at Kristineberg, found that the length at metamorphosis varied from 275μ to 345μ, but at Ven young bottom stages were found with a length of 255μ, and in Copenhagen Harbour he found very small specimens only 150μ long. One of the dangers facing a small larva seeking a place to settle is the possibility that it may be taken in by the inhalent siphon of an adult. Kristensen (1957) observed that where there were dense populations of adults the young were often taken in and then rejected covered in mucus so that they were unable to free themselves and perished.

Although it is normally regarded as an intertidal animal the cockle will live permanently submerged when this is possible, and has been recorded living at a depth of 14 metres. In the Baltic Sea it is always submerged. A factor which may help to restrict the cockle to the intertidal zone near the mouths of estuaries is the presence of subtidal predators. The whelk, *Buccinum undatum* does not normally occur between the tides, but it has been shown experimentally to feed on cockles if the opportunity presents itself. Hancock (1960) found that the maximum rate of feeding was about two cockles a week, although each meal lasted only about one hour. The whelk can crawl slowly up to an unsuspecting cockle and insert the lip of its shell between the open valves, wedging them open so that the soft body of the cockle is accessible to the proboscis of the whelk.

In a review of the British marine lamellibranchs, Tebble (1966) has placed *C. edule* in the genus *Cerastoderma*, and has put the closely allied *C. lamarcki* in the same genus. It seems likely that in the past these two species have been confused, and records of *C. edule* from salinities below 20‰ probably refer to *C. lamarcki*. Peterson (1958) has studied the distribution of these species in Danish waters, and found that *C. edule* did not occur in salinities below 20‰, while *C. lamarcki* was found over a range from 5 to 25‰. A third species, *C. exiguum*, was found where the salinity was about 10‰. Peterson noted that *C. lamarcki* was often found among vegetation, not always buried in the sand, and he observed that young specimens were capable of climbing,

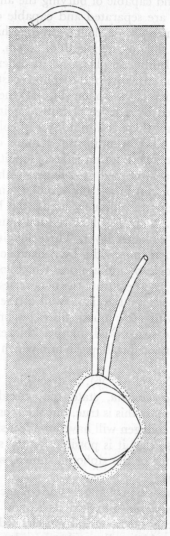

Figure 43. Scrobicularia plana showing the positions of the siphons when feeding. The lighter area around the shell indicates a region of aerated mud. Actual length of the shell about 4 cm

using bysus threads in the manner already described for young *Mytilus* p. 149).

Scrobicularia plana is well adapted to live in deep soft mud. The foot is large and capable of burying the animal in a short time. The siphons are separated and capable of considerable extension. A specimen with a shell 4 cm long has been seen to extend its inhalent siphon to a length of 28 cm. The process of extension is remarkably complex, and in order to understand it some details of the structure of the siphon must be considered. In a transverse section of the siphon (Fig. 44) the outer covering is seen to be formed by a layer of cubical epithelium. Inside this outer epithelium there is a thin layer of collagen, containing a few muscle fibres running in a more or less circular direction. The next layer is a thin series of longitudinal muscles, and inside this is a large bed of collagen with muscle fibres. These fibres do not run in a truly circular direction, but lie in two series which run obliquely across the bed of collagen and form a lattice. A layer of longitudinal muscles lies on the inner edge of the collagen layer. These longitudinal muscles are bounded on their inner sides by a haemocoelic space and are broken into blocks by radial fibres which extend across the haemocoelic space and penetrate the next series of longitudinal muscles which lie nearer the inner wall of the siphon. A thin layer of collagen with a few muscle fibres lies on the inner edge of the haemocoelic space, and another thicker layer lies just underneath the inner cubical epithelium which lines the lumen of the siphon.

Chapman and Newell (1956), in describing this structure of the siphons of *Scrobicularia*, state that circular muscles are absent. In a strict sense this is true, but the muscle fibres which run in the beds of collagen will have an effect similar to that of circular muscles, so that it is possible for the siphonal wall to reduce the diameter of the siphonal lumen, and to close the ends of the siphons.

The complex disposition of muscle and collagen in the siphonal wall enables *Scrobicularia* to extend its siphon without closing the end. Contraction of the radial muscles will pull the inner layers towards the thick collagen layer and exert pressure on the haemocoelic space. This pressure is translated into a longitudinal pressure which extends the siphon. Contraction of the siphon is effected by the longitudinal muscles.

The mechanism described above is quite adequate for extending the siphon into sea water, but if the siphon has to penetrate through a substratum a different mechanism is

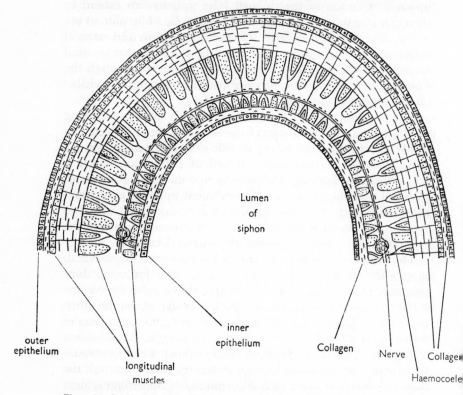

Lumen
of
siphon

outer
epithelium

longitudinal
muscles

inner
epithelium

Collagen

Nerve

Collagen

Haemocoele

Figure 44. Scrobicularia plana, diagram of a transverse section of the inhalent siphon, only half the section is shown

brought into play (Green, 1967). The end of the siphon closes and makes small circling movements which soften the substratum prior to penetration. The siphon then elongates as if being blown up by increased pressure in the lumen. The details of this mechanism are as yet unknown, but the siphons of *Scrobicularia* are most effective in penetrating fine sand or mud.

If *Scrobicularia* is propped in a natural position in sea water the siphons can only extend upwards to a distance of about 12 cm, but if the animal is covered with a layer of fine sand the inhalent siphon can extend upwards to a height of 28 cm. It seems as if the support of the sand is necessary for maximum upward extension of the siphons. The siphons can extend to their full length if the animal is allowed to lie on its side in sea water, but then the siphons also lie horizontally. In natural conditions the siphons are supported by the fine sand or mud through which they pass. If the substratum is firm enough the siphons lie in permanent burrows and the openings are visible on the surface of the mud.

The inhalent siphon of *Scrobicularia* is very active in sucking up deposits from the surface of the mud. Material collected in this way is transported to the mantle cavity and sorted on the gills. Material for rejection is carried by ciliary currents to a point near the base of the inhalent siphon. Expulsion of this material is a violent process. The exhalent siphon is closed and the shell valves are strongly adducted so that a jet of water spurts from the open end of the inhalent siphon and carries the rejected material some distance away from the feeding area.

The inhalent siphon is not only a feeding organ, but serves to supply the mantle cavity with aerated water for respiratory purposes. This ability to draw water down from above the surface of the mud enables *Scrobicularia* to live at depths down to 30 cm in black mud which allows little or no circulation of water in its interstices and is lacking in oxygen. Examination of the cavity inhabited by *Scrobicularia* in such a mud indicates that there must be some leakage of aerated water through the mantle lobes. The layer of mud or fine sand immediately next to the shell is usually pale and oxidised in contrast to the general black reduced substratum.

By drawing in water from above the surface of the mud, *Scrobicularia* exposes itself to the variations in salinity which this water undergoes. Freeman and Rigler (1957) have shown that *S. plana* does not osmoregulate except when the salinity falls below 30 per cent sea water. This means that over a considerable range of salinities the osmotic pressure of the blood varies with that of the water above the surface of the mud.

But *Scrobicularia* does not have to keep its siphons extended into the water, it can retract them and close its shell. This ability enables *Scrobicularia* to survive in areas where the salinity may fall to a very low level for a short period. In the Gwendraeth Estuary a dense population of over 500 per sq m was found thriving in a mud flat where the salinity of the overlaying water in winter fell as low as 2‰ for a short period as the tide ebbed (Green, 1957a).

Scrobicularia plana may live as long as 18 years, and reach a shell length of 54 mm. In the first 3 or 4 years of life growth is fairly rapid, with the shell increasing in length by about 6 mm per year. In later years the rate of growth declines, and once an age of 10 years has been reached the length of the shell increases by only a millimetre each year.

A feature of the biology of *Scrobicularia plana* is that the optimum conditions for the settlement of spat do not always coincide with optimum conditions for survival and growth. Spooner and Moore (1940) found that in St John's Lake, near the mouth of the Tamar, there were very heavy settlements of spat, but most of the specimens died at an early age. They refer to a similar phenomenon found in the Fleet, which is a brackish area enclosed by a shingle spit in Dorset. Here vast numbers of very small individuals, less than a millimetre in length, were found in a muddy area where *Zostera hornemanniana* and *Ruppia* were growing. This particular area was permanently covered with water, and as *Scrobicularia* prefers intertidal conditions, was not suitable for permanent settlement. Further, the substratum in the intertidal zone was mainly gravel and also unsuitable for settlement, so that although there was an abundant supply of spat none of the specimens succeeded in establishing themselves and growing.

Once an area has been heavily and successfully settled by *Scrobicularia*, and the individuals have reached a shell length of about 30 mm, so that they can live at a depth beyond the reach of most predators, the comparatively long life of this bivalve means that new spat attempting settlement will encounter difficulties. The major difficulty is the activity of the inhalent siphons of large individuals. These continuously remove the soft surface layers on which the spat settle. The long duration of occupation by individuals means that an area suitable for growth may be fully occupied and rendered unsuitable for spat settlement for periods of several years at a stretch.

The northern range of *Scrobicularia plana* appears to be limited by low winter temperatures. In the severe winter of 1963 many populations around the British Isles were wiped out, but a few populations in sheltered areas seem to have escaped the most severe effects so that the repopulation of the areas which were depopulated is probably only a matter of time.

Macoma balthica, which is similar in basic structure to *Scrobicularia*, is much more tolerant of cold, and was not much affected by the severe weather which decimated *S. plana*. The more northerly distribution of *Macoma balthica* and its occurrence in the Gulfs of Finland and Bothnia, where the sea freezes for several months of the year, are indications of this great resistance to low temperature.

Although similar in structure to *Scrobicularia*, *Macoma* has somewhat different habits. It does not burrow so deeply, it lies on its left-side (at least *M. nasuta* does), and it moves about frequently. Brafield and Newell (1961) found that *M. balthica* moved along U-shaped tracks, and that the first phase of movement was generally towards the sun. The later phases were generally away from the sun. These movements are made by means of the foot, and leave a furrow about 5 mm wide on the sand surface. A looped track of this type enables *Macoma* to cover a wider area in search of food than would be possible if the search was made by the inhalent siphon alone. At the same time the form of the track prevents the bivalve from wandering away from its station on the shore.

The detailed distribution of *Macoma balthica* on a shore has been discussed by Beanland (1940), who concludes that the abundance of this species depends on the amount of food available, and the amount of time available for feeding. The latter factor will vary with tidal level, and will be least near high water mark. The relationship between food and type of substratum has been subjected to further analysis by Newell (1965) who found that populations of *M. balthica* in the Thames Estuary were denser where the grade of deposit was finer. The higher populations in fine deposits were attributed to increased densities of micro-organisms. The main food of *Macoma* seems to be derived from the digestion of micro-organisms in the deposits that are sucked up by the inhalent siphon, and the abundance of this food is related to the surface area of the deposits (see Chapter 13).

Competition for space may occur between *Scrobicularia* and *Macoma*, particularly when the population density is high. *Scrobicularia* has been found in densities of just over a thousand per sq m. At this density near the mouth of the Tamar the majority of the individuals were newly settled spat, with a shell length under 0·6 mm (Spooner and Moore, 1940), but in the Gwendraeth Estuary the mean length of the shells was 30 mm in a region where the density was 1,024 per sq m. *Macoma* has been found in populations with densities up to 5,900 per sq m in the Mersey Estuary (Fraser, 1932). At these high densities one of the species is usually dominant, and very few or no specimens of the other species are present. At lower densities the two species can coexist in almost equal numbers. Table 18 shows some of the population densities that have been recorded from European shores.

TABLE 18

NUMBERS OF 'SCROBICULARIA PLANA' AND 'MACOMA BALTHICA' ON
EUROPEAN SHORES

Locality	Authority	No. of S. plana per sq m	No. of M. balthica per sq m
Mersey	Fraser (1932)	0	2,000–5,900
Skalling	Thamdrup (1935)	20–60	300–1,000 (6,300)
Forth and Clyde	Stephen (1929, 1930, 1931, 1932)	136	150–200
Kyle Scotnish	Raymont (1955)	150–330	250–428
Tamar	Spooner and Moore (1940)	100–280 (1094)	36–76
Gwendraeth	Green (1957a & unpubl.)	500–1024	0–10.

Towards the mouth of an estuary, where the salinity is higher and the deposits tend to be coarser and cleaner, *Scrobicularia* and *Macoma* are often replaced by *Tellina tenuis*. Perkins (1956) found indications of an inverse relation between the numbers of *Macoma balthica* and *Tellina tenuis* in a traverse of a sandbank near the mouth of the Dee Estuary. Near the mouth of the Exe Estuary the numbers of *T. tenuis* were greatest in regions where the other two species were not abundant. This was particularly noticeable on a traverse ranging from high to low water. *Tellina tenuis* was abundant near low water while the other two species were more abundant higher on the shore (Holme, 1949).

Tellina tenuis lies on its side, with the right valve uppermost (Holme, 1961). The two siphons are capable of great extension, similar to that observed in *Scrobicularia*, and the general method of feeding is similar to that genus, so that competition between the two forms is to be expected. In some parts of Kames Bay the numbers of *T. tenuis* may be as high as 7,588 per sq m (Stephen, 1929). The spacing of *T. tenuis* has been studied by Holme (1950) who found that some populations were not randomly distributed in the sand but were fairly evenly spaced. He suggests that siphonal contact may be the clue that *Tellina* uses to estimate the distance from its nearest neighbour.

The gaper or soft clam, *Mya arenaria*, has its two siphons joined together. Each siphon keeps its own lumen, but the walls are joined together and surrounded by a complex series of muscles. The combined siphons are capable of great elongation, so that the clam can lie nearly 2 ft below the surface of the muddy sand and extend the siphonal tip up to the surface.

The method of extension of the siphons is of interest because it forms a contrast with that found in *Scrobicularia* (p. 157). It has been shown by Chapman and Newell (1956) that the combined siphons do not extend smoothly, but in a series of jerks, and each jerk is accompanied by an adduction of the shell valves. The mantle edges are fused through most of their length, and the small foot opening is capable of being closed by a muscular flap. This means that the pressure inside the mantle cavity is increased when the valves are drawn together by the adductor muscles. Now the lumen of each siphon is continuous with the mantle cavity, so that if the tip of the siphon is closed the increase in pressure will tend to elongate the siphons. Shortening of the siphons is effected by a series of longitudinal muscles.

As is usual in lamellibranchs the inhalent siphon carries

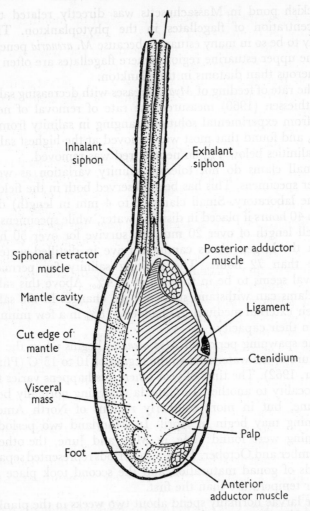

Figure 45. Diagram of the structure of *Mya arenaria*. The right shell valve has been removed and the siphons are shown only partly extended. The shell reaches a length of 15 cm

water for respiratory purposes and passes it through the gills where any food particles in suspension are trapped and transported to the labial palps and then to the mouth. Matthiessen (1960) has shown that the growth rate of young clams in a brackish pond in Massachusetts was directly related to the concentration of flagellates in the phytoplankton. This is likely to be so in many estuaries, because *M. arenaria* penetrates to the upper estuarine regions where flagellates are often more numerous than diatoms in the plankton.

The rate of feeding of *Mya* decreases with decreasing salinity. Matthiessen (1960) measured the rate of removal of neutral red from experimental solutions ranging in salinity from 0 to 31‰ and found that most was removed at the highest salinity. At salinities below 4‰ no neutral red was removed.

Small clams do not tolerate salinity variation as well as larger specimens, This has been observed both in the field and in the laboratory. Small clams (2 to 4 mm in length) die in 30 to 40 hours if placed in distilled water, while specimens with a shell length of over 20 mm can survive for over 50 hours. Even the largest clams cannot survive in distilled water for more than 72 hours. The limiting salinity for permanent survival seems to be in the region of 4‰. Above this salinity the clams can withstand sudden large changes in the salinity of their external medium. A change of 18‰ in a few minutes is within their capacity.

The spawning period of *Mya arenaria* begins when the temperature of the water rises to the region of 10 to 15°C (Pfitzenmeyer, 1962). The time of year when this happens varies from one locality to another. In Canada spawning normally begins in June, but in more southerly regions of North America spawning may begin in April. In Maryland two periods of spawning were found: one in May and June, the other in September and October. These two periods represented separate periods of gonad maturation and the second took place at a higher temperature than the first.

The larvae normally spend about two weeks in the plankton before settling. As with other estuarine lamellibranchs the size at metamorphosis is very variable (Jørgensen, 1946). A newly settled *Mya* may have a shell varying in length from 200μ to 320μ. The larva attaches itself by means of a byssus thread to sand grains or plants in the surface layers of the substratum. This byssus may be retained until the young *Mya* reaches a shell length of 7 mm, but sooner or later the byssus is lost and the young *Mya* takes up the adult position with the shell buried in a vertical position and the siphons reaching to the surface of the substratum. In the adult *Mya* the foot is greatly reduced and the animal is not capable of reburrowing when removed from its deep burrow.

Mya arenaria grows comparatively rapidly and may reach a length of 30 mm by its first winter. By the time it is fully grown the shell may be 15 cm long and the animal more than 8 years old.

A single female *Mya* may produce three million eggs in a year, but enormous numbers are lost by being flushed out to sea before they can settle. Ayers (1956) estimates that of those that do succeed in settling about one per cent survive to sexual maturity, which is reached after a period of 5 years. For an adult population to be maintained at a constant level there must be a total settlement of 200 spat over 5 years, or 40 settled spat per year to replace two adults.

Estuarine Barnacles

A typical sessile barnacle, such as *Balanus*, has a mantle composed of a tough membrane and calcareous plates into which it can retract the rest of the body and seal it off from the outside world by means of the scuto-tergal valves. (Fig. 46). Such an arrangement provides many advantages for a intertidal estuarine animal. The valves may be closed when the tide is out, or when the salinity falls to a very low level.

The opening of the valves of *Balanus balanoides* after a period of closure has been studied by Barnes and Barnes (1958). If this species is exposed to air for several hours, and then immersed in sea water, the valves open and a small amount of gas escapes from the mantle cavity. The cirri then begin to beat, and a full feeding rhythm may be established in about two minutes. If the barnacle is flooded with distilled water the valves begin to open, a small amount of gas may escape, but then the valves close tightly again. At salinities above 17‰ normal cirral activity is usually resumed, although near to this value there may be some hesitation. At salinities below 17‰, using specimens taken from truly marine conditions, the valves may open

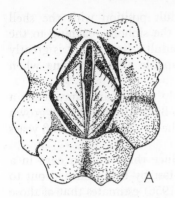 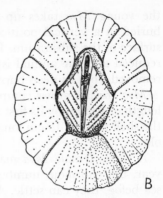

Figure 46. Sessile barnacles
A, *Eliminius modestus.* B, *Balanus improvisus.* The scuto-tergal valves are drawn slightly more open in *Elminius*

slightly, but the cirri do not resume normal activity. If sea water is replaced by pure sodium chloride in solution of equivalent concentration to sea water the cirri may make a few attempts at activity, but normal activity is never resumed. Now if calcium salts are added to the sodium chloride solution the cirri will resume normal activity. Magnesium salts produce a similar effect, but are not so effective as calcium salts in inducing activity of the cirri. In dilutions of sodium chloride less than sea water the calcium and magnesium salts are effective only if the total salinity is above 50 per cent of sea water. In more dilute solutions the closure of the valves was not prevented by the calcium and magnesium salts.

When the tide ebbs from a shore the behaviour of the barnacles varies according to the shelter and humidity of their surroundings. If the mantle cavity remained full of water and the valves remained closed there would be enough oxygen dissolved in the water to enable the barnacle to respire aerobically for 10 minutes (Barnes, Finlayson and Piatigorsky, 1963). But an intertidal barnacle does not keep its mantle cavity full of water when the tide goes out. A considerable proportion of the water is expelled by movements of the cirri so that an air space is formed in the mantle cavity. This air space often remains in communication with the outside air by means of a small gap, or pneumatophore, between the scuto-tergal valves.

Small movements of the barnacle within the mantle cavity serve to move air through the pneumatophore and so renew the supply of air in the mantle cavity. In this way the barnacle avoids the necessity of relying on the oxygen dissolved in the water in the mantle cavity and can continue to respire aerobically throughout the tidal cycle. If there is a real danger of severe desiccation, as for instance if the barnacle is on a bare rock exposed to the sun and wind, the scuto-tergal valves can be more tightly closed. The barnacle can then respire anaerobically, with the production of lactic acid. In water of low salinity the barnacle can also close its valves and respire anaerobically. When conditions improve and the water surrounding the barnacle is suitable the lactic acid is rapidly oxidised and the oxygen debt repaid.

The eggs of barnacles are retained in the mantle cavity until they hatch as nauplii. These larvae swim in the plankton and pass through six naupliar stages before changing into cypris larvae. In *Balanus eburneus* the first naupliar stage lasts between 15 minutes and 4 hours at 26°C, and the later naupliar stages last between 1 and 4 days (Costlow and Bookhout, 1957). The cypris stage of *B. eburneus* lasts between 1 and 14 days, but successful settlement was observed only in those settling in less than 4 days. The larval development of *Balanus amphitrite denticulatus* is very similar in its duration, taking between 7 and 10 days from hatching to settling at 26°C. Similar durations of the naupliar stages have been found in *Balanus balanoides* (10 days from hatching to cypris at 20°C), *Chthamalus stellatus* (10 days) and *Elminius modestus* (6 days) (Moyse, 1963).

The cypris larva resembles an ostracod, having a bivalved carapace, but it has more thoracic legs, and these are all similar to one another instead of having different forms as in the ostracods. This is the stage which attaches itself to a solid substratum by means of a cement gland on the antennule. The cypris then metamorphoses into the adult form. The thoracic legs become longer and transform into the cirri of the adult. The cypris bivalved carapace is lost and a new carapace or mantle develops. Once the mantle, with its calcareous plates, has been formed, the barnacle has established itself in the position

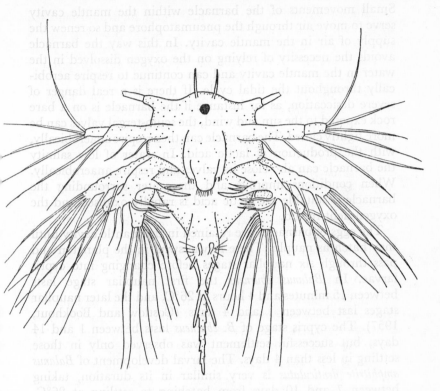

Figure 47. Nauplius larva of *Elminius modestus,* the length of the body is about 0·4 mm

in which it will remain for the rest of its life. It follows from this that the selection of a settling site is a critical phase in the life of a barnacle. If the cypris settles in a place where there is no food the barnacle will starve. In those species which are obligatory cross-fertilisers the distance to the next barnacle is important. If the distance is too great the barnacles will be unable to reproduce. Many species of barnacles settle gregariously in places already colonised by members of their own species. The factor inducing them to settle seems to be the protein arthropodin in the cuticle of the adult barnacles (Crisp and Meadows, 1962). This is detected when the cypris comes into contact with a surface bearing barnacles or the remains of barnacles. Once it has found such a surface the cypris will readily metamorphose, but if it fails to find such a surface it can delay metamorphosis for over a week.

The aggregation of barnacles is an obvious aid to reproduction. Many species are hermaphrodites, but most need to cross fertilise. Spermatozoa are transferred from one individual to another by means of a penis which may extend to a length of several centimetres, but above such a distance the transfer of sperm becomes impossible. *Elminius modestus* is capable of fertilising another individual at a distance of 3 to 5 cm (Crisp, 1958). This species is a slightly protandrous hermaphrodite. Testes and penes are found in specimens with a length of 3 to 5 mm, and ovaries are found in specimens longer than 4 mm. In summer an individual can become mature in about 6 weeks after settling, and adults are capable of reproduction at all sub-lethal temperatures above 6°C. This means that *E. modestus* has an extensive breeding season and has the potential of producing several generations in a year.

The original habitant of *E. modestus* was on the shores of Australasia, but at some time after 1940 it was introduced into European waters. Since that date it has spread rapidly and in many areas of North-western Europe it has replaced *Balanus balanoides*. This is particularly so near the mouths of estuaries, for as well as having a more extensive breeding season and more rapid generation time *Elminius modestus* has a greater tolerance to dilution of sea water than *Balanus balanoides*.

Elminius modestus has not yet entered the Baltic Sea, so that its penetration into stable salinities cannot be compared with *Balanus balanoides*, which in regions of stable salinities penetrates down to a salinity of 12‰, but its congener, *B. improvisus*, penetrates to lower salinities of only 2 or 3‰. This is the only barnacle in the Gulfs of Finland and Bothnia, and has been recorded attached to the wing cases of the beetle *Hydrophilus piceus*.

The distribution of a barnacle in an estuary may be determined not only by the salinity tolerance of the adult, but also by the resistance to dilution of the medium shown by the larva. The nauplii of *Balanus balanoides*, *Elminius modestus* and *Chthamalus stellatus* are immobilised by salinities below 12‰. Above this salinity they are active and continue to be so even in salinities above that of sea water, until the salinity reaches 50‰, when they are again immobilised (Bhatnagar and Crisp, 1965). There is considerable interaction betweeen salinity tolerance and temperature. *Elminius modestus* nauplii show a greater resistance to dilution of the medium than do the nauplii of the other two species. At a temperature of 16°C and a salinity of 5‰ the nauplii of *E. modestus* survive for 30 minutes, the nauplii of *C. stellatus* survive for 10 minutes, while the nauplii of *B. balanoides* survive for less than 5 minutes. At 31°C in sea water the nauplii of *B. balanoides* survive for only 4 hours, *E. modestus* survive for 8 hours, and those of *C. stellatus* survive for over 48 hours. These results can be related to the known distributions of these species. *Balanus balanoides* is a northern species extending from the Arctic as far south as the north of Spain. *Elminius modestus* was introduced into Europe from Australasia, and now extends from Scotland to the south of Portugal, while *Chthamalus stellatus* is a southern species, extending from Scotland to West Africa. The temperatures at which the eggs of these three species will develop also reflect their geographical distribution. The eggs of *Balanus balanoides* develop normally at all temperatures between 2 and 16°C, while those of *Chthamalus stellatus* need temperatures above 9°C, and preferably above 16°C before they develop, and the upper limit for development is correspondingly high, reaching 31°C. *Elminius modestus* produces remarkably eurythermic eggs, which are capable of development at all temperatures between 3 and 32°C (Patel and Crisp, 1960).

The wide tolerance of variation in environmental conditions shown by *Elminius modestus* is also reflected in the foods which are suitable for rearing the larvae. In a series of experiments made by Moyse (1963) the nauplii of *Balanus balanoides* flourished only when fed on diatoms, but the nauplii of *Elminius modestus* grew well on a diet of certain flagellates as well as on diatoms. *Chthamalus stellatus* nauplii grew well on a diet of flagellates, but did not get beyond the second naupliar stage when fed on diatoms. A form such as *Elminius modestus* which is capable of utilising both diatoms and flagellates is clearly going to be at an advantage in estuaries, where diatoms are often frequent, but where flagellates are also often abundant, and may dominate the phytoplankton.

The life cycle of a barnacle in a tropical estuary has been described by Sandison (1966a), who studied *Balanus pallidus stutsburi* in Lagos Harbour. This barnacle appears to be restricted to brackish areas where the salinity is appreciably less than sea water. Optimal growth and survival were found to occur at salinities between 4 and 10‰. The salinity regime in Lagos Harbour shows great seasonal variation, with low values between June and October, and high values from January to May. During the high salinity period *B. pallidus stutsburi* is absent from the mouth of the harbour (East Mole in Fig. 48A), but present in the upper harbour. Conversely,

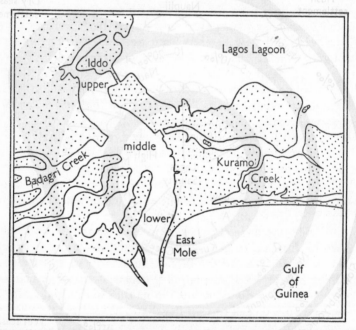

Figure 48. Life cycle of *Balanus pallidus stutsburi* in Lagos Harbour. A (above) diagram of the harbour and lagoon to show localities. B (opposite), annual cycle of events. The widths of the circles at Iddo and East Mole are an indication of the relative sizes of the populations. In Kuramo Creek variations in population numbers are not known. Seasons of high and low salinity in Lagos Harbour are shown in the centre of the figure (after Sandison, 1966)

during the low salinity period the barnacle is abundant at the harbour mouth, but rare in the upper harbour (Iddo in Fig. 48A). A population is present throughout the year in the nearby Kuramo Creek.

Barnacle nauplii were found in Lagos Harbour most abundantly in January and February. Cyprids were found settling in large numbers in the upper harbour at Iddo from February to April. These larvae presumably came from the permanent population in Kuramo Creek. This large settlement survived rather poorly, and by May about 75 per cent had died. The survivors produced nauplii in May. The cyprids from these nauplii settled near the mouth of the harbour as the salinity

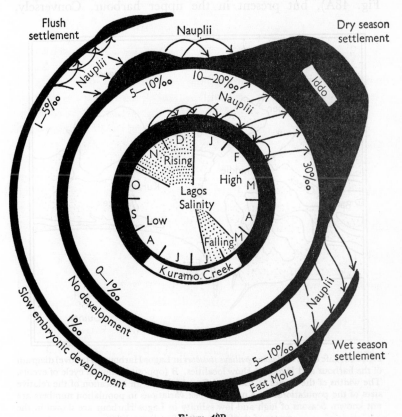

Figure 48B

was falling. The population thus established near the mouth survived until the following December, and produced some nauplii which were carried up into the harbour as the salinity rose at the end of the wet season. When the salinity rose above 30‰ the population at the harbour mouth disappeared.

The presence of *Balanus pallidus stutsburi* in Lagos Harbour thus depends on a supply of larvae from neighbouring creeks, and the population shows a seasonal redistribution that is governed by the wet and dry seasons. The barnacle is sensitive to both low and high salinities, with the optimum well below the salinity of sea water. The life cycle is summarised in Fig. 48B.

Estuarine Isopods

Marine isopods are in many ways preadapted for an estuarine or a semi-terrestrial mode of life. They have well-developed walking legs, biting mouthparts, capable in some cases of dealing with material as tough as wood, and they carry their young in a brood pouch. When the young are liberated from the brood pouch they are miniatures of their parents, so that a planktonic larval stage can be omitted from the life cycle.

The Asellota are generally regarded as being the most primitive isopods. They have a free leg segment which in other isopods is incorporated into the body wall. Members of this group are found in fresh water, brackish water, and in the sea down to the greatest depths. Species of the genus *Jaera* are often abundant in estuaries. In Britain two species were recognised until recently: *J. albifrons* (= *J. marina*) and *J. nordmanni*, but recent work by Bocquet (1953) and Naylor and Haahtela (1966) indicates that the former species should be regarded as a superspecies, comprising four species. An additional species, *J. hopeana*, has been found living on the underside of *Sphaeroma serratum* (Haahtela and Naylor, 1965).

All these species are euryhaline to some extent, but *J. nordmanni* penetrates further into estuaries and higher up the shore than the others (Naylor, Slinn and Spooner, 1960). The genus as a whole is usually found in stony areas, either under

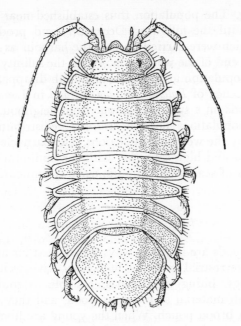

Figure 49. Jaera albifrons (Peracarida: Isopoda), adult female, actual length 5 mm
(after Sars)

stones or under algae attached to stones. The species of the *J. albifrons* group are all very similar morphologically, and can be separated only by certain features of the males. But these small morphological differences are accompanied by some ecological differences. *Jaera albifrons* is most common on sheltered shores, underneath stones in wet areas below high water of neap tides. *Jaera praehirsuta* is more frequently found on algae, but also occurs under stones. *Jaera ischiosetosa* becomes most abundant in fairly strong streams flowing across stony areas, and *J. forsmanni* appears to be somewhat less euryhaline than the other species, being most abundant near the mouth of estuaries under stones in well-drained areas. These ecological preferences are not sufficient to completely isolate the species, so that several species may be found on one shore, and two species may be found under a single stone. In spite of this overlapping occurrence the species remain distinct and hybrids are rare.

On British shores females of *Jaera* may be found with eggs or embryos in the brood pouch throughout the year, although the peak breeding season occurs in the spring and summer.

There are a few species of *Jaera* in fresh water in southern and eastern Europe. They probably encounter severe competition from the widespread and abundant species of the genus *Asellus*. This genus is abundant in the fresh water of the Northern Hemisphere, but few species penetrate into estuaries. One of the exceptions is *Asellus aquaticus*, which is considered in detail in Chapter 8.

In contrast to the Asellota, which are basically crawlers, many members of the Flabellifera can swim vigorously. On looking into a salt marsh pool one frequently sees a member of the genus *Sphaeroma* detach from the marginal vegetation and swim across the pool back downwards with the pleopods beating rapidly. The name of this genus derives from the ability to roll into a ball so that the head touches the telson and the ventral surface is completely hidden. This ability may serve as a protection against small predators, but it may also help to retard desiccation if the salt marsh pool dries out. In summer the cracks in the sides and floors of dry salt marsh pools are often found to contain *Sphaeroma* rolled into a ball awaiting the next spring tide.

Sphaeroma does not carry its young in the external brood pouch, but has four internal pouches which open on the ventral surface inside the external pouch. The eggs are first laid into the external pouch, but the female then bends the head ventrally towards the telson and the eggs are forced into the internal pouches (Kinne, 1954). The possession of these internal pouches is probably an adaptation that allows the female to continue rolling into a ball when carrying young. If the eggs were carried in the external pouch they would prevent the head from reaching the telson, and the female would then be at a disadvantage if attacked by a small predator.

A sequence of *Sphaeroma* species may be found along the length of an estuary. In marine conditions at the mouths of European estuaries, and on the open shore, *Sphaeroma serratum* is the common species. Inside the estuary *Sphaeroma monodi* is most frequent in the outer half, and occurs under any available

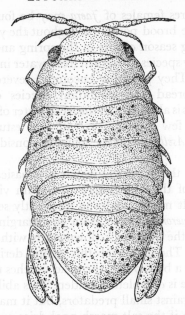

gure 50. Sphaeroma rugicauda (Peracarida: Isopoda), actual length 7 mm

shelter such as empty cockle shells. In salt marsh pools with a considerable range of salinities *S. rugicauda* becomes dominant. There may be some overlap between *S. monodi* and *S. rugicauda*. Even further into the estuary *S. hookeri* is often found in brackish ditches at somewhat lower salinities than *S. rugicauda*. In the Baltic *S. hookeri* is often found among stones covered with the barnacle *Balanus improvisus*.

The life cycle of *Sphaeroma hookeri* has been studied in detail at two localities: at Kiel (Kinne, 1954), and in the harbour at Copenhagen (Jensen, 1955).

At Kiel there are two main broods in a year. The first brood is produced by females that have overwintered. They lay their eggs in May and the young emerge in June. The second brood is produced by the females of the first brood which become mature in a month, and lay their eggs in early August. In the harbour at Copenhagen there is only one main brood in a year. This appears somewhat later than the first brood at Kiel, and the females do not become mature in the year of their birth.

The number of eggs per brood varies with the size of the female. At Kiel, *S. hookeri*, with a length of 3·5 mm, produced 10 to 16 young per brood, females with a length of 5 mm produced 41 to 62, while the largest females with a length of 7 mm produced 88 to 115 young in each brood. The last group was exceptional; Kinne (1954) estimates that the mean total number of eggs produced by a female in the normal two broods is about 71.

When a female *Sphaeroma* has liberated her young she moults and loses the oostegites which form the external brood pouch. This means that before she can produce another brood she must moult again and redevelop the oostegites. The females of *S. hookeri* at Kiel lose their external pouches in September and reacquire them in the following April or May.

Another flabelliferan that can swim vigorously is *Eurydice pulchra* (Fig. 51). This is a marine species which penetrates a long way into estuaries, and may be seen swimming swiftly in the rising tide. It is sometimes found in salt marsh pools, but this is often a case of stranding. Normally it appears to prefer open, moving water, and it will burrow into sand. According to

Figure 51. Eurydice pulchra (Peracarida: Isopoda), actual length 6 mm (after Sars)

Soika (1955) this species spends most of its time buried in sand, and emerges to lead a pelagic life at night.

Cyathura carinata is an elongated isopod, reaching a length of about 14 mm and forming burrows in intertidal estuarine muds. Spooner and Moore (1940) found that this species was a dominant form in the middle reaches of the Tamar Estuary, reaching a density of 378 per sq m in some areas. The maximum density was consistently found in the upper part of the intertidal zone, above mid-tide level. This species is a protogynic hermaphrodite, each individual passes through a female phase before changing to a male (Legrand and Juchault, 1963). Populations in France appear to lack males in the period from October to March, but if the gonads are examined it is found that although the external characteristics are female some of the individuals collected in December and January have testes. Later in the year the males are easily identifiable by external characteristics.

The ecology of *Cyathura polita* on the eastern coasts of N. America has been studied in detail by Burbanck (1962). This species seems to be more tolerant of fresh water than the European *C. carinata*. In some areas *C. polita* has been found living in the same water as fresh-water leeches and may-fly nymphs. On Cape Cod this isopod is frequently found in regions where *Spartina* marshes merge with *Typha* marshes. This is a transitional zone with fresh water beginning to dominate. Further south along the Gulf of Mexico, *C. polita* is frequently found in stable sandy areas matted with roots of the tape grass *Vallisneria*.

The tolerance of *C. polita* to fresh water seems to be greater in the southern part of its range, and it seems as if the species is slowly spreading northwards along the eastern coast of N. America. There is evidence that the species has spread further north during the last century. An interesting parallel to this is found in the appearance of *C. carinata* in Sweden during the 1930s in an area further north than it had been recorded previously. Burbanck (1959) has suggested that these two species once formed a single species that was driven southwards by the last Ice Age. The two populations, one on each side of the Atlantic, have now diverged sufficiently to be regarded as separate species, and are now spreading northwards again.

Cyathura polita makes burrows which are simple and unlined, extending downwards to a depth of about 8 cm. The animal generally lies between 5 and 7 cm below the surface, but very young specimens may inhabit the surface layers when an algal mat is present. The food of *C. polita* seems to be a mixture of organic detritus and diatoms, although it also eats dead fish and has been seen in the laboratory to attack and kill small gammarids.

When attempting to define the ecological niche of *C. polita*, Burbanck found that there was no one macroscopic species or group of species that was consistently associated with this isopod. Towards the head of an estuary the associated fauna included *Chiridotea almyra, Gammarus tigrinus, Leptocheirus plumulosus, Leptochelia dubia, Corophium lacustre, Almyrocuma proximoculi*, and the polychaete *Scolecolepides viridis*. But nearer the mouths of estuaries the following associates were found: *Crassostrea virginica, Macoma balthica, Mya arenaria, Venus mercenaria, Nassa obsoleta, Balanus eburneus, Callinectes sapidus* and *Nereis limbata*.

Some physiological aspects of the ecology of *C. polita* were studied by Frankenberg and Burbanck (1963). They studied two widely separated populations, from Georgia and Massachusetts, but found that both had similar physiological responses. The body fluid was maintained hypertonic to the external medium at salinities between 1 and 28‰, but was isotonic between 28 and 43‰. Oxygen consumption by *C. polita* did not appear to be influenced by salinity, and remained at the same level in all salinities between 1 and 37‰.

The breeding season of *C. polita* on Cape Cod extends from May to late August. There may be some aggregation of the population at the beginning of this period. The highest population density recorded by Burbanck (1962) was 4,000 per sq m in a sample from the Pocasset River in May. At other times of the year the maximum density recorded was 1908 per sq m, which was exceptional; the average normally lay between 100 and 200 per sq m.

A point of general interest concerning the genus *Cyathura* is the existence of three blind species living in subterranean fresh waters. These species presumably originated from estuarine ancestors, and one of them, *C. milloti*, from Madagascar, was found in a fresh water resurgence in the intertidal zone. The species most recently discovered, *C. specus*, was taken from an underground lake in a cave in Cuba (Bowman, 1965). The valviferous isopods have the uropods modified to form a pair of door-like structures which can close over the pleopods (Fig. 52). Swimming is effected by the pleopods with the uropods held open so that a current of water can flow out at

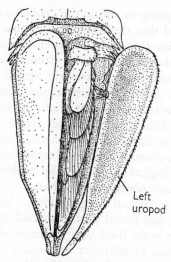

Figure 52. Ventral view of the posterior end of *Mesidotea entomon* (Peracarida: Isopoda). The left uropod has been bent back to reveal the pleopods

the hind end. The legs have well-developed claws so that the animal can cling on to algae or barnacle covered rocks. In British estuaries *Idotea chelipes* (= *I. viridis*) is the species most tolerant of dilution of sea water.

Idotea has the first pair of legs modified as gnathopods, which are used in grasping food, but the members of the subfamily Glyptonotinae have the first three pairs of legs modified in this way. This subfamily includes the huge Antarctic forms of the genus *Glyptonotus*, and the smaller forms *Chiridotea* and *Mesidotea* (= *Saduria*).

The genus *Chiridotea* includes a small number of species which occur on the eastern coast of North America, from Florida to

Nova Scotia. Two species, *C. almyra* and *C. nigrescens*, live in estuaries.

The species of *Chiridotea* are small (5 to 6 mm long) scavengers feeding mainly on animal material. The gnathopods are used to push food into the mouth with a rhythmic packing action at a rate of about once per second. At the same time the mandibles clip away the food at a rate of about four or five bites per second (Tait, 1927). This rhythmic use of the gnathopods does not seem to have been observed in any other isopod.

Perhaps the most interesting valviferan dwelling in brackish water is *Mesidotea entomon*. This large species (Fig. 53) is found in the Baltic, Caspian and Arctic Seas, as well as in certain

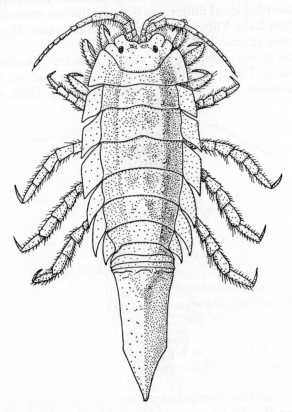

Figure 53. Mesidotea entomon (Peracarida: Isopoda). Actual length 5 cm

fresh-water lakes bordering the Baltic. It lives as a scavenger, but is also capable of catching and killing small animals such as chironomid larvae (Green, 1957b). Specimens taken from fresh-water lakes have the remarkable ability to live in full strength sea water provided that they are acclimatised by stages (Lockwood and Croghan, 1957). The mean chloride content of the haemolymph of specimens from Lake Vättern in Sweden was found to be 239 mM/1. The water of lake Vättern contained only 0·2 mM/1. Specimens from the brackish waters of the Baltic had a haemolymph chloride content of 335 mM/1, and were found to be incapable of surviving in fresh water. It thus appears as if there are two physiological races. There are some slight morphological differences between the two races, and the form from Lake Vättern has been given the name *M. entomon* f. *vetterensis* (Ekman, 1919).

A final group of isopods that may be found in estuaries is the Oniscoidea. This group includes the woodlice and sea slaters. The common sea slater of Western Europe, *Ligia oceanica*

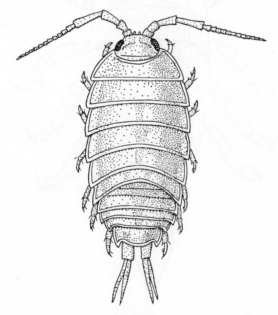

Figure 54. Ligia oceanica (Peracarida: Isopoda), actual length 28 mm (after Sars)

G

(Fig. 54) is found mainly on rocky shores, but also penetrates along the high water mark in estuaries when sufficient stony ground is available. Parry (1953) found that this species could tolerate considerable changes in the concentration of its blood, from 775 to 1,870 mOsm/Kg water. The main problem that *Ligia* faces is one of desiccation, for it spends most of its time out of water. The cuticle of *Ligia* is fairly permeable so that it loses water easily and the blood becomes concentrated. The ability to tolerate an increase in the concentration of the blood seems to be advantageous when *Ligia* is exposed to the sun. The evaporation of water cools the animal, keeping the body temperature several degrees below that of the surrounding air (Edney, 1957). This could be a life-saving mechanism when the temperatures approach the lethal limit for *Ligia*. Of course this loss of water cannot continue for long periods, but for short periods it enables *Ligia* to inhabit areas that might otherwise be uninhabitable. Sooner or later the water that has been lost must be replaced. *Ligia* shows an interesting technique of water uptake, using both oral and anal drinking, The latter process is the more frequent in *Ligia,* and involves dilation of the rectum and antiperistaltic movements which transport water to the mid gut where it is absorbed. Anal uptake of water is a common feature in many aquatic Crustacea (Fox, 1952).

Some species of woodlice, such as the common *Porcellio scaber* and *Oniscus asellus,* are sometimes found in the upper reaches of salt marshes, but these occurrences can be viewed as more or less accidental penetrations from the neighbouring terrestrial communities.

Estuarine Amphipods

Amphipods form an important element in the epifauna of estuaries, and predominant among the estuarine amphipods are the members of the family Gammaridae. The genus *Gammarus* for instance has a large series of species which between them occupy niches with salinities ranging from higher than sea water down to fresh water. In British estuaries the following species are commonly encountered as one proceeds inland from the sea. *Gammarus locusta, G. salinus, G.*

zaddachi, G. duebeni and *G. pulex*. The last species is a fresh-water form, but *G. duebeni* is remarkable for being both a fresh-water form and a marine-brackish form. The distribution of *G. duebeni* in British fresh waters seems to be strongly influenced by the presence or absence of *G. pulex*. This last species appears to

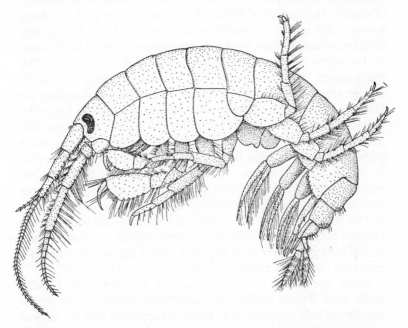

Figure 55. Gammarus duebeni (Peracarida: Amphipoda), actual length 15 mm

have invaded Britain after the Ice Age, and Hynes (1955) has shown that in fresh water it has a higher reproductive potential than *G. duebeni*. As a result, *G. duebeni* is confined to brackish waters in estuaries where *G. pulex* is present at the fresh-water end.

At the seaward end of its distribution *G. duebeni* overlaps the distributions of *G. zaddachi* and *G. salinus* (Spooner, 1947). These two species have been shown by Kinne (1960, 1961a) to have a higher reproductive potential than *G. duebeni* at salinities of 5‰, 10‰ and 30‰, and at temperatures ranging from 12 to 24°C. The rates of growth, moulting and heart beat were also faster in the other two species. At first sight it seems difficult to see how *G. duebeni* manages to survive in the face of competition at both ends of its range. It may be that *G. duebeni* is particularly well adapted to those regions of an estuary where the salinity changes are particularly large and rapid. The distribution of this species in the Gwendraeth Estuary supports this view. Here *G. duebeni* is found in pools where the salinity ranges from below 1 to 30‰ and the change from low to high salinity may take place during the course of a single spring tide.

Gammarus duebeni has the ability to crawl out of pools and to move across damp ground. Lagerspetz (1963) found that when out of water this species has a well-marked humidity response, and can choose the higher of two relative humidities when presented with the choice under experimental conditions. Another species, *G. oceanicus*, also showed a similar ability, but did not move as rapidly on land as *G. duebeni*. A third amphipod, *Pontoporeia affinis*, which does not normally leave the water, showed no ability to distinguish the higher of two humidities. As *Pontoporeia* is always submerged, such an ability would not be of use to it in its normal environment. On the other hand, in the two species of *Gammarus*, and particularly in *G. duebeni*, this ability would be of survival value in their normal habitat. In damp air *G. duebeni* survived better than the other two amphipods, and this resistance to desiccation may also be of importance in the competition of this species with other estuarine gammarids.

The osmotic and ionic regulatory powers of *G. duebeni* have recently been studied in some detail by Shaw and Sutcliffe (1961) and Lockwood (1961, 1964, 1965). Field and laboratory studies have shown that this species can tolerate salinities from 0·2 mM/l up to over twice the concentration of sea water. When *G. duebeni* is in sea water the blood is slightly hyperosmotic to the medium, and can be maintained well above the concentration of the external medium if this is reduced below that of sea water. If the concentration of the external medium is reduced slowly, *G. duebeni* from brackish water can be adapted to live in solutions down to 0·2 mM/l. A comparison with the fresh-water species *G. pulex* shows that the latter can survive in concentrations down to ·006 mM/l. Both species have a mechanism for the uptake of sodium, which is revealed by tracer experiments using ^{24}Na. The uptake of sodium by G. pulex in low external concentrations is more rapid than that of G. duebeni, but the mechanism of G. pulex quickly becomes saturated and the rate of uptake does not then increase with increasing external concentrations. The rate of sodium uptake by G. duebeni continues to increase with increasing external concentrations, so that although the rate is lower at an external concentration of 0·2 mM/1, by the time the external concentration has reached 3 mM/1 the rate is twice as high as that of G. pulex. It is clear that G. pulex is the better fresh-water animal and that G. duebeni is better adapted to dilute brackish water.

When in sea water G. duebeni produces urine isotonic with the blood, but when in fresh water the urine is hypotonic to the blood. This is important in relation to the conservation of salt when the animal is in a dilute external medium. The volume of urine produced can be taken as roughly proportional to the difference between the concentration of the blood and the external medium (Werntz, 1962). If the urine remained isotonic with the blood the total salt loss would be much greater in the more dilute medium. This would then place an extra load on the salt uptake mechanism. The relative importance of the salt loss via urine and via the body surface has a direct bearing here. Lockwood (1964) has found that when G. duebeni is producing isotonic urine about 80 per cent of the total sodium loss is via the urine. This is in marked contrast to the larger Crustacea living in dilute media. For instance, in Eriocheir sinensis only 14 per cent of the total salt loss takes place via the urine, and in the African river crab, Potamon niloticus, the figure is reduced to 1 per cent.

The eggs of Gammarus duebeni are laid into a brood pouch at the bases of the thoracic limbs, and hatch as miniature adults after a period which varies according to the temperature. At 18°C the incubation period is 14 days, but at 4·7°C the time is extended to 54 days (Hynes, 1954). The time taken to reach maturity at a temperature of 15 to 20°C varies from 23 to 30 weeks. Specimens kept in captivity have lived as long as 25 months, but in nature they probably live for a maximum of 15 to 17 months.

If females of *G. duebeni* are kept at a constant salinity of 10‰ the sex of their offspring depends on the temperature during a critical period of a few days before oviposition. If the temperature is below 5°C the young are males, if the temperature is above 6°C the young are females. Between 5 and 6°C the young are produced in mixed broods. The critical temperature for sex determination may be changed if the animals are kept at a constant high temperature (19–20°C), or if they are kept in higher salinities (Kinne, 1952, 1953). The exact significance of these findings is uncertain, but Kinne (1961b) has made a number of suggestions concerning their ecological importance. It may well be that at high temperatures the sexual activity of a small number of males born early in the year will be sufficient to fertilise the eggs of a larger number of females. At low temperatures the rate of moulting, and so the frequency of egg laying, by the females will be lower, and the statistical chances of a male finding a female in the right condition for fertilisation are reduced. A higher number of males would then be necessary to ensure the fertilisation of all the females.

In addition to the genus *Gammarus* there are several other genera of the Gammaridae with representatives in estuaries. We cannot mention them all here, but the genus *Melita* deserves mention because it shows a sequence of species along the length of an estuary. In British estuaries *Melita palmata* occupies the outer part, while *M. pellucida* penetrates further towards fresh water (Goodhart, 1941; Jones, 1948). The genus *Marinogammarus* contains about half-a-dozen species (Sexton and Spooner, 1940). The commonest species in British estuaries is *M. marinus*. The other species penetrate little or not at all into estuaries, but several of them occur on rocky shores, particularly where a small stream of fresh water crosses the intertidal zone. *Marinogammarus pirloti*, for instance, is often found under stones in company with the flatworm *Procerodes ulvae* (see p. 116).

The family Talitridae has penetrated into estuaries by taking a route high on the shore. The common sand hopper, *Talitrus saltator*, lives on sandy shores, and burrows just above the high water mark in well-drained sand. During the daytime *Talitrus* remains in the burrow that it made early in the morning, but at night it emerges to scavenge on the shore. *Orchestia gammarella,* which belongs to the same family, is often abundant under wrack cast high up on rocky shores. This species also penetrates well into estuaries and is common under stones in the upper parts of salt marshes. The process of hopping by a sudden extension of the normally flexed tail is a very effective protection against predators. It is a common experience to turn over a stone and see several dozen *Orchestia,* then to see them hop away so rapidly that they all disappear in a few seconds.

Corophium volutator is the best known of the amphipods which form burrows in fine sand and mud, but it is only one member of a large genus which has about 45 species found in a wide range of salinities. Most of these species live in marine or brackish habitats, but a few have extended into fresh water. Before discussing the other species *C. volutator* will be considered in some detail to provide a basis for comparison.

The general form of *Corophium volutator* is shown in Fig. 56. The second antennae are very large and are carried in front of the animal as it moves about. The fifth walking legs are longer than the others and are used to brace the animal inside its burrow. The first two pairs of thoracic limbs are modified to help in feeding and are called gnathopods. The second pair of gnathopods carries a large series of filtering setae.

The food of *Corophium* seems to be organic detritus and its associated micro-organisms. Particles are sorted out from the

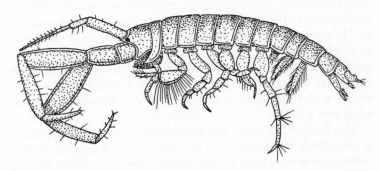

Figure 56. Corophium volutator (Peracarida: Amphipoda), actual length of body
8 mm

substratum by means of the gnathopods, and then passed to the other mouthparts. This method is used when the animal is partly or wholly out of its burrow. When in its burrow the pleopods on the underside of the abdomen are moved backwards and forwards creating a current which flows through the U-shaped burrow. This current passes in over the head and out past the abdomen, and serves to draw in aerated water for respiratory purposes. It also carries small particles which are trapped by the fringe of filter setae on the edges of the second gnathopods. Food collected in this way is transferred forwards to the mouth by a complex series of movements of the mouthparts.

In a suitable mud the population of *Corophium volutator* can rise to a very high level. Spooner and Moore (1940) recorded 11,000 per sq m in parts of the Tamar Estuary, but this is a relatively low figure when compared to the 63,000 young per sq m recorded by Watkin (1941) in a sample taken from the Dovey Estuary in May.

According to Watkin the population in the Dovey Estuary overwinters in two groups: large specimens, over 7 mm in length which begin breeding in February, and smaller specimens which do not become mature until March or April. The young born early in the year become mature and start to breed in July and August, so that there are two generations in a year. An individual *C. volutator* does not live for much more than a year, usually somewhat less. In the Gulf of Finland the breeding season lasts from May until the end of September (Segerstråle, 1940). The young are carried in the brood pouch for about four weeks, and some of the females may produce four or five broods in a season.

The young remain in the maternal brood pouch until they are miniatures of the adults. In general the young appear to be liberated in the burrow, and Thamdrup (1935) has described small side burrows branching from the main parental burrow. At first these small burrows remain functionally connected with the parental burrow, but as the young increase in size their burrows become progressively more independent.

Although the burrow is a fairly permanent structure, *C. volutator* does not restrict its life to such a confined space. This amphipod is capable of walking over the surface of the mud, either when the tide is in or when the tide has just ebbed and left the surface of the mud wet. *Corophium* can also swim, and the males are thought to leave their burrows and swim in search of females. Local movements of whole populations have been recorded. The details of these movements have not been worked out, but Watkin (1941) found that during the course of a year the centres of populations shifted. He also found that in some parts of the Dovey Estuary the population of *C. volutator* diminished when the soil became gradually drier, and this diminution coincided with colonisation of the area by the glass-wort *Salicornia europaea* and a tendency for *Corophium arenarium* to replace *C. volutator*.

Differences in ecology and substrate selection between *Corophium volutator* and *C. arenarium* have been studied by Gee (1961) and Meadows (1964a and b). In general *C. volutator* is found in areas with a high silt and clay content combined with a high water content. If given the choice of burrowing into a fine sand or a coarse sand it will choose the finer. *Corophium arenarium* is found in drier areas with a lower silt and clay content. A further difference is that *C. volutator* shows a distinct preference for muds with a reduced oxygen content, while *C. arenarium* prefers better aerated sands.

It is not only the physical structure of the substratum that renders it attractive or otherwise to *Corophium*. The surfaces of sand grains and silt particles are normally covered with a film of bacteria and other micro-organism. Meadows (1964a) has shown that destruction of this film by boiling with acid removes the attraction of the specific substrates of both *C. volutator* and *C. arenarium*, even though all the acid was washed away most scrupulously. Treatment of the sands with salt solutions which cause bacteria to detach from the particles does not necessarily cause the sand grains to become unattractive to *Corophium*. This may indicate that some substance secreted by the micro-organisms remains on the particles and is at least partly responsible for rendering the sand attractive to *Corophium*.

Apart from their different substratum preferences, *C. volutator* and *C. arenarium* differ in their tolerance of dilution of sea water. In the extreme case *C. arenarium* dies in less than an hour when placed in tap water, while *C. volutator* can survive as long as 16 days.

At the seaward end of an estuary *Corophium volutator* tends to be replaced by other species of the genus. In the canal that runs from Amsterdam to the North Sea *C. sextoni* and *C. insidiosum* were found near the mouth (Stock, 1952). The latter species normally builds tubes of fine particles cemented by a secretion from the second pair of legs, and attached to some solid object or enmeshed in seaweeds. It can also burrow in mud, and Stock found it in large numbers together with *C. volutator*.

In beds of *Zostera marina* lying in shallow water near the mouths of estuaries, another species, *C. crassicorne*, is often found burrowing into fine sand. A closely allied species, *C. bonelli*, builds tubes on hydroids and weeds in full strength sea water. The last species is unusual among amphipods in that it appears to be parthenogenetic (Crawford, 1937). Although thousands of females have been found the male remains unknown.

Another species, *C. multisetosum*, has been found burrowing in clay and in tubes attached to a hard substratum in almost fresh water. This species overlaps yet another, *C. lacustre*, which also builds tubes, particularly on the hydroid *Cordylophora*. In spite of its name, *C. lacustre* does not seem to be a truly freshwater species, although it does seem to be on the threshold of becoming one. A species which has crossed this threshold is found in many rivers in Europe. This is *C. curvispinum*, which is thought to have spread from the Caspian and Black Seas into the Volga and Danube and then into other rivers, so that it is now found in rivers draining into the Baltic and North Seas. There is a single record from England, in the River Avon at Tewkesbury.

ESTUARINE SHRIMPS AND PRAWNS

The Decapoda Natantia includes those decapods which swim by means of their abdominal appendages. It should be made clear that the terms 'shrimp' and 'prawn' have no scientific validity. In England the term prawn is usually applied to members of the genus *Palaemon*, and shrimp is applied to *Crangon vulgaris*. Members of other families are called either shrimps or prawns. In N. America the term shrimp is applied to members of the Penaeidae, but English authors often call these prawns.

Within the Decapoda Natantia there is a wide range of families with representatives penetrating into estuarine waters, and, particularly in tropical regions, into fresh water. For instance, in a review of the brackish water prawns of Malaya, Johnson (1965) lists representatives of five different families. It is not possible to discuss all the different estuarine shrimps and prawns here, but examples have been chosen to illustrate certain points.

The Penaeidae form the basis of the Texas shrimp fishery. The young shrimps thrive best in waters of reduced salinity, but the adults migrate seawards to breed. The eggs are shed freely in the sea, and give rise to nauplius larvae. After passing through several moults the nauplius gives rise to a protozoea, which has stalked eyes and distinct thoracic segments, but it still swims in the same manner as the nauplius, using its antennae. After three moults the protozoea gives rise to a zoea; the number of appendages increases and the general shrimp-like form is well established. The zoeal stage swims by means of its thoracic limbs. When the abdominal limbs have developed sufficiently to take over the role of swimming the shrimp has entered its post-larval stages. At subsequent moults additions are made to the segments and limbs, until after about 14 stages the adult form is assumed.

The young white shrimp, *Penaeus setiferus*, enters the coastal waters of Texas and Louisiana, and penetrates into salinities as low as 0·4‰. This species is also capable of living at a salinity of 41‰, but in these conditions it does not grow well (Gunter, 1961). The young of the brown shrimp, *Penaeus aztecus*, can also penetrate to low salinities, reaching 0·8‰ in some parts of its range. The third important species inhabiting the same general area, *P. duorarum*, the pink shrimp, is not as euryhaline as the other two and does not penetrate into salinities less than 2·7‰.

Similar migrations of young penaeids into low salinities have been studied in Malaya by Hall (1962) who recorded 11 species in the mangrove swamps near Singapore.

Most of the other families of shrimps do not shed their eggs but carry them attached to the abdominal appendages of the female. The larval form emerging from the egg is usually at a more advanced stage than the nauplius of the Penaeidae. In some families the eggs are relatively large and the form that emerges is at an advanced zoeal stage, or even post-larval.

The family Palaemonidae has a considerable number of genera which have species in brackish or fresh water. *Palaemon*, *Macrobrachium*, *Leptocarpus* and *Palaemonetes* are examples of such genera.

The genus *Palaemonetes* contains about 20 species, of which 3 are marine in habit, 5 are found in brackish water and 12 are found in fresh water (Holthuis, 1950). In Europe *P. varians* is abundant in brackish drainage ditches and ponds, often in company with fresh-water animals at salinities down to 0·5‰. In southern Europe *P. antennarius* occurs in true fresh-water conditions. The osmoregulation of both these species has been studied (Panikkar, 1941; Parry, 1955, 1957, 1961).

When *Palaemonetes varians* is in sea water it maintains the concentration of its blood below that of the external medium. In 60 to 70 per cent sea water the blood is isotonic with the medium, but at lower salinities the blood is kept hypertonic to the medium. When the external medium is reduced experimentally to a salinity of 0·1‰ the blood concentration is

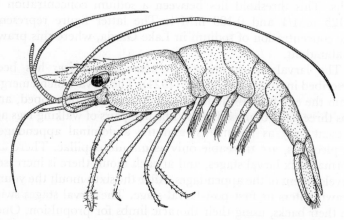

Figure 57. Palaemonetes varians (Decapoda: Natantia), length about 4 cm

maintained at about 18·9‰. The osmotic pressure and chloride content of the urine produced by *P. varians* does not differ much from the values found in the blood, but as the salinity of the external medium decreases from about 60 per cent sea water the rate of urine production increases. At low salinities the production of urine may reach 1·8 per cent of the body weight per hour. When the blood is isotonic with the external medium the rate of urine flow is at its lowest, being about 0·2 per cent of the body weight per hour.

Palaemonetes antennarius also produces urine which is isosmotic with the blood, but the concentration of the blood is lower than that of *P. varians*. This reduction of the osmotic concentration of the blood will relieve the stress of living in fresh water to some extent, but *P. antennarius* is remarkable in that it produces a copius flow of urine, which reaches about 2 per cent of the body weight per hour. This is much higher than the rate of flow recorded from fresh water crayfish, which produce urine at only one-tenth of this rate.

A high rate of flow of urine isotonic with the blood must drain a considerable amount of salt from the body. This means that there must be a considerable expenditure of energy in replacing this salt by active uptake from the surrounding water. There appears to be a threshold concentration of the external medium below which *P. antennarius* is incapable of taking up salts. This threshold lies between a sodium concentration of 0·125 mM/l and 0·183 mM/l. The latter figure represents the concentration of sodium in Lake Garda, where this prawn is abundant.

The larval development of *Palaemonetes varians* has been described in detail by Gurney (1924). The larva which emerges from the egg has all the head appendages well developed, and has three pairs of maxillipeds. The five pairs of walking legs are present only as rudiments, and the abdominal appendages, or pleopods, are traceable only as minute papillae. There are normally six larval stages, and at each moult there is increased development of the appendages. After the sixth moult the young prawn enters its first post-larval stage. The larval stages swim on their backs, using their thoracic limbs for propulsion. Once the post-larval stages are entered the prawns swim the right way up, using their abdominal appendages. The assumption of the adult posture coincides with the completion of development of the otocyst at the base of the antennule. The larval stages lack a functional otocyst, although traces of its development can be seen as early as the third larval stage. In the first post-larval stage the young prawn is 7 or 8 mm long. To become sexually mature it must pass through a series of about 15 moults until it reaches a length of about 25 mm. Each larval stage and the early post-larval stages last about 4 or 5 days, but the interval between moults in the later stages increases up to 10 days or more.

Another important natant decapod in European estuaries is the common shrimp, *Crangon vulgaris*. This species moves out of the estuaries in winter, but returns in the spring. The first pair of legs is equipped with strong pincers, and *Crangon* is capable of preying on *Nereis* and other worms. The annual

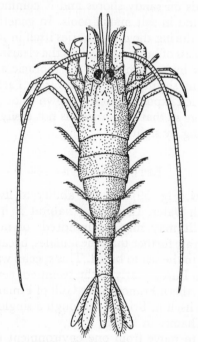

Figure 58. Crangon vulgaris (Decapoda: Natantia), length about 6 cm

cycle of *Crangon vulgaris* in the Bristol Channel has been studied by Lloyd and Yonge (1947). In the more saline areas the breeding season extends from January to August, but in the Severn Estuary there is a shorter season, from March to June. Under estuarine conditions a single brood is produced in each year, but in marine conditions there are two broods. The females in the estuary become mature after three years, but those that remain in the Bristol Channel produce a single brood in their second year and two broods in each of the third and fourth years. The normal length of life of a female is four or five years; the males live for about three years. Breeding by *C. vulgaris* appears to be inhibited by combined low temperatures and low salinity. The migration of the adults out of the estuary in winter indicates that they can tolerate low salinity only when the water is warm.

Crangon vulgaris is not very particular in its choice of substratum. It abounds on sandy shores and is common on mud; it may also be found in salt marsh pools. In general *C. vulgaris* is nocturnal, and during the day it buries itself in sand in shallow water. The process of burying involves the clearing of sand from under the body by movements of the thoracic and abdominal limbs so that the shrimp sinks downwards. Part of the dorsal surface is often left exposed, but the brown colouration provides good camouflage so that the shrimp is not easily detected by a predator hunting by sight.

ESTUARINE CRABS

Eriocheir sinensis (Fig. 59) has the ability to live in both sea water and fresh water. Its normal habitat is in the rivers of China, where it may migrate hundreds of miles upstream. The males migrate further than the females, because the females have to return to the sea to breed. This species was accidentally introduced into Europe early in the twentieth century and now ranges from northern France to the Gulf of Finland, but has not yet established itself in Britain, although a single specimen was found in the Thames in 1935.

The ability to move from one environment to another has made *Eriocheir* a subject suitable for the study of mechanisms of

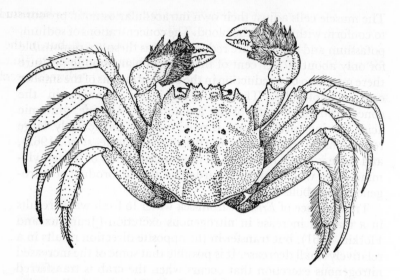

Figure 59. Eriocheir sinensis (Decapoda: Brachyura), width of carapace about 6 cm

osmoregulation. It has been found that this crab can keep the osmotic concentration of its blood above that of dilute external media and below concentrated external media. The gills of *Eriocheir* can actively take up sodium and chloride ions from the surrounding medium (Koch, Evans and Schicks, 1954); this is a vital part of the adaptation to living in fresh water. Another feature of *Eriocheir* is that its body surface is much less permeable to salts than are the surfaces of marine crabs. This reduces the amount of osmotic stress on the animal in fresh water and brings the loss of salts down to a level at which the uptake mechanism can provide adequate compensation.

In fresh water the blood of *Eriocheir* has a freezing point depression (Δ) of about $1 \cdot 1°C$ when the Δ of the water is $0 \cdot 02°C$. In sea water the blood is isosmotic with the external medium ($\Delta = 2 \cdot 09°C$). This means that although the crab is regulating well in fresh water and keeping its blood well above the concentration of the external medium, the blood concentration is reduced considerably from the concentration found when the crab is in sea water. The cells bathed by the blood must be subjected to a considerable osmotic change in their medium.

The muscle cells adjust their own intracellular osmotic pressure to conform with that of the blood. The concentrations of sodium, potassium and chloride are reduced, but as these are responsible for only about 40 per cent of the intracellular osmotic pressure there must also be a reduction in the concentrations of the smaller organic molecules. There are significant changes in the concentrations of some of the amino acids. Alanine, glutamic acid, glycine, proline and threonine show particularly large reductions in concentration (Bricteux-Gregorie *et al.*, 1962), and there is also a reduction in the concentration of trimethylamine oxide which can be regarded as an end product of nitrogenous metabolism.

Transference of *Eriocheir* from sea water to fresh water results in a marked increase in nitrogenous excretion (Jeuniaux and Florkin, 1961), but transfer in the opposite direction results in a relatively small decrease. It is possible that some of the increased nitrogenous excretion that occurs when the crab is transferred to fresh water is a result of the reduction in osmotically active nitrogenous compounds in the muscles. A study of the proteins in the muscles of *Eriocheir* during transfer from fresh water to sea water revealed no differences detectable by electrophoresis (Florkin and Schoffeniels, 1965), so that the main regulators of intracellular osmotic pressure seem to be the smaller nitrogenous molecules, particularly the non-essential amino acids such as glycine, glutamic acid and alanine.

The only crab which penetrates well into estuaries in Britain is the shore crab, *Carcinus maenas*. This species inhabits a wide range of habitats from open rocky shores to sheltered mud flats and salt marsh pools. It is most numerous where shelter is available, and large numbers of small specimens may be found under stones in the middle reaches of estuaries. In habits *C. maenas* appears to be a generalised predator, eating anything that it finds or catches. This wide range of feeding habits is coupled with an ability to osmoregulate at salinities lower than the blood concentration, so that the crab is well adapted to live in estuaries.

The osmotic and ionic regulation of *Carcinus* has been studied by a number of workers (Duval, 1925; Schlieper, 1929; Nagel, 1934; Webb, 1940; Shaw, 1955a and b, 1958, 1961;

Gilbert, 1959; Robertson, 1960; Bryan, 1961). The concentration of the blood can be maintained well above that of the external medium in concentrations lower than sea water down to about 6‰, which is the lowest salinity at which this crab will survive permanently. At concentrations higher than sea water the blood concentration increases with that of the medium. In low concentrations of the external medium there is an

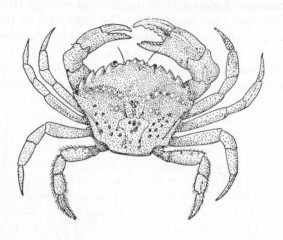

Figure 60. Carcinus maenas (Decapoda: Brachyura), width of carapace about 6 cm

active uptake of salts which compensates for any loss through the body surface. Urine production increases in the more dilute media. In sea water the amount of urine produced is equal to 3·6 per cent of the body weight per day, but in a salinity of 14‰ the urine production increases to 30 per cent of the body weight per day. The loss of salts through the urine does not increase proportionately; Shaw (1961) found that the amount of sodium lost through the urine was never more than 20 per cent of that lost through the body surface.

Carcinus maenas exhibits rhythmic patterns of locomotory activity with peaks of activity coinciding with high tide and with darkness (Naylor, 1958, 1960). This crab also moves up and down the shore. The larger specimens move further than the smaller specimens, and show some seasonal variation in this movement. In summer many of the crabs move up the shore with the advancing tide and remain on the shore when the tide ebbs, provided that there is sufficient shelter on the shore, but in winter a much higher proportion of the population moves down again with the tide. During the coldest months the crabs show no signs of moving upshore with the tide (Naylor, 1962). There is also evidence that estuarine individuals of *C. maenas* move seawards during the cold weather (Broekhuysen, 1936), and that females carrying eggs tend to remain at the seaward ends of estuaries until the larvae hatch (Rasmussen, 1959).

Females with eggs attached to the abdominal pleopods are found throughout the year at Plymouth, but are most abundant

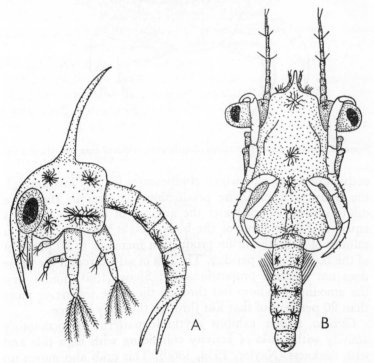

Figure 61. Carcinus maenas (Decapoda: Brachyura)
A, Zoea larva, length 1·4 mm, lateral view. B, Megalopa larva, dorsal view, length about 3 mm (B, after Atkins)

in February and March. The larva which hatches from the egg is a zoea (Fig. 61A). Such larvae are most abundant in the plankton in the spring and early summer. The number of zoeal stages is probably about seven, and then there is a metamorphosis into a megalopa, which resembles a small crab with a thin extended abdomen.

The megalopa can both swim and walk. When swimming the third and fourth walking legs are held in a remarkable manner, which has been fully described by Atkins (1954c). The third leg curves round the side of the body and extends forwards so that the claw loops over the eye, while the fourth leg is folded partly on to the dorsal surface of the carapace (Fig. 61B). This removes the last legs from the field of activity of the pleopods which propel the megalopa when it is swimming. The two anterior walking legs and the cheliped are held close against the side of the body giving the overall shape a somewhat more streamlined effect than if they were extended.

The megalopa metamorphoses into a small crab. The life span of *C. maenas* is about 3 or 4 years (Broekhuysen, 1936), and the young crab continues moulting until it reaches a carapace width of 70 to 85 mm, after this it stops moulting. This cessation of moulting usually occurs about ten moults after the crab has become sexually mature.

The moulting process is under hormonal control, and it has been shown that the activities of the Y-organ, which lies in the head, initiate the onset of moulting. In some crabs the Y-organ degenerates when the final size has been reached, but in *Carcinus* the Y-organ persists. It has been possible to take a crab which has ceased to moult and cause it to moult again by removing the eyestalks (Carlisle, 1957). This suggests that the eyestalks produce a substance which inhibits the moult inducing activity of the Y-organ. Removal of the eyestalks removes this inhibition, allowing an extra moult. In the spider crab, *Maia squinado*, the Y-organ degenerates, and removing the eyestalks of a full-grown specimen does not induce an extra moult.

The gills of *Carcinus* are housed in two chambers, one on each side of the body. Water is drawn into these chambers through gaps near the bases of the legs, and particularly near the bases of the chelipeds. Passage of water through the gill chamber is caused by movements of the scaphognathite. This is a flap attached to the base of the maxilla, lying in a plane parallel to the direction of current flow. After passing over the gills the water leaves the gill chamber via openings between the antennae. About 80 per cent of the water entering the gill chamber does so through Milne Edward's openings at the bases of the chelipeds (Arudpragasam and Naylor, 1964b).

The direction of flow through the gill chamber can be reversed, so that water enters between the antennae and leaves via the openings at the bases of the legs. Such a reversal is a regular feature in the normal crab (Arudpragasam and Naylor, 1964a), and serves to pass water over the more posterior gills which do not receive as good a supply of water as the anterior gills which are bathed by the large volume of water entering through the openings of Milne Edwards.

Carcinus often remains on the shore after the tide has receded, and so may find itself in a small pool which quickly becomes depleted of oxygen. In such a situation the openings between the antennae can be pushed out of the water and bubbles of air taken into the gill chamber and passed through the water bathing the gills by the action of the scaphognathite. A stream of bubbles may then be seen emerging through the openings of Milne Edwards. In this way the crab can aerate the water in its gill chamber and remain active in conditions under which a purely aquatic method of respiration would be inadequate.

In contrast with European estuaries the Brisbane river estuary in Australia has an abundance of crab species. Snelling (1959) found 23 species, though not all of these were common. The common species exhibited a definite zonation, both along the estuary and vertically at any one locality within the estuary. The two species that penetrated furthest towards fresh water were *Halicarcinus australios* and a *Sesarma* species. The former also occurred at the mouth of the estuary, but the latter was found only in brackish water and was replaced at the seaward end of the estuary by *Sesarma erythrodactyla*. A general pattern of vertical zonation was found to include four main zones:

1. Sesarmine zone, an upper zone, above high water of neap tides, usually with a mixture of mud and gravel. *Helice haswellianus* and *Sesarma erythrodactyla* occupy this zone.

2. Ocypodine zone, extending from high water of neap tides to mid-tide level. *Uca longidigitum* and *Heloecius cordiformis* occupy this zone. When mangroves are present the two are codominant, but in the absence of mangroves *Uca* becomes dominant.

3. Upper Macrophthalmine zone, extending from mid-tide level to low water neaps. In the lower reaches of the estuary *Euplax tridentata* was dominant, but further upstream *Cleistostoma mcneilli* was the dominant crab in this zone.

4. Lower Macrophthalmine zone, lying below low water of neap tides. In this zone the mud is usually very soft with numerous pools. Large numbers of *Macrophthalmus setosus* burrow in this zone.

These zones are only well developed in the lower and middle reaches of the estuary. As one progresses further upstream fewer and fewer species are found until at a point forty miles from the mouth only two species are found: *Halicarcinus australis* on the lower shore, and a *Sesarma* sp. on the upper shore.

A study of the gill areas of a variety of American crabs has been made by Gray (1957), who found that there was a tendency towards a reduction in gill area per gram body weight in the more terrestrial species. The wharf crab, *Sesarma cinereum*, spends most of its time on land, and has an average gill area of 638 mm² per gram body weight. This is a low figure when compared with an active aquatic species such as *Callinectes sapidus*, which has an average gill area of 1367 mm² per gram body weight. There was considerable variation in the gill area in relation to body weight within one species. The younger, smaller crabs had higher gill areas per unit body weight. Within the species living in water there was a marked difference between the sluggish spider crabs and the active swimmers such as *Callinectes* and other members of the Portunidae. For each unit of body weight the spider crabs had only half the gill area of the portunids.

Among the fiddler crabs of the genus *Uca* three species were

Figure 62. Sesarma cinereum (Decapoda: Brachyura), width of carapace 18 mm

examined by Gray (1957). *Uca pugilator* was found on sandy beaches near the seaward end of estuaries, *U. minax* was found a long way upstream, where the tidal range was small and *U. pugnax* occupied the salt marshes intermediate between the ranges of the other two species. There was some difference in the sizes of the specimens examined for each species, but if the highest gill area per gram body weight is taken for each species (Table 19) it is seen that *U. minax* has the highest value. This

TABLE 19

BODY WEIGHTS AND GILL AREAS IN THE GENUS 'UCA'
(After Gray, 1957)

Species	Weight of smallest crab examined	Maximum gill area per gram weight (mm²)
U. pugilator	1·6	817
U. pugnax	1·0	889
U. minax	3·8	904

species is exposed to the air for shorter periods than the other two and so conforms to the general rule of the more aquatic species having larger gill areas. The difference between *U. minax* and the other two is probably greater than is shown in Table 19 because the smallest specimens examined of this species were considerably larger than the other two species, and the gill area per gram body weight is higher in smaller specimens.

The larval development of *Callinectes sapidus* has been followed in the laboratory by Costlow and Bookhout (1959),

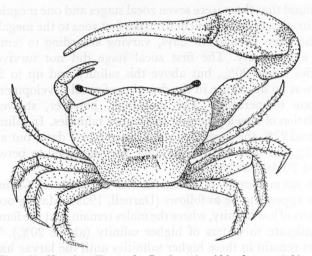

Figure 63. Uca minax (Decapoda: Brachyura), width of carapace 34 mm

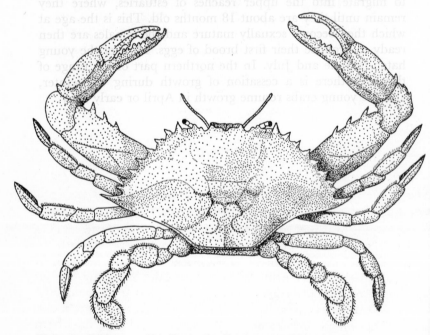

Figure 64. Callinectes sapidus (Decapoda: Brachyura), width of carapace 14 cm

who found that there were seven zoeal stages and one megalopa. The duration of development from the first zoea to the megalopa took between 31 and 49 days, varying according to temperature and salinity. The first zoeal stage did not survive at salinities below 20‰, but above this salinity and up to 32‰ there was no difference in the duration of zoeal development at any one temperature. The megalopa, however, showed a retardation of development at the higher salinities. In salinities of 20 and 27‰ the megalopa stage lasted 6 or 7 days, but at 31 and 32‰ the duration of this stage was increased to between 10 and 20 days at 25°C.

The normal cycle of events in the life cycle of *Callinectes sapidus* appears to be as follows (Darnell, 1959). Mating occurs in waters of low salinity, where the males remain, but the females then migrate to waters of higher salinity (above 20‰). The females remain in these higher salinities until the larvae hatch. Once the larval stages have been passed the young crabs begin to migrate into the upper reaches of estuaries, where they remain until they are about 18 months old. This is the age at which they become sexually mature and the females are then ready to produce their first brood of eggs. Most of the young hatch in June and July. In the northern part of the range of this crab there is a cessation of growth during the winter, and the young crabs resume growth in April or early May.

CHAPTER 7

ESTUARINE MICROBENTHOS

THE microfauna has been defined as consisting of those animals passing through a 1 mm sieve (Smidt, 1951). This is a fairly loose definition, because many elongated forms, such as nematodes, may be several mm in length, and still pass through a 1 mm sieve. There is no advantage to be gained in trying to make the definition more precise. The small annelids and nematodes which exceed 1 mm in length are often very slender and are similar in width to some of the larger ciliates. To separate the nematodes on the basis of a single dimension would clearly be artificial and unworkable. Grouping small animals together in this way is to some extent a matter of convenience, but there are certain problems encountered in the study of small animals which are different from those encountered in the study of larger forms. The time scale of events and the durations of life cycles are greatly compressed, and surface phenomena at interfaces between particles and water, or water and air, become important.

The most important elements of the microfauna of estuaries are the Protozoa, Nematoda and Crustacea. Other groups, such as the Turbellaria, Coelenterata, Rotifera, Gastrotricha, Annelida and Acarina may assume local importance. The young stages of some members of the macrofauna may temporarily become members of the microfauna. Newly settled spat of bivalves and young gastropods are often well within the size limits of the microfauna and share the same environment.

The environment is influenced very strongly by the size of the particles forming the substratum. In estuaries one is not usually concerned with very coarse particles, but with sands and muds. Sands offer the advantages of high permeability, good aeration and comparatively large interstices through which small forms can travel without moving the sand grains.

Muds offer a vast particulate surface for the growth of bacteria which can be utilised as food, but the interstices are small and aeration is sometimes restricted to the top few millimetres. The small size of the interstices in mud may be compensated for by the property of thixotropy (see p. 27), which permits the passage of burrowing forms by a reduction in resistance when the mud is agitated.

ESTUARINE CILIATES

Some of the microhabitats available to the microfauna are illustrated by the studies of ciliates made by Fauré-Fremiet (1950, 1951) and Webb (1956). The term mesoporal is applied to those ciliates which inhabit sands with a mean particle size between 0·4 and 1·8 mm, and microporal to those living in sand with a mean particle size between 0.1 and 0.3 mm. Euryporal forms may be found living in both ranges of particle size. Some typical euryporal ciliates are: *Trachelocerca*

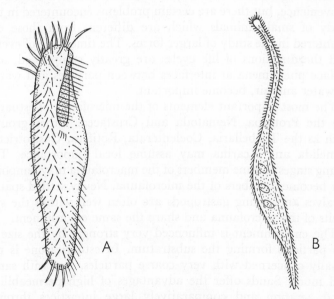

Figure 65. Euryporal ciliates
A, *Holosticha kessleri*, length 150μ. B, *Trachelocerca phoenicopterus*, length 500μ

phoenicopterus, *Holosticha kessleri*, *Condylostoma patulum* and *Micromitra retractilis*.

It must not be thought that the ciliates are restricted to an interstitial life. An even more important habitat is formed by the film of bacteria, diatoms and other algae which forms at the interface between the substratum and the water. This film often exhibits considerable cohesiveness, binding the surface particles together. Part of this cohesion is due to the mucilaginous secretions of micro-organisms and part is due to water surface tension and adsorptive properties associated with small particles. The ciliates themselves secrete mucus which may add to the general cohesion of the film. Some ciliates are particularly associated with the diatom film: *Chilodon calkinis*, *Cryptopharynx setigerus* and *Chlamydodon mnemosyne* are examples.

Free standing water in sheltered situations may also develop a surface film which is often composed of protein or lipoprotein (Goldacre, 1949). Diatoms and bacteria are often found on the underside of the surface film, and ciliates may also enter this community, which is called the neuston. Webb (1956) records the following ciliates from the neuston of a small pool in the Dee Estuary: *Chlamydodon triquetrus*, *Euplotes charon*, *Lacrymaria coronata*, *Holosticha kessleri*, *Loxophyllum helus* var. *multiverrucosum*, *Placus socialis* and *Uronychia transfuga*. These ciliates were accompanied by the flagellates *Oxyrris marina* and *Euglena limosa*.

The films developed on mud and sand surfaces form the basis of the trophic structure of the ciliate communities in estuaries. Bacteria, diatoms and organic detritus form the main foods of many ciliates, some of which are in turn eaten by larger ciliates. The foods of the common marsh ciliates from the Dee Estuary are given in Table 20.

Many ciliates take a mixed diet, but some are more specialised. *Nassula citraea* feeds on filaments of the blue-green alga *Oscillatoria*. The filaments are taken in through the strongly ciliated cytopharynx and may be seen coiled within the cytoplasm. The diatom feeders often show preference for diatoms of a particular size. *Chilodon calkinsi*, which is about 50μ long, takes smaller diatoms than *Chlamydodon triquetra*, which reaches a length of 120μ. Two or three species of diatom

TABLE 20

FOOD OF THE COMMON MARSH CILIATES IN THE DEE ESTUARY

(After Webb, 1956) (subsidiary food sources are indicated in parentheses)

GROUP I *Diatoms*
Chilodon calkinsi (bacteria)
Chlamydodon mnemosyne
C. triquetrus
Condylostoma patens (detritus)
C. patulum (bacteria?)
Frontonia marina (bacteria)
Strombidium elegans (bacteria)
S. styliferum (bacteria)
GROUP II *Bacteria*
Cohnilembus pusillus
Euplotes charon (detritus)
Mesodinium acarus
M. pulex
Micromitra retractilis
Pleuronema crassum
P. marinum
Spirostomum loxodes (diatoms)
S. teres (diatoms)
Stylonychia pustulata (diatoms)
Vorticella microstoma (detritus? diatoms?)
GROUP III *Bacteria and detritus*
Aspidisca crenata (diatoms)
Blepharisma salinarum (diatoms)

GROUP IV *Diatoms and bacteria*
Chilodontopsis elongata
Crytopharynx setigerus
Diophrys appendiculatus (detritus)
D. scutum (detritus)
Placus socialis
GROUP V *Bacteria, diatoms and detritus*
Euplotes harpa
Holosticha kessleri
Opisthotricha parallelis
Uronychia transfuga (protozoa)
GROUP VI *Protozoa*
Lacrymaria coronata
L. lagenula
L. olor var. marina
Lionotus fasciola
L. folium
Loxophyllum fasciolatum
L. helus
L. meleagris
L. setigerum
L. undulatum
Trachelocerca fusca
T. phoenicopterus
T. subviridis

feeders may be found together, but they are usually feeding on diatoms of different sizes, thus reducing competition for food. *Chilodon calkinsi* and *Chlamydodon triquetra* may occur abundantly together, but the latter species rarely occurs with *Frontonia marinus*, which also feeds on large diatoms.

Most of the ciliates which feed on bacteria in estuaries are discontinuous feeders. They swim actively for a period, then stop and feed. When food is abundant they remain still for long periods. When food is scarce the periods of swimming may become longer so that the chances of finding a suitable feeding area are increased.

Ciliates which prey on other protozoa are generally less numerous than the bacterial and diatom feeders. *Lacrymaria olor* var. *maritima* is an estuarine predatory ciliate which usually occurs in small numbers in field collections. If this species is placed in a rich culture of the bacterial feeder *Cohnilembus pusillus* the latter is rapidly decimated, as a flourishing population of *Lacrymaria* develops.

It is probable that in nature there is a continuous shifting of ciliate populations. Aggregations of bacteria and diatoms are exploited by species such as *Cohnilembus pusillus* and *Chilodon calkinsi*. When the populations of these species are sufficiently dense they provide the opportunity for the development of predatory populations.

When the food supply is exhausted many ciliates encyst, but this does not safeguard them from the attacks of other ciliates. *Loxophyllum setigerum* has been observed ingesting the cysts of *Diophrys appendiculatus*, and *Lacrymaria coronata* has been seen feeding on the cysts of *Nassula citraea* formed when the food supply of the latter has been exhausted.

The vertical zonation of ciliates in the sediments of a brackish-water beach on Asko island in the Baltic Sea has recently been studied by Fenchel and Jansson (1966). The salinity of the water in the region where this study was made lay between 6.1 and 6.5‰, and the availability of oxygen below the sand surface was very limited. In some of the samples, at one centimetre below the sand surface the oxygen available was only 2 per cent of that available in the water above the sand. Most of the ciliates were confined to the top 2 cm of sand, but there were some species which were consistently found in the deeper layers. *Mesodinium pupula* was found in its greatest numbers at least 3 cm below the sand surface, and in some samples penetrated as far as 6 or 7 cm below the surface. *Sonderia vorax* also penetrated well below the sand surface and was often accompanied by *S. schizostoma*, but the vertical range of the former species was greater. This difference between closely allied species could be related to their feeding habits. *Sonderia vorax* feeds on flagellates, diatoms and bacteria, and might be expected to find food over a wide vertical range. *Sonderia schizostoma* feeds almost exclusively on sulphur bacteria of the genus *Beggiatoa*. This habit restricts its vertical range, and *S. schizostoma* was often absent from the top centimetre of sand when its congener was present.

A general correlation was found between the distribution of Eh and the zonation of ciliates. Where the Eh fell rapidly below the surface the vertical zonation of the ciliates was compressed. Thus in one locality where the Eh fell to -100 mv in the top 2 cm *Mesodinium pupula* was most abundant between 3 and 4 cm from the surface, and in another locality where the Eh barely fell below Omv in the top 8 cm the maximum abundance of *M. pupula* was found between 5 and 6 cm.

The internal osmotic pressure of ciliates is regulated by the activities of contractile vacuoles. The evidence that these vacuoles regulated the water content of the cell body has been reviewed by Kitching (1938, 1952, 1954). The rate of vacuolar output by a fresh-water ciliate can be decreased by increasing the salinity of the external medium. That is in accord with the idea that the vacuole is concerned with eliminating water which enters the cell body when the internal osmotic pressure is higher than that of the medium. The active nature of this process is demonstrated by the fact that very dilute cyanide solutions inhibit vacuolar output and increase the cell volume of fresh-water peritrichs. If sublethal concentrations of cyanide are used the effect is reversible. Low concentrations of cyanide inhibit some of the respiratory enzymes, so that the output of water by the contractile vacuole is dependent on energy released in cellular respiration.

Estuarine Foraminifera

The Foraminifera are predominantly a marine group with a few aberrant fresh-water forms. The living Foraminiferan has long delicate pseudopodia which ramify in all directions and unite to form a network at some distance from the main cell body. These elongated pseudopodia are used to collect food, and in some species they appear to be capable of producing a toxic substance which can paralyse small animals such as copepods (Sandon, 1957). A shell of varying complexity is present. Sometimes this shell is formed of minute sand grains, or it may be a simple sac of a chitin-like substance, but most frequently the shell is calcareous and consists of several chambers. The arrangement of these chambers varies greatly from one genus to another. In highly evolved forms there is a dimorphic alternation of generations. Forms with a large primary chamber (megascleric individuals) give rise to gametes which fuse and develop into forms with a small primary chamber (microscleric individuals). These microscleric forms give rise to small amoeboid masses, each of which may develop into a megasceleric individual. In some species it appears that several generations of microscleric individuals may occur between megascleric generations.

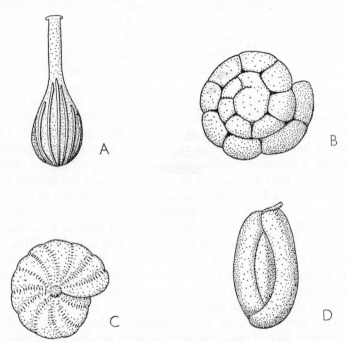

Figure 66 Estuarine Foraminifera
A, *Lagena sulcata*. B, *Ammonia beccarii*. C, *Elphidium striato-punctata*. D, *Milliammina fusca*. All are slightly less than 1 mm in maximum dimension

Many of the marine genera have species which penetrate at least a short distance into estuaries, and some species are known which can be regarded as euryhaline. The classical work dealing with estuarine Foraminifera is that of Brady and Robertson (1870). Recent work has been concerned with the analysis of the effects of salinity on growth and survival (Bradshaw, 1955, 1957, 1961; Murray, 1963, 1965), although there have been some notable studies on ecological distribution (e.g. Phleger, 1960, 1965).

The survey made by Brady and Robertson revealed that a considerable number of species penetrated into British estuaries, but five were more widespread than others and reached lower salinities. They listed (modern names are given here):

> Miliammina fusca
> Trochammina inflata
> Ammonia beccarii
> Nonion depressulus
> Elphidium striato-punctata

Of these the last is probably the commonest of all the British brackish-water Foraminifera. Where the water is almost fresh the shell is very thin and often has a pale green colour. This species has sometimes been regarded as a variety of *Elphidium crispum*, a common marine species. The work of Murray (1963) has shown that a reduction in salinity reduces the feeding and growth rate of *E. crispum*. The survival of this species in reduced salinity varies with the temperature of the water. At a salinity of 15‰ *E. crispum* survives for at least 15 days at 8°C, but at 16°C the cultures died. If the salinity was raised to 20‰ the animals survived at both temperatures for 38 days. A reduction in salinity will reduce the calcium content of the water, and a deficiency of calcium may be the cause of the delicate shells of some of the estuarine species.

The distribution of Foraminifera in the salt marshes of Galveston Bay, Texas, has been studied by Phleger (1965). A point of general interest is that his list of common species is very similar to that of Brady and Robertson. Phleger found for instance that *Trochammina inflata* was abundant in the *Salicornia* zone, but less abundant in the *Spartina* zone and in other parts of the marsh. *Ammonia beccarii* was abundant alongside the marsh channels, and *Miliammina fusca* was another common species. *Trochammina macrescens* is widespread in the marshes of the Gulf of Mexico, particularly in those with a lowered salinity, and has been recorded from similar salinities in Massachussetts. This species has a very thin shell which tends to collapse when

H

dried, and was originally described as a variety of *T. inflata*. This thin-shelled form is widespread in Europe.

PROTOHYDRA

Protohydra leuckarti is a small hydroid reaching a length of about one millimetre and lacking tentacles. This minute predator occurs in numbers exceeding 50,000 per sq m in some Danish estuaries and lagoons (Muus, 1966), and is known from an area extending from the Gulf of Finland to the Bay of Arcachon (Nyholm, 1951) in salinities from 3.8 to 30‰ (Remane, 1958).

Protohydra has been found living in both sand and mud, but prefers a mixed bottom in shallow water. A recent account of the biology of the species has been given by Muus (1966) who found that *P. leuckarti* fed mainly on harpacticoid copepods and on nematodes, but would rarely take halacarid mites or ostracods. When fed on a single *Tachidius* adult copepod per day the hydroid divided once every 4 days at room temperature. The normal method of reproduction is a transverse division. This process took 12 to 24 hours at 20°C and about 9 days at 4 or 5°C. The details of sexual reproduction have not been fully worked out, but there seems to be a period during the autumn when eggs are produced. As far as is known a single relatively large egg is produced at a time, and the process from the first appearance of the egg to spawning takes about 4 weeks.

MICRO-TURBELLARIANS IN ESTUARIES

The smaller turbellarians in estuaries belong to three orders: the Acoela, Rhabdocoela and Alloeocoela. The first of these orders contains simple worms lacking a digestive cavity and excretory system. The acoeles are all basically marine forms, but several species are known from brackish waters in the Baltic and Caspian Seas. The latter sea has the distinction of containing the largest known acoele, *Anaperus sulcatus*, which reaches a length of 12 mm.

The relationship between temperature and salinity resistance in the acoele *Convoluta roscoffensis* has been studied by Gompel and Legendre (1928). They found that the worms could survive at temperatures up to 38°C in sea water and in salinities down to 19‰, but when the salinity fell to 17 they died at 35°C and gyrated abnormally at temperatures above 23°C. At a salinity of 10‰ they died at 23°C. The upper limit of salinity tolerance at temperatures below 23°C lay in the region of 60‰. *Convoluta roscoffensis* is a marine form, and the above results show that its thermal tolerance is reduced by osmotic stress.

The Rhabdocoela are more complex than the acoeles. They have a simple gut without diverticula. Numerous species are known from brackish water. The best known region is the northern part of Germany and the Baltic, where a group of active workers (Ax, Karling, Luther, Meixner) have made detailed studies of their systematics and distribution. Some species, such as *Macrostomum appendiculatum* and *Gyratrix hermaphroditus*, can live in both fresh water and the sea. An interesting feature of the latter species is that the excretory system is less well developed in individuals found living in the sea (Kromhout, 1943). Presumably the greater osmotic inflow when the animal is in fresh water necessitates the better development of the excretory system to cope with the danger of excessive swelling.

One of the most abundant of the sub-groups of rhabdocoeles in brackish water is the sub-order Kalyptorhyncha. These animals have a protrusible proboscis as well as a muscular pharynx. The proboscis is muscular and glandular, and is capable of being shot out at high speed when the rhabdocoele contacts a small copepod. The prey adheres to the proboscis, which is then bent backwards to transport the food to the pharynx which can also be protruded. *Gyratrix* is a member of this group.

Within the species *Gyratrix hermaphroditus* there are several races, some of which have been found to differ in chromosome numbers. On the coasts of Finland there are populations with a diploid chromosome number of 4, others with 6, and yet others with 8. Reuter (1961) found that the forms with a diploid chromosome number of 6 were parthenogenetic, and each individual produced an average of 8 or 9 eggs in a lifetime. Specimens with a diploid number of 8 were not parthenogenetic and produced an average of 7 eggs in a lifetime. Both forms had

Figure 67. *Gyratrix hermaphroditus* (Rhabdocoela: Kalyptorhyncha), actual length
2 mm

an upper salinity tolerance of about 12‰ and could tolerate temperatures up to 36°C. These results, coupled with the variation in the structure of the excretory system, indicate that although as a species *G. hermaphroditus* is euryhaline there are genetical sub-groups within the species that have a more restricted salinity tolerance, and each individual animal is not holoeuryhaline.

The rhabdocoeles are hermaphrodites, and a few have been shown to fertilise their own eggs, but the general rule is for an exchange of sperm to take place with another individual. Some of the rhabdocoeles can produce two types of egg. Thin-shelled eggs develop quickly and give rise to miniature worms after a short time. Thick-shelled eggs may lie dormant for several months. These resting eggs are capable of withstanding freezing and drying, so they that provide a means of surviving in severe environmental conditions and may also act as a dispersal phase.

Within the Alloeocoela there is a considerable range of organisation. Some are not much more complex than the acoeles, but others are more complex than the rhabdocoeles, and have diverticula from their intestines. Some of the brackish-water species have complex batteries of sense organs, including ciliated pits, eyes and a single median statocyst. The genus *Monocelis* has a considerable number of marine species, and several in estuaries. The posterior end of *Monocelis* is modified to form an adhesive disc, and the worm may be observed to stop moving from time to time and remain attached in one position.

All the small flatworms are basically carnivores, although some species include diatoms in their diet, and a few may live almost entirely on these small plants. Some of the marine acoeles of the genus *Convoluta* have a symbiotic relationship with unicellular green algae.

ESTUARINE NEMATODA

Nematodes are among the most ubiquitous of animals, but the identification of species in this group is not easy, so that there are comparatively few studies of their occurrence in estuaries. The only detailed study of the nematodes in a British estuary is that of Capstick (1959), who identified 37 species from the Blyth Estuary. The distribution of nematodes in this estuary showed several characteristics which may be applicable to other areas.

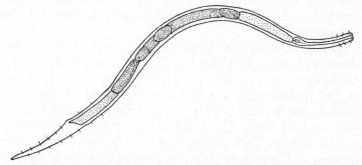

Figure 68. Prochromadorella bioculata, a nematode capable of living in fresh and brackish water. Actual length 0·7 mm

Most nematodes were found in the top 2 cm of the substratum. Below a depth of 4 or 5 cm very few nematodes were found. The richest populations were found in mud, where densities of 28,000 in 100 sq cm were recorded. In sandy areas the populations were sparser, about 3,000 in 100 sq cm.

The overall abundance of nematodes decreased as the head of the estuary was approached from the middle reaches. There were some specific exceptions to this general trend. For instance, *Thalassoalaimus tardus* and *Theristus setosus* became more abundant towards the head of the estuary, and 5 other species were found only in samples taken near the head. The most abundant species, such as *Spirina parasitifera* and *Anaplostoma viviparum*, decreased markedly in abundance towards the head of the estuary, and this decrease was responsible for the overall decrease in total nematodes.

The intertidal distribution of the various species showed considerable variation. Some species, such as *Anaplostoma viviparum*, *Sabateria* spp. and *Theristus* spp. had a fairly continuous distribution throughout the intertidal range. Others, such as *Spirina parasitifera*, showed maximum population densities on the upper part of the shore, with a gradual decrease to the upper and lower ends.

Free living marine nematodes have been divided into four groups by Wieser (1952). These groups represent morphological types which differ in feeding habits.

In the first group, which contains 97 genera, there is no distinct mouth cavity. The oesophagus is suctorial and the food taken by these nematodes appears to be semi-liquid, or at least it never contains any large particles. These nematodes may feed selectively on small particles and bacteria in deposits. *Rhabditis marina* appears to feed on bacteria and small flagellates.

The second group contains 73 genera with unarmed mouth cavities. The lips appear to play an active part in the ingestion of food, which includes larger particles than those taken by the first group. Some species in this group feed on diatoms. *Zygonemella striata*, for instance, feeds on relatively large diatoms, *Theristus setosus* feeds on small diatoms and euglenoids, and *Desmolaimus zeelandicus* has been recorded as eating the diatom *Pleurosigma*.

In the third group the mouth cavity contains small projections which can pierce algal cells, or scrape such cells from larger surfaces. These nematodes, which belong to 104 genera, may be termed epigrowth feeders. Examples include *Paracanthonchus mortenseni* which feeds on diatoms, and *Spirina similis* which has been recorded containing 'green vegetable matter'.

Wieser's fourth group contains 87 genera with large powerful armature in the mouth cavity. Many of these are predators which either swallow their prey whole or else pierce and suck, but some have also been recorded feeding on plant materials. The estuarine species *Sphaerolaimus gracilis* feeds on smaller nematodes of the genera *Theristus* and *Axonolaimus*. Marine species have been found preying on small polychaetes: *Pontonema californicum* is an example. An even more remarkable habit is exhibited by *Oncholaimus dujardini*, which has been observed feeding on the polyzoan *Zoobotryon pellucidum*.

The four morphological groups defined by Wieser (1952) show a differential distribution in substrata of various grades. In fine sand rich in deposits of organic detritus the second group tends to be dominant. In clean fine sand lacking much organic debris the predatory nematodes form about 48 per cent of the total population. Solid poorly aerated muds also tend to have a fauna of predatory nematodes, particularly members of the genus *Sphaerolaimus*. However, Perkins (1958) found that the anaerobic black layer at Whitstable was inhabited by *Monohystera filicaudata*, *Araeolaimus villosus* and *Euchromadora vulgaris*, and all of these were feeding on sulphur bacteria, despite the fact that they each belong to a different group in Wieser's classification. It seems that the structure of the mouth may provide an indication of the method of obtaining food, but does not give a precise guide to the nature of the food taken by a nematode.

MICRO ANNELIDS

The smaller annelids provide the clearest demonstration of the artificiality of the classification into macro and microfauna. Because of their elongate shape they often pass through the sieves used to separate out the macrofauna from sand and mud and so they are neglected in studies that are restricted to the larger forms. On the other hand their length is often such that including them in the microfauna cannot be done without misgivings. In spite of this difficulty some of the annelids are best considered under the heading of microbenthos.

One of the most delightful of all estuarine annelids is a small fan worm, *Manayunkia aestuarina.* This is a sabellid, related to

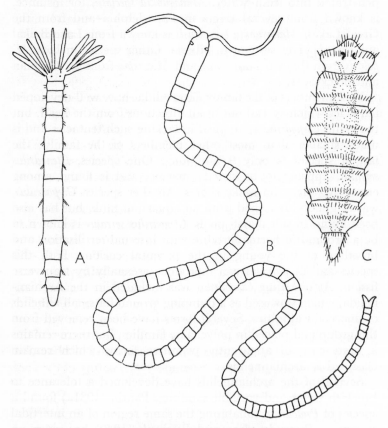

Figure 69. Estuarine micro-annelids

A, *Manayunkia aestuarina* (Polychaeta: Sabellidae), actual length 6 mm. B, *Protodrilus* sp. (Archiannelida), actual length 10 mm; this is an example of the extreme elongation found in certain members of the interstitial fauna. C, *Dinophilus* sp. (Archiannelida), actual length 2 mm such large forms as *Sabella* and *Bispira*, and yet it is so small that the largest specimens are only 6 or 7 mm. long. Like *Sabella*, this small polychaete has green blood containing the respiratory pigment chlorocruorin.

Manayunkia aestuarina is common in the estuaries of western Europe, and occurs in regions where the salinity may fall to 2‰ for short periods. With this ability to tolerate low salinities it is not surprising that some members of the genus have penetrated into fresh water. *Manayunkia speciosa*, for instance, is known from several rivers in Philadelphia and from the Great Lakes. *Manayunkia baicalensis* is known from Lake Baikal and several river systems in Siberia. Other species include *M. polaris* from the Murman coast and *M. caspia* from the Caspian and Black Seas.

Most members of the family Cirratulidae have well-developed tentacles which spread out in all directions from the head, but the genus *Ctenodrilus* is atypical in lacking such tentacles and is much smaller than most other members of the family; the largest species is only 9 mm long. One species, *Ctenodrilus serratus*, appears to be strictly marine, and is found among encrusting algae on rocky shores. Another species, *C. parvulus*, was originally described from an aquarium tank, but has also been found in salt marsh pools. *Ctenodrilus serratus* is known to be a protandric hermaphrodite with internal fertilisation and brooding of the young in the parental coelom. Both this species and *C. parvulus* can reproduce asexually by transverse fission. At one time *Ctenodrilus* was included in the Archiannelida, which was used as a dumping ground for small annelids of uncertain affinities. Several forms have been removed from this group and placed in polychaete families, but there remains a hard core of apparently primitive forms which remain classified as archiannelids.

Several of the archiannelids have developed a tolerance to brackish water and variable salinities. Pantin (1931a) found a species of *Protodrilus* inhabiting the same region of an intertidal stream as *Procerodes ulvae*, and Ruebush (1940) has given an account of the biology of *Dinophilus gardineri*. This species was found in brackish pools, and in 'Lillie's Ditch', near the Woods Hole Biological Station. The colour of *D. gardineri* is normally bright orange, but this colour fades if the animal is starved of its normal food, which consists mainly of diatoms, although other algae and a few Protozoa are also eaten.

In unfavourable conditions *Dinophilus gardineri* can enclose itself in a gelatinous capsule which appears to be relatively impermeable. The eggs are also laid in gelatinous capsules formed by glands in the epidermis of the parent worm. Up to 16 eggs may be laid in a single capsule from which the young emerge 6 or 7 days after laying.

The sexes of *D. gardineri* are separate, and the males practice hypodermal impregnation. The muscular penis penetrates the body wall of the female anywhere, except on the head, and the sperms are injected into the female body cavity. This method of fertilisation is advantageous to an estuarine animal in that it avoids subjecting the sperms to osmotic stress from the external medium.

The lower limit of salinity tolerance of *D. gardineri* is about 1.3‰, and it is capable of living in full-strength sea water. Specimens from low salinities have body volumes 80–90 per cent greater than those living in sea water. The nephridia are also larger and more conspicuous in low salinities. This may be interpreted as an indication that these organs are concerned with osmoregulation, and is a parallel to the situation described in the rhabdocoele *Gyratrix hermaphroditus* (p. 217).

ESTUARINE OSTRACODS

The ostracods are characterised by the possession of a bivalved carapace which can be adducted to enclose all the limbs. The number of limbs is greatly reduced in comparison with other Crustacea; only three pairs are present on the thorax. The body is usually terminated by a caudal furca, consisting of two rami which often bear spines and setae. The carapace is often heavily calcified and may be ornamented with ridges and tubercles as well as spines around the free margins. Dorsally the two carapace valves meet to form a hinge. Some species can swim freely in the water, and have long setae on their antennae. Species which do not swim lack these setae, or have them present in a reduced form. Many of the bottom dwelling ostracods appear to be general scavengers, capable of feeding on living algae and on detritus. Some are also capable of preying on smaller members of the microfauna.

The two most important families in brackish water are the Cypridae and the Cytheridae. The former is basically a fresh-water group, and the latter is basically marine. For instance, in a monograph of the German species (Klie, 1938) there are 92 species of the family Cypridae; two of these live in brackish water, and the rest are fresh-water forms. Of the German Cytherids, Klie describes 75 species belonging to 26 genera; only three of these genera are found in fresh water and the remainder are marine.

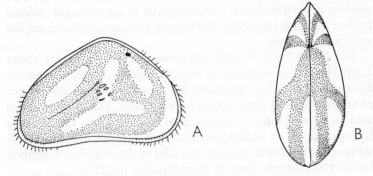

Figure 70. Heterocypris salina (Ostracoda: Cypridae)
A, lateral view. B, Dorsal view, length about 1·2 mm (after Sars)

Heterocypris salina (Fig. 70) is a brackish-water species, occurring in salt marsh pools with a fairly low salinity (usually under 10‰). In good conditions this species will produce two generations in a year. The closely allied *H. incongruens* may also be found in pools with a low salinity. This species is also remarkably tolerant to poorly aerated conditions, and can survive for two weeks in water smelling of hydrogen sulphide (Green, 1959). The duration of development from hatching to maturity takes about 16 days at 22°C, and in this time the ostracod passes through seven moults. In spring time, at low temperatures, the time to reach maturity may be about 5 weeks, and the duration of life may be 3 months. The eggs are capable of resisting desiccation so that the species can persist in temporary pools. In the laboratory *H. incongruens* can be reared very successfully on pure cultures of algae, including *Chlorella*, *Anabaena* and *Oscillatoria* (Green, 1959).

Theisen (1966) has studied the life histories of 7 species in a shallow bay on the Danish Coast, where the salinity fluctuated between 9 and 20‰. The fauna of this area was typically estuarine, including *Nereis diversicolor*, *Corophium volutator*, *Cardium lamarcki* and *Gobius microps*. Three of the ostracod species (*Cyprideis torosa*, *Leptocythere lacertosa*, and *Loxoconcha elliptica*) persisted as adults throughout the winter, but the other 4 species (*Elofsonia baltica*, *Cytherura gibba*, *Cytherois fischeri* and *Cytherois arenicola*) overwintered as eggs. Within the 7 species there was considerable variation in the duration of development. The eggs of *Elofsonia baltica* hatched about 4 days after being laid, but those of *Cyprideis torosa* took about 30 days to hatch. The latter species also took a very long time to reach maturity—63 days at temperatures between 18 and 24°C, while *Elofsonia baltica* took only 11 days to pass through its immature stages.

The salinity tolerances of a large number of species have been summarised by Neale (1965). About 30 species may be found at salinities between 0.5‰ and 20‰ and some of these are capable of living in higher salinities.

The salinity tolerance of *Hemicythere conradi* has been studied experimentally by Kornicker and Wise (1960). They found that this species when acclimated to a salinity of 33‰ could tolerate and remain active in any salinity between 6‰ and 65‰. In field collections they found *H. conradi* in salinities ranging from 8‰ to 51‰, so that the laboratory tolerance was somewhat greater than the natural range. This ostracod is one of the species found in the Upper Laguna Madre which undergoes considerable changes in salinity according to weather conditions. In dry years the salinity in some areas may rise above 65‰ and the ostracod presumably does not survive, but in wet years the salinity remains below the upper limit of tolerance, and the ostracod is found burrowing in silty sand.

As we have seen with other groups, the salinity tolerance of adults may be greater than that of the eggs or young larvae, and the occurrence of a species may be restricted more by the salinity tolerance of the young than by that of the older stages. A study of the natural occurrence of 7 species of brackish-water ostracods and a parallel laboratory study of the salinities in which their larvae hatched has been made by Theisen (1966). His results are summarised in Table 21. There is good agreement between the two sets of data, and it is clear that the salinity range over which larvae are capable of hatching is generally somewhat less than the range of natural occurrence of adults.

TABLE 21

SALINITY TOLERANCE OF OSTRACODS FROM A SHALLOW BAY ON THE DANISH COAST
(After Thiesen, 1966)

Species	Cyprideis torosa	Leptocythere lacertosa	Loxoconcha elliptica	Elofsonia baltica
Salinities of areas where adults have been recorded	0–60‰	5–<32‰	0·5–30‰	<1–32‰
Salinities of cultures in which larvae hatched	1–40‰	5–30‰	8–30‰	1–30‰

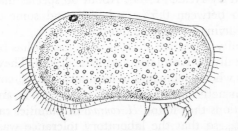

Figure 71. Leptocythere castanea (Ostracoda: Cytheridae), lateral view, actual length about 0·65 mm

Apart from exerting an effect on the survival of ostracods the salinity of the medium may also affect growth. Barker (1963) has shown that two euryhaline cytherids, *Leptocythere castanea* and *Loxoconcha impressa*, are smaller when collected from localities with low salinities. In a series of samples collected at points along the Tamar Estuary in August and September there was a gradient of size along the salinity gradient, and a sharp decrease in salinity was paralleled by a similarly sharp decrease in the average size of the ostracods.

BENTHIC COPEPODS

The harpacticoid copepods have been monographed with remarkable thoroughness by Lang (1948, 1966). The total number of pages in these two publications is somewhat in excess of two thousand, and the scope is world wide. In contrast to this vast amount of systematic information there is relatively little biological information available. Few species have been reared through their complete life cycles in controlled conditions, and the durations of the various stages are poorly known.

The majority of harpacticoids have a body form similar to those depicted in Figs 72A and B, but some are more elongated, and others are flattened, either laterally to resemble minute amphipods, or dorso-ventrally to resemble tiny isopods. As far as is known the harpacticoids are general browsers and scavengers, eating small algae or any bits of detritus that they can find. The large number of species and their variety of body form lead one to suspect that there may well be a much greater range in feeding habits, but this aspect of harpacticoid biology has not yet been studied.

It is not possible to consider all the species that have been recorded in estuaries; for instance Wells (1963) recorded 92 species from the Exe Estuary. Instead, a small number of species have been selected for special mention because they occur in the most typical estuarine habitats where the salinity shows large and rapid fluctuations.

The two species of the genus *Platychelipus* shows many of the characteristics of estuarine harpacticoids. They are benthic for the whole of their lives. The nauplii do not attempt to swim, but crawl between the mud particles, using their antennules and antennae (Barnett, 1966). The adults have the swimming setae on the legs modified into stout spines which are more suitable for pushing the animal through mud. These spines are shorter and stouter in *P. littoralis* than in *P. laophontoides*. This difference may be related to the habitats of the two species.

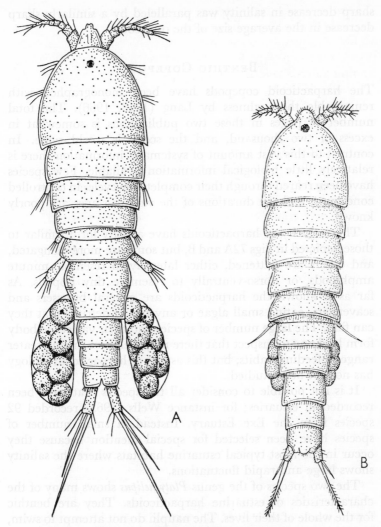

Figure 72. Estuarine harpacticoid copepods
A, *Stenhelia palustris*, adult female, length 0·8 mm. B, *Nannopus palustris*, adult female, length 0·6 mm (after Sars)

Platychelipus littoralis generally occurs higher on the shore where the mud is likely to be stiffer and drier, while *P. laophontoides* is found in softer mud lower on the shore.

Most harpacticoids appear to pass through five nauplius stages and six copepodid stages. The sixth copepodid is the adult. The females often mate while still in their fifth copepodid stage, but they do not lay eggs until they are fully adult. Several batches of fertilised eggs can result from a single mating (Nicholls, 1935; Fraser, 1936).

There is a considerable range of form among the nauplii of the harpacticoids. Some are not very different from those of other copepods, but others are considerably modified. For instance, the nauplius of *Stenhelia palustris* is wider than long, and is incapable of swimming. Bresciani (1960) has described the behaviour of this nauplius. If placed in water on a smooth surface the nauplius is incapable of co-ordinated movement, but if a few grains of sand are added the nauplius can move adroitly between them, using its antennae and mandibles. When moving through the spaces in sand the nauplius travels sideways and rotates about its transverse axis as it moves. Basically this curious movement is a series of forward rolls combined with lateral displacement, so that the resulting path is helicoidal (Fig. 73B).

Stenhelia palustris can tolerate a wide range of temperature and salinity. Its geographical distribution extends from the north of Greenland to France, and it is a common member of the microfauna of salt marsh pools in Britain. Bresciani (1960) found it in a region where the salinity varied between 6 and 16‰ and the temperature ranged from 0 to 26°C.

Some of the harpacticoids are holoeuryhaline, being capable of living in both sea and fresh water: *Horsiella brevicornis* is an example. Noodt (1957) regards this species as characteristic of the region of variable salinity colonised by algae of the genus *Vaucheria*, and he lists seven other European species as holoeuryhaline. However, it has not been established that each of these species is capable of completing its life history over the whole range from sea to fresh water. Fig. 74 shows the ranges of salinity in which some well-known European species have been found under natural conditions. Many other species with overlapping salinity ranges might have been included, but have been omitted for the sake of clarity.

Tigriopus brevicornis (= *T. fulva*) is not strictly speaking an

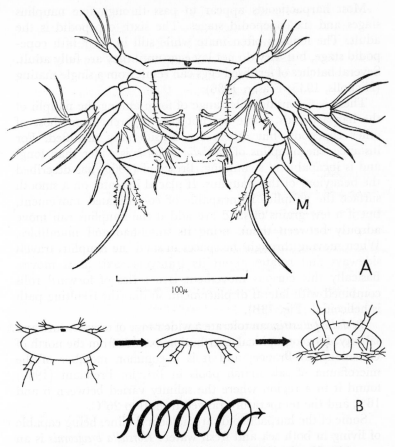

Figure 73. Nauplius of *Stenhelia palustris* (Copepoda: Harpacticoida)
A, week old nauplius, M = mandible. B, diagram to show how the nauplius moves
sideways (after Bresciani, 1960)

estuarine species, but it has such a remarkable resistance to
salinity change that it deserves mention for comparison with
the truly estuarine forms. The normal habitat of *T. brevicornis*
is in pools high on rocky shores. These pools are reached by the
high spring tides, but not by intervening tides. The salinity in
these pools varies according to the weather. If there is heavy
rain after a spring tide the salinity may fall well below that of
sea water, but if there is a spell of hot dry weather there may be

considerable evaporation and the concentration of salt may
rise to over twice the salinity of sea water. Ranade (1953) has
shown that *T. brevicornis* can live normally in salinities ranging
from 4.2 to 90‰. At salinities above 90‰ the copepods become
quiescent, but can resume normal activity if placed in lower
salinities within about 30 hours. If the salinity is increased

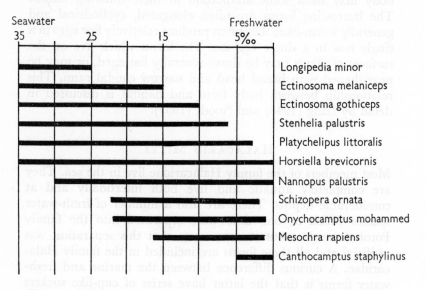

Seawater   Freshwater
35   25   15   5‰

Longipedia minor
Ectinosoma melaniceps
Ectinosoma gothiceps
Stenhelia palustris
Platychelipus littoralis
Horsiella brevicornis
Nannopus palustris
Schizopera ornata
Onychocamptus mohammed
Mesochra rapiens
Canthocamptus staphylinus

Figure 74. Salinities at which some common harpacticoid copepods have been
found under natural conditions (based on the data given by Gurney, 1932 and
Lang, 1948)

beyond 180‰ the copepods must be returned to a lower
salinity within 3 hours, otherwise they die. The lethal temperature for this species varies with salinity: at a salinity of 4.2
the copepods die at 34°C, but at a salinity of 90‰ the lethal
temperature increases to 41.8°C. This appears to be an excep-
tion to the general rule that maximum resistance to one
environmental factor can only be achieved when other factors
are optimal (see p. 113). In the normal habitat of *T. brevicornis*
high temperatures and high salinities will tend to occur together, and this species has developed a resistance to both
simultaneously.

Salinity is not the only factor governing the distribution of harpacticoids in estuaries. The nature of the substratum is also important. Some species, such as *Harpacticus flexus*, live in sandy areas, but instead of burrowing into the sand they skim over the surface. Others, such as *Nitocra typica*, burrow into the sand or wriggle among the strands of filamentous algae. The form of the body may show some adaptation to these differing habits. The burrowing forms are often elongated, cyclindrical and generally worm-like; they often produce relatively few eggs in a single row in a single egg sac. The forms which live on the surfaces of plants may be dorso-ventrally flattened, or may be pear-shaped with broad head and narrow caudal rami. This relationship between body form and habitat is discussed in detail by Lang (1948) and Noodt (1957).

HALACARID MITES

Most members of the family Halacaridae live in the sea. They are completely aquatic and live both intertidally and at considerable depths. There are also a number of fresh-water species which at one time were separated into the family Porohalacaridae, but it now seems that this separation was artificial and all these forms are included in the family Halacaridae. A curious difference between the marine and fresh-water forms is that the latter have series of cup-like suckers around the exterior of the genital aperture; no such suckers are present in the marine forms, although there may be similar structures inside the genital aperture. This might at first sight seem to be a fundamental difference between the two groups, but it seems to have been acquired separately in several different subfamilies. In one estuarine species, *Isobactrus uniscutatus*, which belongs to a marine genus, there are some structures resembling the external suckers of the fresh-water forms. It may well be that the development of external genital suckers is an adaptation to life in fresh water. These suckers presumably help to keep the genital apertures of male and female closely apposed during the transfer of spermatozoa. Greater efficiency in this respect may be necessary in fresh water to avoid subjecting the sperms to osmotic stress. It is noteworthy that numerous species of fresh-water mites belonging to the group Hydrachnellae are equipped with external genital suckers.

During its post-embryonic development a halacarid passes through a single larval phase and two nymphal phases. The larval phase has only 3 pairs of legs; the nymphs have 4 pairs

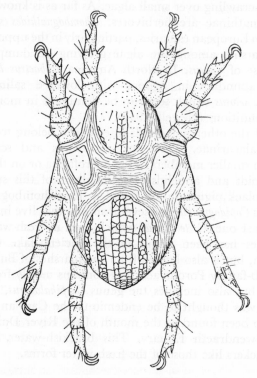

Figure 75. Rhombognathides spinipes (Acari: Halacaridae), dorsal view of nymph, actual length of body 360μ

of legs, but the gonads and genital aperture are not fully developed. In the females the second nymphal stage is the mating stage, and this receives sperms from adult males. This is a curious parallel to the harpacticoid copepods (see p. 230) where the adult males mate with females that have not entered their final adult stage. The duration of life of the halacarids has not been studied in detail, but the whole cycle from newly hatched larva to adult may take as long as a year (cf. Andre, 1946).

A high proportion of the halacarids found in estuaries belong to the sub-family Rhombognathinae. These mites (cf. Fig. 75) usually have a black pigment masking the internal organs and are found crawling over small algae. As far as is known all the Rhombognathinae are herbivores. *Rhombognathides spinipes* is common in European estuaries, particularly in the upper reaches of salt marshes among the algae growing on damp ground at the base of *Juncus*. In North America *Isobactrus hutchinsoni* becomes abundant in regions with lowered salinity, and replaces *I. setosus* which tends to be dominant in more strictly marine conditions (Newell, 1947).

Most of the other halacarids in estuaries belong to the sub-family Halacarinae. These are carnivores and scavengers, feeding on smaller members of the microfauna or on the eggs of harpacticoids and annelids. The members of this sub-family lack the black pigmentation found in the Rhombognathinae. The genus *Copidognathus* has several species that live in estuaries and at least one, *C. tectiporus*, which lives in fresh water. This last species has been found in the ancient lake Ohrid in Jugoslavia, but is also known from the marshes of Bulgaria.

The sub-family Porohalacarinae contains mostly fresh-water dwellers, but also includes the genus *Caspihalacarus*, which at one time was thought to be endemic to the Caspian Sea, but it has also been found at the mouth of the River Dnieper and in the Gwendraeth Estuary. This brackish-water form has genital suckers like those of the fresh-water forms.

CHAPTER 8

THE FRESH-WATER COMPONENT

WHEN a spring tide pushes into an estuary it causes the river level to rise further upstream than brackish water ever penetrates. Frequently the river will overflow its normal channel and extend over its flood plain to an extent which varies with the flow of the river and the height of the tide. When the tide recedes it leaves pools in the flood plain. Upstream these pools will be fresh, lower downstream they will be slightly brackish, and in the region of a salt marsh the salinity will be much nearer to the concentration of sea water. Pools such as these may dry out in a hot summer, but in the cooler seasons will persist from one spring tide to the next. A series of pools with salinities ranging from fresh water to sea water will provide opportunities for animals to penetrate from fresh water into the estuary, and will allow estuarine animals of marine origin the opportunity of progressive adaptation to fresh water.

In many estuaries the flood plains are now drained by man-made ditches, but these may also provide a range of salinities similar to the naturally occurring pools.

Some species belonging to groups which are predominantly fresh-water dwellers have taken the opportunities offered by riverside pools. Many examples are found among the insects. The variety of insects penetrating into brackish water is shown well by the work of Butler and Popham (1958), who studied pools and dykes near the mouth of the River Humber. They found 35 species living in water with a salinity of 7‰ or more, and of these the following 11 were found in salinities of 17‰ or more.

Hemiptera
Sigara selecta
,, stagnalis
,, falleni

236

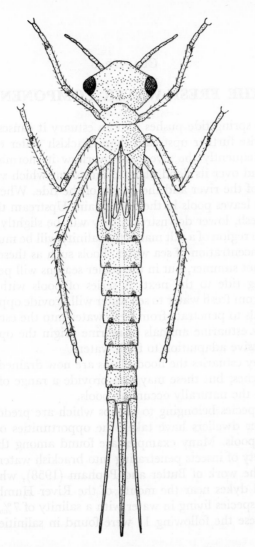

Figure 76. Larva of *Ischnura elegans* (Odonata). Actual length about 17 mm.
Several species of this genus have euryhaline larvae

Coleoptera
Hydrobius fuscipes
Helophorus brevipalpus
,,   flavipes
Ochthebius marinus
Enochrus bicolor
Haliplus obliquus
Agabus conspersus
Odonata
Ischnura elegans

An interesting feature of the lists given by Butler and Popham is that of the 35 species recorded in salinities over 7‰, about half were not found at salinities above 8·5‰. They suggest that this salinity, which is equal to approximately a quarter of sea water, is a critical level for many fresh-water insects. This suggestion agrees well with the dividing line between the second and third grades of euryhaline limnobionts suggested by Remane (1958) (cf. p. 67).

AN ESTUARINE CADDIS LARVA

The different orders of insects differ greatly in the numbers of species which have penetrated into brackish water from fresh water. Among the 188 species of British Trichoptera, for instance, there is only one species, *Limnephilus affinis*, that can be regarded as a well-established brackish-water animal. This species has been studied in some detail by Sutcliffe (1960a, 1961), who worked with a population which was known to have been established in a salt marsh for at least 30 years. The larvae of *L. affinis* were found to be capable of surviving several months in a salinity of 26‰ at 15°C, and could complete their life history in a salinity of 17‰.

This particular population fed mainly on dead leaves which blew into the salt marsh pools from a nearby copse. The life cycle could be completed on this food, but the diet was not necessarily restricted to fallen leaves. It was observed that if the opportunity arose the caddis larvae would devour the larvae of two diptera, *Chironomus aprilinus* and *Ephydra riparia*, which occurred in the same pools. If crowded together the larvae of *L. affinis* would consume one another.

Larvae of *Limnephilus affinis* can regulate the sodium content of the blood when the external medium is somewhat stronger than the blood. When the external concentration reaches approximately 27‰ the regulatory mechanism loses its efficiency and the sodium concentration in the blood rises rapidly towards that of the external medium. The body wall of this caddis larva is very permeable to water, so that the animal will be subjected to considerable osmotic stress when in strong external media. This stress is combated to some extent by increasing the non-electrolyte concentration in the blood. When the concentration of the external medium is high the chief mechanism to regulate the concentration of salt in the blood is the uptake of water from the rectal fluid. This enables the larva to drink salt water and to eliminate salt after taking up some of the water.

ESTUARINE DIPTERA

The dipterous family Chironomidae has numerous species with larvae that dwell in fresh water, and fewer species that live on the sea shore or in estuaries. *Chironomus aprilinus* is a common estuarine species, the larvae of which may be found in salt marsh pools, where they can complete their life cycles in salinities up to 17‰, and can survive short periods in salinities up to 30‰. *Cricotopus vitripennis* is even better adapted to marine and estuarine conditions. This species occurs in two forms: the larvae of the typical form live in rock pools near high water mark, and the variety *halophilus* lives in estuarine salt marsh pools. The rock pool form is better able to withstand large changes in salinity. Sutcliffe (1960b) transferred larvae direct from tap water to double-strength sea water, and found that they not only survived but completed their life cycles. Larvae of the estuarine form died in about four days when placed in concentrations above 150 per cent sea water, but were capable of completing their life cycles at all lower concentrations. The larvae of both forms were found to be capable of regulating the osmotic pressure of their blood. When larvae were placed in media with concentrations over one-third of sea water they kept the blood concentration well below that of the external medium. Both forms maintained the same blood concentration, but the rock pool form could continue the regulation at concentrations higher than those in which the estuarine form could survive.

Some chironomid larvae construct mucous tubes, which may be U-shaped when made in mud. These tubes are irrigated by undulatory movements of the body. A current of water is drawn in over the head of the larva and expelled over the tail. Some species use this respiratory current also as part of a feeding mechanism. A mucous net, shaped like an inverted cone, is secreted near the mouth of the tube and filters small particles from the incoming current. The net and the material trapped on it are eaten at fairly regular intervals, and then the cycle is restarted with the secretion of a new net.

Mosquitoes (family Culicidae) have also invaded brackish water. Among the 30 British species there are four which may be found breeding in salt marshes (Marshall, 1936). Three of these belong to the genus *Aedes*, and of these *A. detritus* and *A. caspius* are particularly common and widespread. The larvae of both species live in salt marsh pools and can tolerate wide fluctuations in salinity. The areas particularly favoured by these mosquitoes lie at the upper limits of the spring tides. Hollows and pools lying below the lowest spring tides are not suitable for breeding because the time between hatching of eggs and emergence of the adults is normally two or three weeks in an English summer, and if the breeding pool was subjected to every spring tide the larvae and pupae would be washed out to sea before development was completed. Ideal breeding conditions are often found in pools formed by seepage through a sea wall built at the back of a salt marsh. Such pools may be more or less permanent and are not subject to washing out by the tide. In such a pool *Aedes detritus* may breed throughout the year, but *Aedes caspius* seems to require higher temperatures for the eggs to hatch, so that in Britain no larvae hatch in a period extending from December to March, although they do so in the warmer conditions found in Greece. The eggs of these species are not laid in the water, but are laid in a shaded situation on the ground, and hatch when the ground is flooded. These eggs can remain viable for months, and then hatch within a few hours when covered with water.

The larva of *Aedes detritus* can maintain the concentration of solutes in its blood within fairly narrow limits over a wide range of concentrations in the external medium. In fresh water the blood is more concentrated than the medium, and in double-strength sea water the blood is kept well below the concentration of the external medium (Beadle, 1939). This is achieved

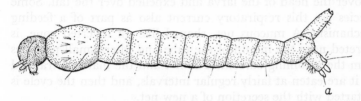

Figure 77. Larva of *Aedes caspius* (Diptera: Culicidae). Length 6 mm. a = anal papillae

in two ways: the body wall is not very permeable, and the rectum is capable of removing water from the contents of its lumen. This results in a concentration of the rectal fluid so that it may be three times as concentrated as the blood and the fluid in the mid-gut (Ramsay, 1950). This is only necessary in saline water. If the larva of *A. detritus* is kept in fresh water there is no concentration of the rectal fluid, which may then be less concentrated than the mid-gut fluid and the blood. Species of *Aedes* which live only in fresh water are more permeable than *A. detritus* and lack the ability to regulate the concentration of the blood when the external medium is more concentrated than the blood. An interesting anatomical difference is found between the salt-marsh species and the fresh-water species of *Aedes*. The anal papillae (Fig. 77) are much larger in the fresh-water forms. These papillae are known to play a part in the salt balance of the fresh-water forms by secreting salt from the external medium into the body. The need for this active uptake of salt is apparently reduced in the larvae living in saline media. The larvae of *A. detritus* probably drink the saline medium in which they live so that they have no problem in obtaining salts, but have to eliminate excess salts by producing a concentrated fluid in the rectum.

The malarial mosquito, *Anopheles maculipennis*, occurs in Britain in two forms, one of which (var. *messeae*) breeds in fresh water and the other (var. *atroparvus*) breeds in brackish water. The two varieties also differ in adult habits. The females of *messeae* hibernate in cold outbuildings, and rarely bite man, but the females of *atroparvus* enter warm, dark and poorly ventilated dwellings and frequently attack man. These habits make *atroparvus* more effective as a vector. Malaria was common in some coastal districts of England during the nineteenth century, but has now disappeared as an indigenous disease, probably because houses are now better designed from the viewpoint of lighting and ventilation, and the breeding areas of *atroparvus* have been reduced by drainage and the building of sea defences. The occasional cases of malaria contracted in England are usually traced to a traveller bringing back the disease and visiting a marshland where *A. maculipennis* var. *atroparvus* can transmit the disease to a new host.

The differentiation of subspecific categories of mosquitoes with larvae able to live in brackish water is not confined to Europe. In tropical Australia *Anopheles amictus* is found as two subspecies. One of these is rarely found in brackish water, but the other *A. amictus hilli* is most frequent in coastal areas and has been recorded in pools with a salinity of 42‰ (Lee and Woodhill, 1944).

In the southern states of the United States a number of mosquitoes breed in brackish or full-strength sea water (Carpenter, Middlekauff and Chamberlin, 1946). *Aedes taeniorhynchus* is widespread, living in salt marshes on both coasts from California to Peru and New England to Brazil. Members of the genus *Deinocerites* live on open mud flats; their larvae are found in crab holes, and the adults may be seen resting on the sides of the upper parts of the holes. In Florida *Culex opisthopus* has similar habits.

The family Ceratopogonidae contains a large number of species with varying feeding habits. The females of these minute flies are often blood suckers. Some attack invertebrates, such as caterpillars, others attack birds and mammals including man. They are commonly known as biting midges, and in North America they are called punkies. Many species have aquatic larvae, and some are particularly abundant in salt marshes. The life history of one salt marsh species has recently been studied by Becker (1961). The larvae of *Culicoides circumscriptus* were found to be capable of completing their life cycles in fresh water, and in water with a salinity one and a half times that of sea water. This species has two generations in a year. The adults lay their eggs in damp mud near the edges of pools in the salt marsh. The eggs are banana-shaped, and a little under half a millimetre in length. A female *C. circumscriptus* may lay up to 400 in a batch. The larvae emerge 3 to 8 days after the eggs have been laid. Under the microscope the larva resembles and moves like a minute snake, with a pale brown head and a long, smooth dull-white translucent body. The duration of larval development varies greatly with the temperature and with the amount of food available; in good conditions it may be completed in 14 days.

In British salt marshes it is possible to find *C. circumscriptus* accompanied by 3 other species: *C. maritimus*, *C. salinarius* and *C. riethi*. The larvae of all 4 species may be found in the regions where *Nereis diversicolor* is the dominant predator, and Becker (1958) has shown that this polychaete will eat the larvae. Out of 60 larvae introduced into a dish containing 4 *Nereis*, only 3 remained alive after 18 hours.

As a final example of salinity tolerance by Diptera with aquatic larvae one may take the flies of the genus *Ephydra*. Several species of *Ephydra* are known from estuarine conditions, and the larva of *E. cinerea* lives in the Great Salt Lake, where the osmotic pressure of the lake water is approximately 250 atmospheres (sea water has an osmotic pressure of about 22 atmospheres). In these conditions the larvae maintain their haemolymph at a concentration equivalent to 20 atmospheres ($\Delta = -1\cdot89°C$). A similar concentration of the haemolymph is maintained in tap water when temperatures are low (3°C), but at 21°C the osmotic pressure of the haemolymph falls to about 16 atmospheres ($\Delta = -1\cdot31°C$) in tap water (Nemenz, 1960). These figures for haemolymph concentration are much higher than those reported for the estuarine species of *Ephydra*. For instance, Beyer (1939) found that *E. riparia* had haemolymph with a concentration equivalent to 5 or 6 atmospheres. It is probable that the exceptionally high concentration of the haemolymph of *E. cinerea* is an adaptation to the extreme salinity of its habitat. The estuarine species of *Ephydra* maintain a lower haemolymph concentration, and can withstand the variations of salinity in their natural habitat without any necessity to increase the haemolymph concentration up to the level found in *E. cinerea*.

ESTUARINE WATER BEETLES

The beetles of the family Dytiscidae form an important element of the fauna of fresh waters, where they often abound and prey upon other invertebrates. A considerable number of species extend their activities a short way into brackish water, and a few penetrate above a salinity of 8‰. One of the largest species to penetrate into brackish water in Britain is *Colymbetes fuscus*. This species is widespread in fresh water, but is also found in salt marsh pools. Sutcliffe (1961) observed some adults that remained in pools with salinities between 18 and 23‰ for several weeks, and found larvae in situations where they must have been subjected to salinities of about 30‰ for short periods.

Another common fresh-water dytiscid, *Agabus bipustulatus*, also penetrates into brackish water, but in general it is not found living permanently in salinities above 10‰. The related species *A. conspersus* appears to be a true brackish-water species, rarely if ever found in fresh water.

Some of the smaller dytiscids, belonging to the tribe Hydroporini, also show tendencies towards becoming brackish-water species. The salinities at which a number of these species are found in the Baltic have been ascertained by Lindberg (1944, 1948). *Coelambus parallelogrammus* is one of the commoner brackish-water species, often abundant in salinities of 6 or 7‰. This species is also found in fresh water. In Britain the closely allied *C. impressopunctata* is rarely found in fresh water, but much more frequently in brackish water. This has led Balfour-Browne (1940) to suggest that *C. impressopunctata* is a comparatively recent arrival in Britain, capable of holding its own in brackish water, but only beginning to get a hold in fresh

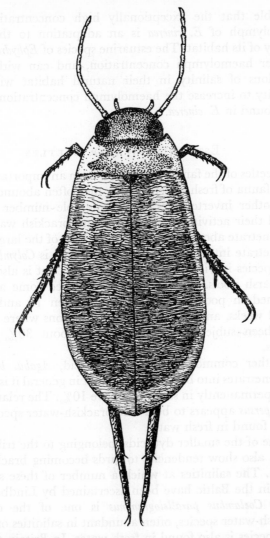

Figure 78. *Colymbetes fuscus* (Coleoptera: Dytiscidae). Length 16 mm

water, where competition with the numerous other small dytiscids is more severe.

The species of *Coelambus* are often accompanied in brackish water by *Hygrotus inaequalis*, and by some species of the family

Haliplidae, such as *Haliplus apicalis*. The haliplids are closely allied to the dytiscids, but they are all small, and appear to have two generations in a year, while the dytiscids have a single generation each year.

Another large family of water beetles, the Hydrophilidae, has numerous examples which have spread from fresh water into brackish water. Some typical fresh-water species, such as the large *Hydrophilus piceus*, are occasionally found in brackish water. This species has been found in the more dilute parts of the Baltic, and, remarkably for a beetle, has been found with barnacles of the species *Balanus improvisus* attached to its wing cases.

Some of the smaller hydrophilids are permanent and regular inhabitants of estuarine waters, particularly in salt marsh pools. Within the genus *Helophorus* for instance there are several species, such as *H. alternans*, *H. avernicus* and *H. aequalis*, which may be found in this habitat. These beetles feed on filamentous green algae and lay their eggs in oothecae which contain about twelve eggs. There are three larval stages before pupation.

Members of the hydrophilid genus *Ochthebius* are all small beetles, about 1 or 2 mm in length. They live in water, or sometimes on damp ground close to water. Most of the species can fly well, so that if their pool becomes dry they can quickly leave and find another. Some of the species are found in fresh water, others in brackish water and others are more or less marine in habit. In Britain there are 15 species (Balfour-Browne, 1958), and a high proportion are found in brackish or marine habitats. In brackish pools one may find *O. marinus*, *O. impressicollis*, *O. viridis* and *O. punctatus*, while on the surface of the salt marsh around such pools *O. auriculatus* may be found. This last species seems to be somewhat more terrestrial in its habits than its congeners. At the seaward end of an estuary, in small rock pools on the upper shore, *O. subinteger* var. *lejolisi* represents the final stage of penetration of the marine habit as far as the British species are concerned.

On the shores of Madeira *Ochthebius subinteger* is found in pools high on the shore and is replaced by *O. heeri* lower on the shore. But the species which has developed the greatest tolerance for high salinities is probably *O. quadricollis*, the biology of which has been studied by Jacquin (1956). This beetle was found in small splash pools in depressions on a stone jetty near Algiers. The water in these depressions showed considerable fluctuations in salinity according to the local weather and the occurrence of storms which carried sea water over the jetty. The adults and larvae were capable of living for three months in distilled water, but in this medium the adults did not breed. Eggs placed in distilled water did not hatch. In sea water and in a salinity of 94‰ the life cycle was completed normally, but at a salinity of 112‰ the adults survived for only 8 days, and at 162‰ survival fell to just over a day. If taken out of water the adults died in about 4 hours in dry air, but if the air was saturated with water vapour they survived for 20 days. The larvae also survived well in damp air, and could pupate successfully in relative humidities over 80 per cent.

ESTUARINE WATER BUGS

The aquatic Hemiptera Heteroptera, or water bugs, have also made some progress in penetrating the brackish environment from the fresh-water end. The most successful species in Britain in this respect is *Sigara stagnalis* (Fig. 79). This species may be found in salt marsh pools with salinities up to 25‰, but it probably cannot breed in salinities higher than 18‰. In external media with concentrations less than 20‰ the freezing point depression (Δ) of the blood is maintained above that of the medium. In fresh water the Δ of the blood is about 0·7°C, and in a salinity of 18‰ the Δ of the blood is 0·9°C (Claus, 1937). It is clear that *S. stagnalis* has an efficient regulatory system that enables it to cope with variations in the salinity of its external medium. The oxygen consumption of *S. stagnalis* in different salinities shows a minimum at 6‰. Both above and below this value the rate of oxygen consumption increases. This may be an indication that *S. stagnalis* is in an optimum salinity at 6‰.

In the spring the adults of *S. stagnalis* lay eggs which give rise to a generation that becomes mature in July. This summer generation then lays eggs to produce another generation that becomes mature in September, but overwinters before laying eggs in the following spring. There are thus two generations in

I

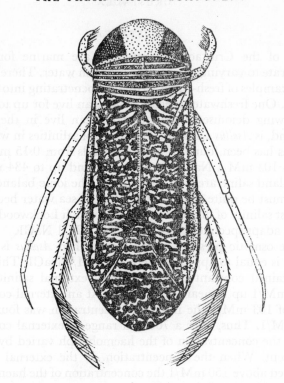

Figure 79. Sigara stagnalis (Hemiptera: Corixidae). Length 6 mm

a year, and the adults remain active over a wide range of temperatures. The geographical range of this species is wide, extending from the Canary Islands to the Baltic Coast of Sweden.

Sigara selecta may also be found in salt marsh pools with *S. stagnalis*, but other water bugs are not found as frequently in brackish water as the *Sigara* species. *Notonecta viridis* is found in both fresh and brackish water, and is more common in the latter. The common pond skater, *Gerris thoracicus*, is sometimes found on the surface of salt marsh pools. In this case the insect is not submerged and is not subjected to the osmotic stresses experienced by fully aquatic insects.

CRUSTACEA

Most of the Crustacea in estuaries are marine forms that penetrate to varying extents into brackish water. There are very few examples of fresh-water Crustacea penetrating into brackish water. One freshwater species, which can live for up to 16 days in flowing deionised water, and yet can live in the Gulf of Finland, is *Asellus aquaticus*. The range of salinities in which this species has been found in nature extends from 0·15 m equiv/l Na to 103 mM/l NaCl in the Baltic, and up to 434 mM/l in the inland saline areas of Westphalia. The ionic balance in this area must be quite different from that of sea water because the highest salinity of diluted sea water to which Lockwood (1959a) could adapt specimens was about 200 mM/l NaCl.

The osmotic pressure of the blood when *Asellus* is in fresh water is equal to approximately 150 mM/l NaCl. This level is maintained constant over a range of external salinities from 0·08 mM/l up to nearly 100 mM/l. At an external concentration of 103 mM/l the internal concentration was found to be 170mM/l. Thus, over a 700-fold range of external concentrations, the concentration of the haemolymph varied by only 15 per cent. When the concentration of the external medium is raised above 150 mM/l the concentration of the haemolymph varies with that of the medium.

The ionic regulatory mechanism of *Asellus* is sensitive to temperature changes (Lockwood, 1960). The rate of active uptake of sodium increases with increasing temperature, but the rate of loss of sodium does not increase at higher temperatures. In very dilute external media *Asellus* is better able to maintain its internal sodium concentration when the temperature is high rather than low.

The smaller Crustacea that penetrate from fresh water into brackish conditions have not received much attention from a physiological point of view. Some of the cyclopoid copepods that are widespread in fresh water can live in areas near the heads of estuaries. *Cyclops viridis* was found throughout the year in the plankton at the head of the Delaware River Estuary (Cronin, Daiber and Hulbert, 1962), and was most abundant in spring and summer. In European estuaries *Cyclops bisetosus* is found in small pools, often of a temporary nature near the head of an estuary. This species occurs in a wide range of fresh-water habitats, including springs, caves, tree holes, and it has also been recorded in salinities as high as 49‰ (cf. Gurney, 1933).

ESTUARINE OLIGOCHAETA

The oligochaete annelids are basically a fresh-water group. Even those which have become dwellers in soil, such as the typical earthworms, retain many of the physiological character-istics of fresh-water animals (Ramsay, 1949). The smaller forms, such as the Naididae and Tubificidae are often dominant members of the benthos. In the case of the members of the latter family they may be overwhelmingly the most numerous animals in polluted areas. Figures of 100,000 in a sq m are sometimes recorded. Penetration from fresh water towards brackish and marine conditions has been effected most frequently by the tubificids, and less often by the naidids. Members of the latter family rarely penetrate to salinities above 8‰, and most species remain in salinities below this. Two species which penetrate to higher salinities are *Amphichaeta sannio* and *Paranais litoralis*. The latter is frequently found in sandy areas, and has been seen swimming actively among filaments of *Enteromorpha*.

In good conditions the naidids reproduce asexually by trans-verse fission, producing two or three individuals that may remain attached to each other for several days (cf. Herlant-Meewis, 1958).

The Tubificidae are often impressive in their abundance; parts of the Thames Estuary are coloured red by thousands of these worms in shallow water at low tide. The normal position for a tubificid is head down in the mud, with the tail protruding. The protrusion and waving activity of the tail varies inversely with the oxygen content of the water. The undulatory motion is more rapid when the oxygen content of the water is low. The activities of tubificids are important in aerating the top layers of mud. Undulation of the tail drives a current of water towards the head and circulates aerated water through the mud.

Tubifex costatus can be found at salinities between 2 and 36‰, while *Limnodrilus pseudogaster* lives in salinities up to 24‰. This species does not occur in fresh water, but the lower limit of salinity at which it is found has not yet been established (Dahl, 1960). Another species, *Peloscolex benedeni* seems to require salinities above 12‰. Yet other species have become more or less marine. *Thalassodrilus prostatus* has been found under seaweed in the Menai Straits (Brinkhurst, 1963), and *Clitellio arenarius* is common in sand and gravel between tides, particularly where there is a trickle of fresh water.

The breeding biology of *Tubifex costatus* in a brackish locality has been studied by Brinkhurst (1964). The worms became mature at the end of April and laid their eggs in cocoons in May and June. By July large numbers of small worms had appeared. It seemed that the worms bred once only at an age of about two years, and then they died.

ROTIFERA IN BRACKISH WATER

There are few rotifers that are truly estuarine, although a considerable number of species may be found in low salinities. Most of these species are well known fresh-water forms. Valikangas (1926) has listed the species found in the harbour at Helsinki, and has given the upper limits of salinity tolerance of each species that he found. A similar list of species from the salt marsh pools on the estuary of the River Dee in England has been given by Galliford (1956). Some species were found in salt marsh pools with widely varying salinities. For instance, *Notholca squamula* was found in salinities from 1·5 to 24‰, *Proales reinhardti* from 5·7 to 31‰, and *Testudinella clypeata* from 1·4 to 27·1‰.

The only fresh-water rotifers that have been studied in detail from an osmoregulatory point of view are the members of the genus *Asplanchna*. These are all transparent sac-like forms with an easily visible excretory system including a contractile vesicle. Two valuable contributions to the study of these rotifers have been made recently by Pontin (1964, 1966). The details of the structure of the excretory system vary from one species of *Asplanchna* to another. In general there is a pair of protonephridia which open into a vesicle. This vesicle fills with fluid and contracts at intervals which vary with temperature and the salinity of the external medium. Each protonephridium consists of a single unbranched tubule which receives ducts from a number of flame bulbs. The number of flame bulbs in each nephridium varies from 4 in *Asplanchna priodonta* up to 230 in the largest varieties of *A. amphora*. This variation can be related to the surface areas of the different species; the species with the largest surface area having the most flame bulbs. The output of fluid from the contractile vesicle is easily calculated from the frequency of contraction and the diameter of the vesicle before contraction. Estimates of the output from 3 species of *Asplanchna* showed that when calculated in terms of the output per flame bulb the figure for each species resembled that of the other species. Measurements showed that the flame bulbs of the different species were also very similar in size, so that in this genus at least there appears to be an optimum size for a flame bulb, and any increase in surface area is dealt with by an increase in the number of flame bulbs rather than by any varia-tion in the size of these structures.

The osmoregulatory function of the protonephridia and contractile vesicle of *Asplanchna* has been demonstrated by immersing specimens in different salinities and recording the output from the vesicle. The intervals between contractions are increased when the salinity of the external medium is increased, and the effect is reversed on return to a lower salinity (Green, 1957). The rate of beat of the vibratile flames has also been shown to vary with salinity. In a concentration of 120 mOsm/l the flames of *Asplanchna priodonta* beat about 500 times per minute at 15°C; this is only a quarter of the rate of beat when the external concentration is 40 mOsm/l.

An interesting feature of *Asplanchna* is that it is a turgid animal. The body wall, although thin, is tight like an inflated balloon. Pontin (1966) estimates that the body fluid of *A. priodonta* is isotonic with a solution of 100–160 mOsm sodium chloride/l, and suggests that the turgor pressure is of the order of 40 mOsm sodium chloride/l.

The upper limit of salinity tolerance for the species of *Asplanchna* appears to be in the region of 4‰. There are rotifers of the genus *Synchaeta* which live in much higher salinities, up to 35‰. Unfortunately the osmotic regulation of these species has not yet been analysed.

THE FRESH-WATER COMPONENT IN THE FAUNA OF THE BRACKISH SEAS

We have already seen how some animals of fresh-water origin have penetrated along estuaries and are now capable of living in fully marine conditions. The number of species that have made the full transition is relatively small. In the brackish seas the large areas of stable low salinity facilitate the invasion of fresh-water animals. Generally the seas with the lowest salinities have the greatest proportion of fresh-water animals in their fauna, but the gradient of salinity between the sea and the neighbouring fresh water is also an important factor.

The Black Sea has a higher salinity (17–18‰ at the surface) than the other brackish seas, and has a smaller proportion of fresh-water forms in its fauna. Among the fish, for instance, there are 37 fresh water and migratory species out of a total list of 180 species. In the neighbouring Sea of Azov, which has a mean salinity of 11‰, and a region in the Gulf of Tarangog, which is almost fresh, there are 79 species of fish, but 33 of these are either fresh water or migratory forms. This difference in the ichthyofauna of the two seas is due mainly to a reduction in the numbers of marine species of Mediterranean origin.

In the Caspian Sea there are 78 species of fish, 4 of which have been introduced by man. Of the remaining species 30 belong to the family Gobiidae and represent the development of an endemic group not found elsewhere. A further 15 species are cyprinids of fresh-water origin, including bream (*Abramis brama*), roach (*Rutilus rutilus*) and carp (*Cyprinus carpio*). The trend to a reduction in total species and an increased proportion of fresh-water forms is continued in the Aral Sea. Because it is so isolated from other seas this sea has only 24 species of fish, of which 12 are cyprinids. Other fresh-water fish in the Aral Sea include the pike (*Esox lucius*), and a large cat fish (*Silurus glanis*). The zooplankton of the Aral Sea also has a strong fresh-water component. For instance, *Mesocyclops leuckarti* and *M. hyalinus* are the commonest of the copepods; this pair of species is a common feature of the plankton of many temperate and sub-tropical lakes. The fresh-water Cladocera *Ceriodaphnia reticulata* and *Moina microphthalma* are also numerous in the Aral Sea.

The Baltic Sea offers considerable opportunities for the invasion of fresh-water animals. The extended salinity gradients and the large areas with salinities below 4‰ (see map on p. 99) provide regions suitable for the mass acclimatisation of fresh-

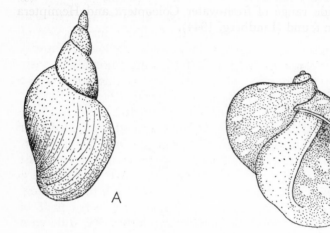

Figure 80. Fresh-water gastropods from the Baltic Sea
A, *Limnaea stagnalis*. B, *Theodoxus fluviatilis*

water animals. The relatively low temperatures in the Baltic may aid the penetration of fresh-water forms by reducing competition from marine forms, because many marine animals cannot osmoregulate well at low temperatures. The Baltic fish fauna includes 58 species, of which 33 are fresh water or migratory forms. Some of the species are the same as those found in the southern brackish seas, for example, the pike (*Esox lucius*), roach (*Rutilus rutilus*) and bream (*Abramis brama*). Among the gastropod molluscs there is also an important group of fresh-water species that have invaded the Baltic, including *Lymnaea stagnalis* and *Theodoxus fluviatilis* (Fig. 80). The fresh-water isopod *Asellus aquaticus* penetrates into salinities as high as 15‰, and similar salinities are tolerated by the larvae of the moth *Acentropus niveus*, which is widespread in European fresh waters south of the Baltic. The rockpools around the shores of the Baltic also contain many fresh-water animals. Some of these pools are in fact fresh, being formed by accumulation of rain in hollows on impervious granite, but others are occasionally flooded by the sea and are then subjected to evaporation so that the salinity may rise above that of the nearby sea. In such pools *Daphnia magna* has been found living in a salinity of 10 or 12‰, and a wide range of fresh-water Coleoptera and Hemiptera have been found (Lindberg, 1944).

CHAPTER 9

THE TERRESTRIAL COMPONENT

IN Chapter 3 it was shown that the vegetation of an estuary, particularly in the region of a salt marsh, shows a transition from aquatic to terrestrial plant communities. In the upper reaches of a salt marsh the plants may be submerged for only a few hours on a few days in a year. Progressive and gradual adaptation to submersion is facilitated by the pattern of spring and neap tides, providing an opportunity for terrestrial animals to penetrate down the shore.

Terrestrial arthropods have taken this opportunity more frequently than other terrestrial animals. For instance, various species of soil-dwelling mites (Acari) have penetrated to differing extents down the shores of estuaries. Those dwelling in salt marshes have been studied in detail by Luxton (1964). In the Juncetum he found that the dominant mite was the oribatid *Platynothrus peltifer*, while in the region where *Festuca rubra* was the dominant plant *Hygroribates schneideri* was most abundant, and was accompanied by *Hermannia pulchella*. Lower still on the shore, in the Puccinellietum, yet another oribatid, *Punctoribates quadrivertex*, became dominant. Several of the mites living in salt marshes were found to be viviparous. *Cheiroseius necorniger*, a mesostigmatid mite with a wide vertical distribution on the marsh, was found to be facultatively ovoviviparous, either retaining the eggs internally until they hatched or else laying eggs which took 8 days to hatch. *Hygroribates schneideri* was found to be capable of surviving more or less permanent submersion, and could reproduce viviparously under these conditions.

In general the detritus eating oribatid mites are found on salt marshes in larger numbers than the predatory mesostigmatid mites. Both groups show zonation in their occurrence, that of the oribatids is described above, but the mesostigmatids show

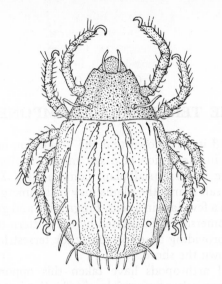

Figure 81. Platynothrus peltifer (Acari: Orbiatei). Actual length 0·9 mm

a parallel zonation. In the *Festuca* zone *Macrocheles subbadius*, *Leioseius salinus* and *Arctoseius cetratus* are the most important forms. *Cheiroseius necorniger* extends from the *Festuca* zone well down into the *Puccinellia* zone, where it is joined by *Digamassellus halophilus* and *Halolaelaps nodosus*.

The predatory mites include Collembola as part of their diet, and in a salt marsh there is often an abundance of these flightless insects. *Archisotoma besselsi* is often found under seaweeds, or associated with felt-like layers of *Vaucheria*, and I have seen dense patches of this species on bare mud. Sometimes this collembolan aggregates into almost solid masses on the surface of mud, and thousands of individuals may be found in a few sq cm. The cause or purpose of this aggregation is unknown.

The common collembolan of rocky shores, *Anurida maritima*, also extends some considerable distance into estuaries. The normal habitat on rocky shores is within crevices, but on estuarine shores this species is often abundant on the salting cliff. The vertical range of *A. maritima* on a salt marsh is thus often restricted to a narrow zone around the edge of the marsh. Higher up on the marsh there are other Collembola, and in regions rarely covered by the tide it is possible to find such typically terrestrial species as *Isotoma viridis*, *Entomobrya lanuginosa* and *Tomocerus longicornis*. On the surface of salt marsh pools it is sometimes possible to find another collembolan,

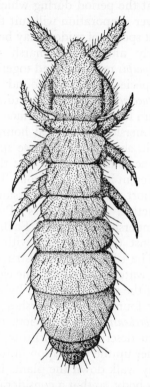

Figure 82. Anurida maritima (Collembola). Actual length 2·5 mm

Xenylla humicola. This species was also one of the dominants among the 33 species of Collembola found by Backlund (1954) in beds of seaweed cast up on the shores of Sweden and Finland.

A curious feature of the occurrence of Collembola is that they often reach their greatest abundance in winter. MacFadyen (1952) quotes several authors, both in Britain and on the Continent, who have found such maxima. This may indicate that the food supply and the relative humidity are most favourable in winter. The breakdown of plant material produced during the summer may take several months to reach a state where it is suitable for certain collembola, and it is well established that most collembolans are very vulnerable to desiccation, so that the period during which there is an excess of precipitation over evaporation will suit them well.

Several different species of spiders may be found on estuarine shores, particularly among salt marsh vegetation. *Erigone arctica* and *E. longipalpis* may be found together in damp places, and the latter species has been found amongst *Halimione portulacoides* and *Puccinellia maritima*. Bristow (1958) found that *E. arcticus* could run freely over the surface of water, but could also withstand immersion for several hours without apparent harm. The smooth, shiny body of these spiders does not trap any air bubbles when submerged, so that the spiders must rely on such air as is present in the lung books. This would probably be adequate for short periods of submersion provided that the spider was not active. The egg sacs of *Erigone* are usually attached to the underside of a stone. Each sac is more or less circular, flat against the stone, but with a raised centre to house the eggs. The silk used to construct these sacs is very closely woven and waterproof, so that the eggs are not subjected to osmotic stress if the stone is covered by a high tide.

Another spider found in salt marshes is one of the wolf spiders, *Lycosa purbeckensis*. The behaviour of this species as the tide rises has been described by Bristow (1923). At first the spiders climb higher up the vegetation in front of the incoming tide, but later they walk down the plants into the water. This species has a hairy body, so that a considerable amount of air is trapped around the spider. This air is sufficient for the spider to survive 10 hours under water. In natural conditions the spider would never be covered for more than 2 or 3 hours.

In the *Spartina alterniflora* zone of North American salt marshes the small spider *Grammonata trivittata* hunts the homopteran *Prokelisia marginata*. This spider behaves like *Lycosa purbeckensis* when faced with the incoming tide. At first it retreats until it reaches the tip of a grass blade, then it turns and walks down into the water, eventually coming to rest at the junction of a grass blade and stem. *Grammonata* has a high resistance to drowning, and can remain submerged for 24 hours without ill effect (Arndt, 1914).

Within the hemipterous family Saldidae there are species which occupy a wide range of habitats. These heteropterans have large eyes and are active predators, so that they are not associated with any particular plant. Some species, such as

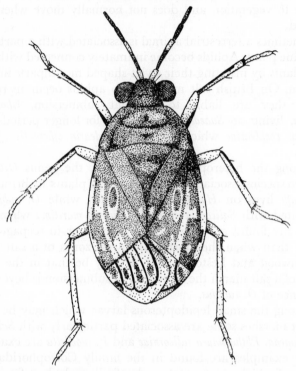

Figure 83. Saldula palustris (Hemiptera: Saldidae). Actual length 4 mm

Saldula orthochila, live on dry land away from water. Others, such as *Saldula scotica* live among shingle at the sides of streams, and yet others, like *Saldula saltatoria*, run over the mud around ponds, both fresh and brackish. Several species inhabit estuaries: *Saldula palustris* and *Halosalda lateralis* are good examples (Brown, 1948). Both are capable of withstanding immersion, but *Saldula palustris* is the more resistant of the two. Consequently it may be found lower down on salt marshes. The young of *S. palustris* live lower on the shore than the adults, and have been shown to be capable of withstanding submersion for 14 hours. This is a much longer period than any to which they would be subjected in their natural habitat. Under natural conditions of submersion the young of *S. palustris* remain active. This contrasts with the behaviour of *Halosalda lateralis* which clings tightly to vegetation and does not normally move when submerged.

Sometimes a terrestrial animal is associated with a particular estuarine plant. Aphids become intimately connected with their host plants by inserting their stylet-shaped mouthparts into the phloem. On British salt marshes two aphids occur in regions where they are liable to frequent submersion. *Chaitaphis suaedae*, living on *Suaeda*, is submerged for longer periods than *Lipaphis cochleariae* which lives on *Cochlearia officinalis* (Jacob, 1956).

Among the heteropterous Hemiptera, the genus *Orthotylus* has two species associated with salt marsh plants in Britain: *O. moncreaffi* lives on *Halimione portulacoides*, while *O. rubidus* is associated with *Salicornia* spp. and *Suaeda maritima* when these plants are found growing around lagoons or in seepage areas rather than when they form the lowest zones of a salt marsh (Southwood and Leston, 1957). It may be that in the lowest zones of a salt marsh the frequency of submersion is beyond the tolerance of *O. rubidus*.

Among the small lepidopterous larvae which may be found on salt marshes some are associated particularly with *Salicornia* and *Suaeda*. *Phthorimaea salicorniae* and *P. suaedella* are examples. Other examples are found in the family Coleophoridae, the larvae of which carry cases made of silk and plant fragments. There is probably differential resistance to submersion among larvae of the genus *Coleophora*, because different species are found associated with different salt marsh plants. *Coleophora salicorniae* occurs on *Salicornia* and *Suaeda*, *C. salinella* lives on *Halimione portulacoides*, and *C. obtusella* lives at the top of the salt marsh on *Juncus maritimus*. The larva of the last species of moth feeds on the seeds of its host plant, and pupates on the stem. The adults are found in July, and their eggs hatch in September.

The larvae overwinter in the dead rush flowers and pupate in the following June. *Coleophora obtusella* thus spends the whole of its life just beyond the reach of the tides, but its congener, *C. salicorniae*, must be submerged at frequent intervals. The larvae live from October to May on the stems of *Salicornia*, and in June they pupate on the ground, without their cases, so that both larvae and pupae are liable to submersion.

Flies of the family Dolichopodidae are predators when adult, and often frequent damp places. Their larvae are found in a variety of situations, such as tree holes, under bark, in cattle dung, and mining in the leaves of various plants. There are also several species that spend their larval lives in estuaries. Dyte (1959) reared *Dolichopus nubilus*, *Porphyrops consobrina*, *Machaerium maritimae* and *Hydrophorus oceanus* from estuarine muds. The last two species were found in areas where the salinity ranged between 15 and 33‰, and *H. oceanus* had a considerable vertical range on the shore, from the zone dominated by *Halimione portulacoides* to the Spartina zone. The adults of *H. oceanus* can often be seen running across the water surface of salt marsh pools.

Near the mouth of an estuary, where a solid substratum provides anchorage for barnacles and seaweeds, other dolichopodids, of the genus *Aphrosyllus* are found. The larva of *A. celtiber* has been found within the barnacle *Balanus balanoides*, but it is uncertain whether or not this was an accidental occurrence. The adults are often abundant in the barnacle zone on open rocky shores, but they have also been seen running over stony estuarine shores coated with *Elminius modestus*.

Members of the order Coleoptera form one of the most conspicuous elements of the terrestrial component of an estuarine fauna, and within this group the predatory family Carabidae is the best represented. Estuarine carabids vary in length from about 3 mm to 30 mm, and different species occupy different niches in the estuarine environment. On loose sandy shores in southern Europe *Eurynebria complanata* (Fig. 84A) is frequently found living just above the tide. This striking species is a fast moving nocturnal predator which ranges down the shore at low tide and hunts the amphipod

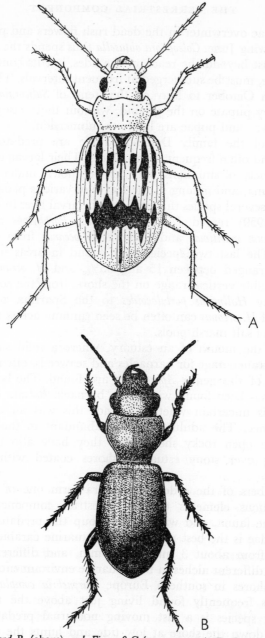

*Figures 84*A *and* B (above) *and Figure 84*C (on next page). Predatory Cleoptera
from sandy areas near the mouths of estuaries. All drawn to the same scale
A, *Eurynebria complanata*; B, *Broscus cephalotes*

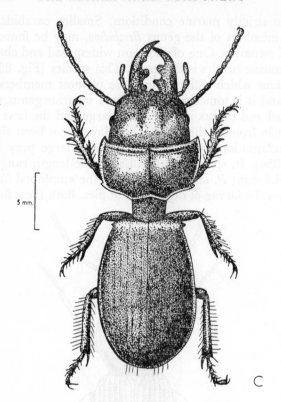

5 mm.

Figure 84C. Scarites buparius (see page 263).

Taltrus saltator. Another species living in similar habitats is *Broscus cephalotes* (Fig. 84B). This species has shorter, stouter legs than *Eurynebria* and is not such a fast runner; the stouter legs may be more efficient for burrowing. The adults of *Broscus* are often abundant on sandy British shores in August, while the larger and more powerful *Scarites buparius* is found on sandy shores in southern Europe. The species listed above are often accompanied by tiger beetles of the genus *Cicindela.* But the tiger beetles are diurnal in habit, racing and flying rapidly across the surface of the sand in bright sunlight.

Because they live in clean sandy areas the beetles mentioned above do not extend very far into estuaries, and they are often found in strictly marine conditions. Smaller carabids, particularly members of the genus *Bembidion*, may be found in all parts of estuaries. One of the most widespread and characteristic estuarine beetles is *B. laterale*. This species (Fig. 85A) has proportions which differ from those of most members of the genus, and it is sometimes placed in a separate genus, *Cillenus*. The head and thorax are relatively large and the jaws project well out in front of the head. These jaws have been shown to be an adaptation to capturing relatively large prey (Green, 1954, 1956). In spite of its small size (the length ranges from 3.5. to 4.2 mm) *B. laterale* preys upon the amphipod *Corophium* and upon the larvae of dolichopodid flies. Both these forms are

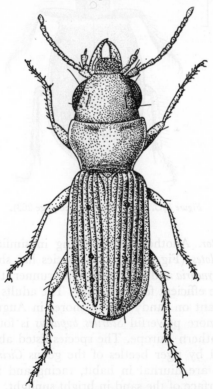

*Figures 85*A (above) *and* B (on next page). Small carabid beetles from estuarine shores. Both are drawn to the same scale.
A, *Bembidion laterale.*

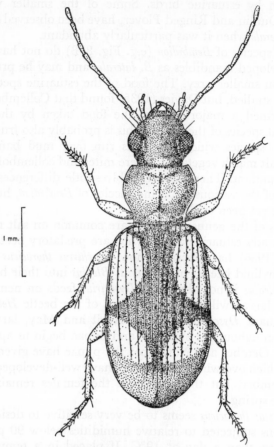

Figure 85B, Bembidion concinnum.

much larger than *B. laterale*, but the long jaws enable the beetle to inflict considerable damage on its larger prey and so quickly incapacitate it. Further, the beetles are often numerous, so that once a *Corophium* is attacked by one beetle others quickly notice the struggle and join in. A single *Corophium* may on occasion be found in the process of being dismembered by up to seven *B. laterale*.

The larvae of *Bembidion laterale* live in burrows in the intertidal zone of estuaries, but the adults run freely across the surface of muddy sand in daylight. This habit exposes them to predation by estuarine birds. Some of the smaller waders, such as Dunlin and Ringed Plover, have been observed feeding on *B. laterale* when it was particularly abundant.

Other species of *Bembidion* (e.g. Fig. 85B) do not have such well-developed mandibles as *B. laterale*, and may be presumed to feed on smaller prey. The food of the estuarine species has not been studied, but Davies (1935) found that Collembola and mites formed a major part of the food taken by the more terrestrial species of the genus. This is probably also true of the estuarine species which as adults run over mud banks and among salt marsh vegetation where mites and collembolans are often abundant. There are probably subtle differences in the ecology of the various estuarine species of *Bembidion*, but these have not yet been explored.

Beetles of the genus *Dyschirius* are common on salt marshes and in sandy estuarine areas. They are predatory beetles. Bro Larsen (1936) has described how *Dyschirius thoracicus* follows the staphylinid beetles of the genus *Bledius* into their burrows. When *Bledius* is not abundant *Dyschirius* feeds on nematodes, and the larvae will feed on the eggs of the beetle *Heterocerus*. In Denmark *Dyschirius* pairs in April and May, larvae are found throughout the summer, and pupae begin to appear in July. By October and November the pupae have given rise to adults which overwinter. The males have well-developed sperm by December, but the oocytes of the females remain small until the spring.

Dyschirius thoracicus seems to be very sensitive to desiccation. Specimens subjected to relative humidities below 90 per cent die in less than a day at 19°C. If placed in a temperature gradient the behaviour of this beetle varies according to whether the substratum is damp or dry (Palmen, 1954). If the substratum is damp the beetles quickly burrow, but if the substratum is dry they move about on the surface for a considerable period. When subjected to gradients ranging from 9 to 31°C and from 12 to 41°C with a dry substratum, the beetles aggregated at the cool end. With a damp substratum and a three-hour experimental period there was no such marked aggregation at the cooler end. Instead the beetles burrowed indiscriminately over the range from 9 to 20°C, and some burrowed where the temperature was as high as 27 to 31°C. If the duration of the experimental period was increased to 6 hours there was a general shift towards the cooler end of the gradient. It is clear that under all conditions the preferred temperature range of this beetle lay under 30°C and was often much lower. If the beetles were kept at a relative humidity of 100 per cent they survived well at 35°C but died quickly at 37°C. The lethal temperature in saturated air is well above that likely to be encountered in nature. Greater danger lies in the possibility of desiccation; this is guarded against by the beetle aways choosing the moister of two humidities when presented with a choice (Perttunen, 1951).

The genus *Bledius* includes several species which are coastal or estuarine in habit. These beetles belong to the family Staphylinidae, and have long bodies with short wing cases which leave most of the abdomen exposed (Fig. 86A). The habitats of the British species which live in salt marshes have been discussed by Steel (1955). *Bledius spectabilis* makes burrows in intertidal sand, usually lower on the shore than the heavily vegetated part of the salt marsh. The allied *B. germanicus* tunnels in the soil of salt marshes, usually above the level of the salting cliff where this is present. This species chooses damp areas but prefers those which are not frequently covered by the tide. *Bledius unicornis* seems to overlap the distributions of both the preceding species, while *B. tricornis* prefers areas in a salt marsh where the surface dries and cracks.

The biology of *Bledius* has been studied in detail by Bro Larsen (1936). *Bledius spectabilis* (Fig. 86A) shows considerable sexual dimorphism. The males have a long forward pointing horn on the anterior border of the prothorax. The females lack this remarkable structure. This species is an algal feeder. It makes burrows in firm sand where the sand grains are coated with algae, or where there are fine networks of small green and blue-green algae. When removing algae from a sand grain the mandibles are held wide apart and the grain is worked over by the other mouthparts.

The tunnels made by *Bledius* are fairly stable structures. They are constructed in firm sand by a technique of excavation in which the beetle uses its mandibles. The small species may carry a single grain at a time, but the larger species may seize up to six grains at a time and hold them between the head and forelegs carrying them to the surface. The tunnels are characterised by the small piles of excavated sand which surround the entrance.

At the beginning of the breeding season, in the latter half of April and at the beginning of May, males and females of *Bledius* may be found together in burrows. *Bledius arenarius*

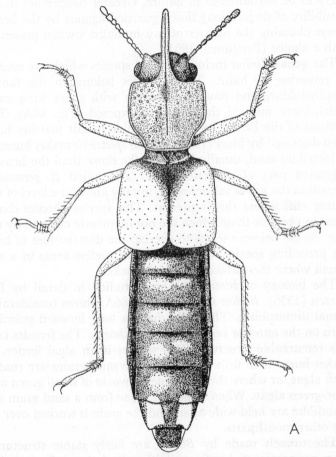

*Figures 86*A (above) *and* B (on next page). Burrowing beetles from estuarine shores. Both are drawn to the same scale. A, *Bledius spectabilis*, male (Family Staphylinidae),

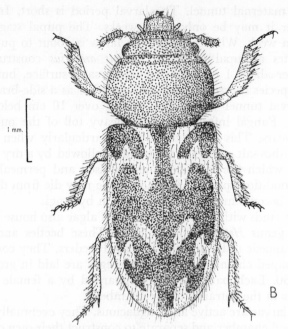

*Figure 86*B, *Heterocerus* sp. (family Heteroceridae).

appears to produce two generations in a year; eggs are abundant in April and again in August. In contrast *B. taurus* and *B. diota*, which do not begin breeding until July, appear to produce a single generation each year.

The eggs of *Bledius arenarius* are laid separately, each in its own isolated chamber about a centimetre below the surface of the sand. Females of the larger species, such as *B. tricornis* and *B. spectabilis*, make tunnels between 5 and 10 cm deep. From the sides of these tunnels open a number (at most 10) of chambers in which the eggs are laid. Again, each egg has its own chamber, and stands on a small pyramid of sand grains with a space between the wall of the chamber and the surface of the egg.

The young larvae of *Bledius* have large heads, strong mandibles, and long thin legs. When the young of *B. tricornis* emerge from their egg chambers they crawl into the maternal tunnel where they remain until they have moulted. It is possible to find 8 or 10 newly emerged larvae clinging to the walls of the maternal tunnel. The larval period is short. In good weather it may be only three weeks. The pupal stage lasts about a week. When a full-grown larva is about to pupate it excavates a pupal chamber. *Bledius arenarius* constructs its chamber about 1 or 2 cm below the sand surface, but some other species construct the pupal chamber as a side-branch of the larval tunnel which may extend over 10 cm below the surface. Fungal infections take a heavy toll of the pupae in some years. This seems to happen particularly when heavy rain washes salt out of the sand and is followed by a dry period during which the sand becomes oxidised and permeable. In these conditions pupae near the surface may die from desiccation or become more liable to infection by fungi.

Firm sands with a surface network of algae also house beetles of the genus *Heterocerus* (Fig. 86B). These beetles and their larvae appear to be detritus and algal feeders. They construct pear-shaped chambers in which the eggs are laid in groups of 20 to 30. Each batch of eggs is guarded by a female which remains in the entrance to the chamber.

The larvae are active and pugnacious. They eventually leave the brood chamber and separate to construct their own cavities in which, when fully grown, they pupate.

The eggs of *Heterocerus* are laid in June in Denmark. Larvae are abundant in July and August, and at the end of the latter month newly emerged adults begin to appear. The adults overwinter, but before they settle down they may disperse from their original breeding ground. The wings of *Heterocerus* are well developed, and swarms of these beetles may be encountered over salt marshes on warm summer evenings.

Not all the beetles living on estuarine shores are ground dwellers; some species are associated with particular salt marsh plants. *Phaedon cochleariae* is a good example. This phytophagus beetle may be found in both adult and larval stages on *Cochlearia officinalis* growing below the high water of spring tides.

Sufficient examples have now been given to establish that estuarine shores are populated by considerable populations of animals of terrestrial origin. Most of the selected examples have been forms which are well adapted to estuarine life, and many of them live in places where they have to withstand regular submersion. In addition to these well-adapted forms there are a large number of terrestrial insects which invade salt marshes when they are uncovered, and then retreat in front of the advancing tide. In the upper parts of salt marshes these insects may have ten or eleven days between spring tides when the area they inhabit is continuously uncovered by the tides.

The insects associated with *Spartina alterniflora* in a North Carolina salt marsh have been studied by Davis and Gray (1966). The ground on which *S. alterniflora* grows is inundated twice daily, but the whole plant is rarely covered, so that the leaves project a foot or more above the water surface when the tide is at its highest. This provides a retreat for those insects which are intolerant of immersion, and enables a considerable range of species to inhabit a plant zone low on the shore.

The insects found associated with *Spartina* in the North Carolina marsh include the following:

Prokelisia marginata (Homoptera: Delphacidae)
Sanctanus aestuarium (Homoptera: Cicadellidae)
Draeculacephala portola (Homoptera: Cicadellidae)
Chaetopsis apicalis (Diptera: Otitidae)
Chaetopsis fulvifrons (Diptera: Otitidae)
Conioscinella infesta (Diptera: Chloropidae)
Ischnodemus badius (Hemiptera: Lygaeidae)
Trigonotylus uhleri (Hemiptera: Miridae)
Orchelimum fidicinum (Orthoptera: Tettigoniidae)
Conocephalus sp. (Orthoptera: Tettigoniidae)
Isohydnocera tabida (Coleoptera: Cleridae)
Mordellistena sp. (Coleoptera: Mordellidae)
Collops nigriceps (Coleoptera: Malachiidae)

In the summer about 30 per cent of the total insects in the *Spartina* association were homopterans, and of these the most important was *Prokelisia marginata*. When this species is submerged it rests head downwards on grass blades and traps bubbles of air under its wings. Arndt (1914) found that submersion for 27 hours did not impair the ability of this species to fly. Higher up the marsh, in the zone dominated by the tall rush, *Juncus roemerianus*, another homopteran, *Keyflana hasta*, was important, and another grasshopper, *Paroxya clavuliger*, was found in this zone. *Juncus roemerianus* reaches a height of 2 metres, and the base of the plant is flooded only on spring tides, so that there is ample opportunity for insects to remain above the tides.

The behaviour of many of the insects in the face of the advancing tide has a parallel in the behaviour of the small terrestrial mammals which invade the upper parts of salt marshes. Voles and field mice often exploit the upper parts of marshes, and their tunnels are sometimes quite abundant. When the spring tides flood their burrows the small rodents move upshore, or climb tall plants until the tide ebbs again. In the salt marshes of Georgia the rice rat, *Oryzomys palustris*, climbs up the stems of *Spartina alterniflora* to escape the high tides. Normally the top two or three feet of the *Spartina* plants remain uncovered by the water. The raccoon (*Procyon lotor*) is a far bulkier animal than the rice rat, but can also remain in the *Spartina* zone during high tide by constructing a ramp. This structure consists of the tops of *Spartina* bent over to form a platform on which the raccoon can lie during the high tide. The mink (*Mustela vison*) is another marsh dweller often found in the *Spartina* zone. This species often makes a nest of dead *Spartina* leaves in a hollow tree trunk washed into the marsh on a high tide.

The ability of so many animals to retreat in front of the tide makes a strict definition of the terrestrial component of an estuarine fauna almost impossible. Nevertheless the validity and importance of this component is illustrated by the large numbers of those species that can live in situations where they are regularly immersed, and where they feed on undoubted members of the estuarine fauna, so playing a part in the general economy of the estuary. If a single example is needed, perhaps *Bembidion laterale*, described on p. 265, is the best.

CHAPTER 10

ESTUARINE FISH

THE fish found in estuaries belong to a wide range of orders and families, but they can also be grouped in various ways according to their habits and the duration of time that they spend in estuaries. The true brackish-water species, such as some of the gobies (*Gobius* spp.), spend all their time in the fluctuating salinities of estuaries. Others, such as the horse mackerel (*Trachurus trachurus*), migrate from the sea into rivers for a short period in late summer and autumn. Yet others, such as some of the whitefish of the genus *Coregonus*, migrate from the fresh regions of rivers into estuaries for the breeding season. In addition to these species which enter estuaries, either from the sea or from fresh water, there are migratory species which pass through estuaries on their way from the sea to fresh water or *vice versa;* well-known examples include the fresh water eels (*Anguilla* spp.) and salmon (*Salmo* spp.).

The three major aspects of fish biology that are of importance in relation to estuarine ecology are osmoregulatory ability, feeding habits, and reproductive habits.

A review of the osmoregulatory abilities of fish has been made by Parry (1966), who distinguishes the following responses to salinity change:

(*a*) Species which survive in conditions of isosmotic constancy and quickly die if transferred to another environment. They may die as a result of changes in the concentration of the body fluids following the movement of water or salts. Marine hagfish, such as *Myxine*, are stenohaline in this way, as are many fish embryos.

(*b*) Species which tolerate some degree of change in the osmotic or ionic levels of their environment by changing and tolerating changes in the concentrations of their body fluids.

274

Some euryhaline fish, particularly elasmobranchs, combine this feature with marked ability to regulate the osmotic concentration of the blood.

(c) Species or stages which may be semi-permeable, allowing free water movement in or out in response to external changes. These species may tolerate changes in body volume or cell volume. Some fish eggs are able to tolerate volume changes of this sort.

(d) Species which reduce the permeability of their surface. This method has its limitations because all fish are faced with the problem of providing a permeable surface for respiration. The fresh water eel (*Anguilla*) is an example that has proceeded in the direction of reducing its permeability by secreting a thick mucous coat.

(e) Species able to compensate actively for movements of water or solutes caused by external changes. This regulation is often brought about by a combination of mechanisms.

The mechanisms available for regulation purposes include: (1) excretion, which may be selective for solutes or water; (2) swallowing the medium, with subsequent selective absorption of ions or water from the gut; and (3) active absorption or excretion of specific ions by the whole epithelium, or at specific sites.

In an estuary the ability to tolerate rapid changes in the external environment must be coupled with the ability to regulate the internal environment if the fish is to become a permanent and successful estuarine dweller. Thus the rate of adaptation to salinity change becomes important. The flounder (*Platichthys flesus*), one of the estuarine fish *par excellence*, can adapt within a few hours when subjected to a sudden change of medium (Motais, 1961).

ESTUARINE TELEOSTS

Fresh-water teleosts maintain the concentration of salts in the blood between 280 and 360 m–Osmole/1, while most marine teleosts keep their blood between 340 and 500 m–Osmole/1. This is well below the concentration of sea water (*ca* 1000 m–Osmole/1). Many marine teleosts can tolerate dilutions of their external environment down to levels equal to the concentration of their own blood. The outer regions of estuaries do not present any major osmotic problems to marine teleosts until the salinity falls to about a third that of sea water.

Marine teleosts utilise four mechanisms to regulate the concentration of salts in the blood: (i) the body surface is relatively impermeable; this effect is brought about by scales, by thickening of the dermis and by the production of mucus; (ii) sea water is swallowed and water is absorbed from the gut together with monovalent ions; (iii) monovalent ions are actively secreted by the gills; (iv) urine is produced in small quantities, so that water is conserved as much as possible. Freshwater teleosts do not have the same problem of water conservation, but they do have the problem of conserving salts to maintain their blood concentration above that of the medium. The gills are involved in this process and have been shown to be capable of taking up salts from the surrounding water (Krogh, 1937; Wikgren, 1953).

When a migratory fish, such as a salmon, enters fresh water from the sea there is a spectacular change in the volume of urine produced; the flow in fresh water may be 100 times as great as in the sea (Holmes, 1961).

The volume of water filtered by the glomeruli of a teleost kidney varies with the environment. A typical fresh-water teleost may filter 200–400 ml/kg fish per day, but in a marine teleost the figure may be as low as 10–15 ml/kg fish per day. The filtrate which passes through the glomeruli has to pass down the renal tubules, and during this process there is resorption of both salts and water. In marine teleosts over 75 per cent of the water filtered through the glomeruli is resorbed. Holmes (1961) found 99 per cent resorption in the rainbow trout (*Salmo gairdneri*) when kept in sea water. In fresh water there is less need for water conservation and the resorption of water by the kidney tubules falls to between 25 and 40 per cent of the glomerular filtrate.

Both marine and fresh water teleosts resorb a high proportion of the filtered chloride, with the figure for fresh-water teleosts reaching 95 per cent, while the marine forms vary between 80 and 90 per cent. It may at first sight seem paradoxical that the kidney of the marine teleost should be so efficient at resorbing salt, but the main route for excretion of monovalent ions is *via* the gills. The efficient resorption by the kidney is a useful feature if the teleost enters water more dilute than the sea.

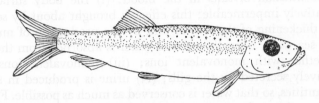

Figure 87. Clupea sprattus, the sprat, length 50 mm

The mackerel (*Scomber scombrus*) may be taken as an example of a marine fish which enters the mouths of estuaries, following prey at certain seasons. Off the south-western coasts of England the mackerel overwinters near the floor of the sea. At this time of year many of the fish that are captured have empty stomachs, and those which have been feeding contain the euphausiid *Nyctiphanes couchi* together with other Crustacea and small teleosts (Steven, 1949). In the early spring some individuals feed on phytoplankton, but by April or May most are feeding on planktonic copepods. In June the mackerel move closer inshore and feed on small fish, particularly *Clupea sprattus,* which is followed into estuaries. In August in the Tamar the mackerel were found to be feeding almost entirely on *C. sprattus* (Hartley, 1940).

Another example from a different continent is the Bombay duck, *Harpodon nehereus.* This species inhabits coastal waters from Zanzibar to China, and is particularly abundant around Bombay and in the estuaries of Bengal and Burma. The ripe adult fish spawn at sea, but immature individuals enter estuaries where they feed on brackish-water shrimps, such as *Acetes indicus* and small teleosts like *Bregmaceros macclellandi.* The Bombay duck does not penetrate into fresh water, but is attracted to regions of lowered salinity where its prey is abundant (Hora, 1934).

The herring, *Clupea harengus*, provides another example of a marine fish entering estuaries. This species has also penetrated the Baltic Sea so far that its distribution overlaps that of the fresh-water pike (*Esox lucius*), and the latter species includes the herring in its diet. The eggs of the herring are capable of developing and hatching at all salinities between 6 and 52‰, and the newly hatched larvae can tolerate a salinity as low as 1.4‰ or as high as 60‰ for nearly 24 hours (Holliday and Blaxter, 1960). This remarkable range of salinity tolerance by the larvae is in fact greater than that of the adults. Holliday and Blaxter (1961) found that North Sea herring with lengths between 9 and 24 cm could tolerate salinities from 6 to 40‰, and within this range they could tolerate direct transfer between different salinities. During these transfers the concentration of the blood was actively regulated, but nevertheless showed some variation. The blood concentration sometimes fell as low as 13‰, and sometimes rose to 22‰, but within the tolerable range of external salinities the concentration usually recovered to a normal value of about 15.8‰.

It is not possible to consider all the marine teleosts which enter estuaries without turning this chapter into a systematic list. Some idea of the variety of teleosts in an American estuary can be gained from the trophic spectrum on p. 343.

The greatest efficiency in controlling the concentration of the blood might be expected in fish which migrate from the sea into fresh water. The salmon (*Salmo salar*) maintains its blood at 344 m–Osmole/kg water in the sea, and only drops to 328 m–Osmole/kg water in fresh water (Parry, 1961). The common European eel (*Anguilla anguilla*) shows a somewhat larger change, from 428 m–Osmole/kg water in the sea to 346 in fresh water (Portier and Duval, 1922). These figures indicate a good control over the internal environment.

There is considerable variation in the ability of *Anguilla* to osmoregulate at different stages of its life history. It is now well known that the young of the European eel hatch in the region of the Sargasso Sea (Schmidt, 1924). Here they are found as leptocephalus larvae with a length of about 6 mm. The leptocephalus is flattened from side to side and is shaped like a minute, transparent willow leaf. These larvae are carried towards Europe by the Gulf Stream. By day the majority are found at a depth of 50 metres, but at night they migrate towards the surface and most are found at depths between 20 and 30 metres.

By the time that they have reached the continental shelf the larvae have reached a length of 6 or 8 cm and an age of $2\frac{1}{2}$ years. The full-grown leptocephalus stops feeding and metamorphoses into a transparent elver or glass eel. Intermediate stages, or semi-larvae, are found in the surface waters just beyond the edge of the continental shelf. Metamorphosis involves a variety of processes: the body becomes cylindrical in shape, starting at the two extremities; the vertebral column is formed around the notochord; the body muscles are rearranged, and the blood becomes red with haemoglobin so that the heart and main vessels can be seen easily through the transparent body wall. During metamorphosis there is a considerable loss in weight; a recently metamorphosed glass eel may weigh only one-tenth the weight of a leptocephalus. Much of this loss in weight is a loss of water, but at the same time there is a 30 per cent loss in total dry weight.

It is the glass eels which invade the coastal waters and ascend the estuaries of Europe about three years after hatching in the Sargasso Sea. The precise time of arrival of the glass eels at the coast varies with the extent of the continental shelf. On the north coast of Spain, where the shelf is narrow, the glass eels arrive in October and November. In January they arrive on the west coast of Ireland, in February they arrive in the Severn Estuary, and it is not until March or April that they arrive on the coasts of Denmark and Norway (Schmidt, 1909).

During their estuarine life the glass eels gradually become pigmented, and by the time that stripes of pigment are present along the limits of the muscular segments the young eels begin to feed. The dark elvers migrate up the rivers, and as they grow they change gradually into yellow eels, so called because of the pigmentation of the ventral surface. The yellow eels remain in fresh water for 7 to 12 years. Some of the females reach a length of nearly two metres, but the males are smaller. Neither sex becomes completely mature in fresh water, but they metamorphose into silver eels which migrate down to

K

the sea. The silver eel differs from the yellow eel in various ways. The ventral surface is no longer yellow, but silver, due to increased deposition of guanine. The dorsal pigmentation by melanin is also intensified. There is a general increase in the size of the sense organs: the nostrils dilate, the lateral lines become more conspicuous, and the eyes increase in size so that the total volume may be eight times the volume of the eyes of the yellow eel (D'Ancona, 1927). There is also a change in the visual pigment of the retina from the purple colour characteristic of fresh-water fish to the golden colour found in deep sea fish (Carlisle and Denton, 1959). The gut of the silver eel degenerates so that it cannot feed. At this stage the osmoregulatory powers of the eel are at their greatest. If a small yellow eel is taken from fresh water and placed in sea water the concentration of its blood rises. The freezing point depression (Δ) increases from -0.62 to $-0.90°C$ over the course of 12 days, with no indication of stabilisation. Larger yellow eels are more effective in controlling their blood concentration. Twelve days after transfer to sea water the freezing point depression is stabilised about $0.07°C$ lower than the original level in fresh water. The silver eels are even more efficient; the difference between the freezing point depression of the blood in fresh water and in sea water after twelve days is only $0.01°C$ (Boucher-Firly, 1935).

The mechanism of osmoregulation by the eel involves changes in the lipoprotein content of the blood. When transferred from fresh water to sea water the chloride content of the blood of the eel increases by about 50 per cent, but this is compensated for by a reduction in the bicarbonates and lipoproteins in the blood. The lipoproteins adsorb a certain amount of water, so making it unavailable to other solutes. A reduction in the concentration of lipoproteins has the effect of liberating water for the solution of other substances, and so causes a decrease in osmotic pressure greater than the simple molecular effect of reducing concentration (Nicloux, 1938).

It has not yet been established that the silver eels that migrate down to the sea from the rivers of Europe ever reach the Sargasso Sea. It is possible that the parents of European eels are specimens which have grown in the fresh waters on the eastern side of North America. The American eels are usually known as a separate species, *A. rostrata*, which hatches in the same general area as the European eel, but takes only one year to reach the mouths of the American rivers. The case for the European and American eels belonging to the same species has been strongly argued by Tucker (1959).

The fresh-water eels of Japan, China and temperate Australia resemble the American eels in that they take only a year to migrate from their hatching area to the estuaries. Tropical eels have a life cycle similar to that of temperate eels, involving leptocephalus larvae, penetration of estuaries and rivers by elvers, and a seaward migration by silver forms, but the duration of the leptocephalus stage is relatively short. *Anguilla bicolor* for instance, which extends from India to Malaya and New Guinea, metamorphoses from the leptocephalus two or three months after hatching (Jesperson, 1942).

The flatfish, such as the flounder (*Platichthys flesus*), exhibit considerable osmotic control, but appear to have a greater tolerance of variation in the concentration of the blood than many other fish. In sea water the flounder keeps its blood at a concentration of 528 m–Osmole/kg water, but in brackish water (165 m–Osmole/kg water) the blood concentration falls to 286 m–Osmole/kg water (Henschel, 1936). The flounder does not seem to be inconvenienced by this fall in blood

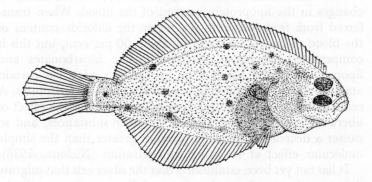

Figure 88. Platichthys flesus, a young specimen of the flounder, taken from the River Crouch in May. Actual length 20 mm, adults reach a length of over 30 cm and a weight of 2 kilograms

concentration, and young flounders can live for several years in fresh water.

The feeding habits of the flounder have received a considerable amount of attention (Scott, 1895; Ascroft, 1900; Murie, 1903; Patterson, 1904, 1906, 1907; Blegvad, 1932; Lubbert and Ehrenbaum, 1936; Larsen, 1936; Stadel, 1936; Hartley, 1940; Bregnballe, 1961), and a considerable amount of variation has been found from one locality to another. In the Tamar Estuary the bulk of the food was made up by three species of Crustacea: *Crangon vulgaris*, *Neomysis vulgaris* and *Schistomysis ornata* (Hartley, 1940). A wide range of other bottom-dwelling animals was also eaten, but few of these were of any importance. Among the polychaetes *Nereis diversicolor* was the species most frequently eaten. In other areas Crustacea also form an important part of the diet. Ascroft (1900) found that *Corophium volutator* was a favourite food of the flounders on the Lancashire coast, and my own observations in the Gwendraeth Estuary show that in summer about 90 per cent of the food taken by flounders is made up by *Corophium* and *Neomysis*. Bregnballe (1961) found that very small flounders (10 mm long) in brackish water on the Danish coast were feeding on harpacticoid copepods and small annelids, including the anterior end of *Pygospio elegans* and whole specimens of the oligochaete *Paranais littoralis*.

In the fresh-water regions of the Elbe the diet of the flounder includes tubificid worms and bivalved molluscs of the fresh-water genus *Sphaerium* (Stadel, 1936). In the River Tweed about 88 per cent of the food taken by flounders consisted of chironomid larvae (Radforth, 1940). A wide range of other insect larvae were also taken, but in no case did the percentage of the total food exceed 1 per cent. In the fresh waters of the rivers Gwendraeth fach and Gwendraeth fawr a considerable proportion of the food taken by young flounders consists of chironomid larvae and oligochaetes.

After feeding and growing in rivers and estuaries the mature flounders migrate seawards to breed. The breeding season off Plymouth appears to be fairly long. A spent female has been captured in February, and a female with a ripe ovary still full of eggs has been found in April. The early post-larval

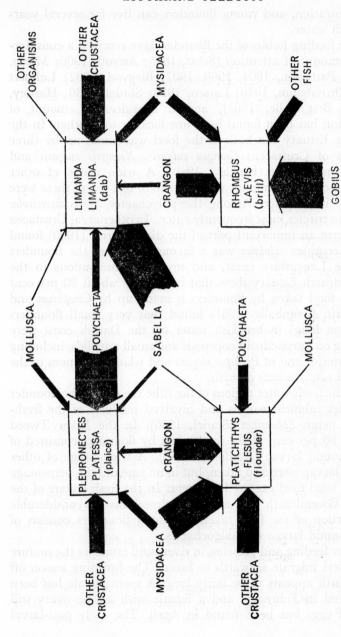

Figure 89. Feeding habits of flatfish in the Tamar Estuary. The arrows point from prey to predator and the thickness of each is proportional to the importance in the diet of the predator. Based on the data of Hartley (1940)

stages can be found in this area from March to June (Russell, 1937).

Hartley (1940) has examined the feeding habits of a range of fish in the Tamar Estuary. Among the flatfish he examined four species in sufficient detail to give a reliable picture of their food. Three of the species: the plaice (*Pleuronectes platessa*), brill (*Rhombus laevis*) and dab (*Limanda limanda*) were essentially marine invaders of the seaward end of the estuary, while the fourth was the flounder. If all the data for each fish are grouped together the mean percentage composition of the food can be calculated. These percentages can then be used to estimate the degree to which the four species compete for food in the Tamar Estuary. Fig. 89 shows that although many items are shared by the four species, each flatfish has its own food preferences. The brill is essentially a fish eater, and a high proportion of its diet consists of *Gobius minutus*. The dab specialises in eating the crowns of the fanworm *Sabella penicillus*. There is not much evidence of competition between these two species, but the flounder and plaice do show some signs of competition. This competition is reduced in several ways. The two fish tend to be separated spatially, with the flounder occupying the fresher parts of the estuary and the plaice remaining near the seaward end. The plaice also includes a much higher proportion of polychaetes in its diet and does not eat as many mysids as the flounder.

A comparison of the type made above is only valid when all the fish have been caught in the same area. We have already seen how the diet of the flounder varies from one locality to another, and the other flatfish might be expected to show similar variations. There are also seasonal variations in diet to be considered when making such comparisons. Hartley's data were gathered over two years, with the samples spread throughout this period so that the mean percentage composition of the diets can be regarded as reliable.

The grey mullets (family Mugilidae) include about 100 species, but the actual number is uncertain because the systematics of the group are somewhat confused. A review of our knowledge of the biology of grey mullet has recently been made by Thomson (1966). A number of species regularly enter estuaries and some penetrate into fresh water, but as far as is known they all breed in the sea. In general the grey mullet are tropical or sub-tropical, but *Liza provensalis* occurs as far north as Iceland. *Mugil cephalus* is the best-known species, and probably has the widest geographical distribution, being found in all seas between latitudes 42°N and 42°S, and extending into very low salinities.

The feeding habits of the *Mugil* species have been studied in some detail. A particularly thorough description of *M. tade* is given by Pillay (1953). The adult fishes are predominantly iliophagous. This means that they obtain their food by straining organic debris and organisms from mud. The mud is sucked into the buccal cavity and worked about between the pharyngeal pads. In some regions of the tropics the local fishermen locate shoals of grey mullet by the sucking noises they make as they take in the mud. The pharyngeal pads are covered with minute denticles and project downwards between the two sets of gill arches. The gill rakers are numerous and each is beset with a series of bristle-like processes which form an efficient sieving apparatus. When the floor of the mouth is raised to pump water over the gills any small particles present in the buccal cavity will be retained on the gill rakers. The pharyngeal pads also contribute to the filter, and the small denticles may help in removing the material trapped on the gill rakers.

Mullet also browse on algae, and Pillay (1953) records *Mugil tade* biting the bark of the mangrove *Avicennia officinalis*. When the bark was closely examined it was found to be covered with a dense growth of the green alga *Protococcus viridis*. Algal browsing by mullet is not completely indiscriminate. Wilson (1951) found that *Mugil chelo* in a tank at the Plymouth aquarium browsed down the algae most thoroughly, but when the red alga *Polysiphonia urceolata* became established it was left untouched and eventually covered all the rockwork in the tank.

Young grey mullet often feed on plankton, including both algae and animals. There are rapid changes in diet as *Mugil tade* grows. Specimens less than 20 mm in length feed mainly on algae. Between 20 and 40 mm in length the young mullet take about 22 per cent of animal food in their diet, including mysids and copepods. Once a length of 40 mm has been reached the proportion of organic debris in the diet increases greatly.

Most of the grey mullet found in estuaries are immature, and in temperate climates they are usually present only during the summer. The mature forms migrate out of the rivers to spawn, but the spawning places have rarely been located. Such evidence as is available indicates that spawning takes place at the surface over deep water. The spawning migration is generally against the prevailing coastal currents. This will result in the eggs and young being carried back towards the estuaries from which the spawning adults have migrated.

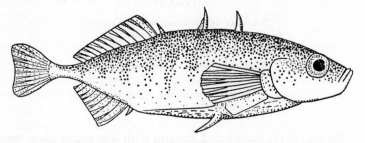

Figure 90. Gasterosteus aculeatus, the three spined stickleback, length 50 mm

The family Gasterosteidae includes a number of species that live in estuaries. The common brackish-water species of North America is *Apeltes quadracus*, the four-spined stickleback. In Britain the commonest species is *Gasterosteus aculeatus*, the three-spined stickleback. The behaviour of this species has been described in great detail (Pelkwijk and Tinbergen, 1937; Tinbergen, 1951; Sevenster-Bol, 1962), but here we can deal only with aspects of the biology of this fish as it relates to the estuarine environment.

Perhaps the most remarkable feature of *G. aculeatus* is that the number of bony plates on the sides of the body varies with both salinity and latitude. This phenomenon has been analysed in detail by Bertin (1925) and Heuts (1947). Populations from brackish water in Belgium have an average of between 20 and 30 plates on each side of the body, while fresh-water populations from the same latitude have an average of 4 or 5 plates. Intermediates between these two averages do occur, for instance the types found in a typical brackish-water population may include a few individuals with as few as 2 plates, but these are rare. As one passes southwards from Belgium there is a reduction in the percentage of *G. aculeatus* with large numbers of plates, and brackish-water populations in the south of France and Algeria have few or no plates. Conversely, on travelling north from Belgium there is an increase in the percentage of *G. aculeatus* with large plate numbers, so that in high latitudes only fish with high plate numbers are found.

TABLE 22

SURVIVAL OF 'GASTEROSTEUS ACULEATUS', TYPE A AT 10°C IN SEA WATER

(After Heuts, 1947)

Number of plates	Mean hours of survival
3 and 4	7·5
5 and 6	11·0
7 and more	13·75

Heuts (1947) designated the fish with low mean plate numbers and small body size as Type A, and those with high plate numbers as Type B. In Belgian populations the Type B individuals live in brackish water and migrate into the sea during the winter. Type A lives its whole life in fresh water in Belgium. These two types differ in their abilities to osmoregulate. If Type B is transferred to salinities below 10‰ it loses the ability to regulate and dies within a few days. Type A can regulate in very low salinities, but also has the ability to live in brackish water, although it cannot withstand sudden transfer to full-strength sea water. Even within Type A there is a relationship between plate number and survival in sea water. Table 22 shows that specimens with 7 or more plates survive for nearly twice as long as specimens with 3 or 4 plates when suddenly transferred to sea water. An effect of temperature is also evident from Table 23. These experimental results are in good accord with the geographical studies of Bertin (1925).

TABLE 23

SURVIVAL OF 'GASTEROSTEUS ACULEATUS',
TYPE A IN FRESH WATER AT 25°–28°C

Number of plates	Mean hours of survival
3 and 4	43
5	39
6	38·5
7	36·75

The hatchability of the eggs produced by the two types of *G. aculeatus* varies with temperature and salinity. At 23°C eggs produced by Type A hatch well at low salinities, but the percentage hatch decreases rapidly at salinities above 12‰. At the same temperature the eggs of Type B do not hatch so well at low salinities, but do hatch well at salinities between 5 and 30‰. At a temperature of 10°C there is a remarkable change: Type A eggs hatch best at a salinity of 16 while Type B eggs hatch best at about 3‰.

The reproductive cycle of *Gasterosteus aculeatus* has been studied by Craig-Bennet (1931), who found that the breeding season in England lasted from April to August, and that the temperature had to rise above 14°C for the young fish to become mature. The habit of the male in building a nest is well known, as is the assumption of a characteristic breeding dress. In the breeding season the male develops a brilliant red coloration of the throat and ventral surface of the body; the iris of the eye and the back of the body become blue in appearance. In this dress the male defends a territory around his nest and eventually guides one or more females to lay eggs in the nest. The musculature of the pectoral fins of the male is larger than that of the female: this feature is associated with the habit of the male of fanning water over the eggs as they develop in the nest. The male also differs from the female in that a large number of the kidney tubules are modified to produce a secretion which is used in sticking the nest together.

In temperate latitudes the time from hatching to maturity is about a year, and in natural populations most individuals live for less than 2 years, although they can live for 4 years in an aquarium. In high latitudes, as in Greenland, *G. aculeatus* does not become mature until it is 3 or 4 years old.

A study of the feeding habits of *Gasterosteus aculeatus* from the coast of Denmark revealed that specimens inhabiting *Zostera* beds fed mainly on Crustaceans; copepods formed a major fraction, but *Idotea* and mysids were also eaten (Blegvad, 1917). The detailed composition of the food varies from one locality to another. Hynes (1950) studied a population in a small stream called the Birkit in Cheshire, and found that Crustacea and insects were the main food, with *Asellus* forming up to 19 per cent of the total food. Both Hynes and Blegvad noted that plant material was taken when animal food was scarce. Hynes also studied the ten-spined stickleback (*Pygosteus pungitius*) and found its diet to be basically the same as that of *Gasterosteus aculeatus*. The two species often occur together, but must presumably have some differences in their ecology. Hynes suggests that they have different breeding habits, and noted that *Pygosteus pungitius* tended to nest in more densely weeded areas than *Gasterosteus aculeatus*.

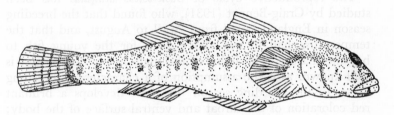

Figure 91. Gobius microps, length 49 mm

In British estuaries the most abundant and widespread fish is probably *Gobius microps*. This species can live in fresh water during the summer, and can also survive in salt marsh pools where the salinity rises above that of sea water. Breeding populations inhabit the lower reaches of estuaries, but the young fish migrate upstream to the fresher regions.

In habits the adult *Gobius microps* is a benthic fish and feeds mainly on benthic animals. The detailed composition of the diet varies from one part of an estuary to another. The feeding habits of *G. microps* in three different regions of the Gwendraeth Estuary are shown in Fig. 92. In fresh water there is competition with the minnow (*Phoxinus*) for chironomid larvae, but

A : GWENDRAETH FACH

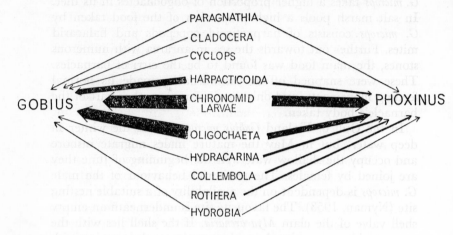

PARAGNATHIA
CLADOCERA
CYCLOPS
HARPACTICOIDA
CHIRONOMID LARVAE
OLIGOCHAETA
HYDRACARINA
COLLEMBOLA
ROTIFERA
HYDROBIA

GOBIUS

PHOXINUS

B : SALT MARSH POOLS

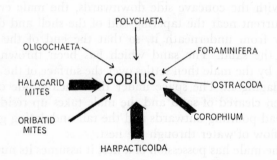

POLYCHAETA
OLIGOCHAETA
FORAMINIFERA
HALACARID MITES
GOBIUS
OSTRACODA
ORIBATID MITES
COROPHIUM
HARPACTICOIDA

C : PASTOUN SCAR

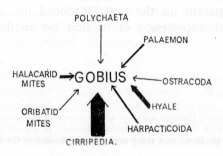

POLYCHAETA
PALAEMON
HALACARID MITES
GOBIUS
OSTRACODA
ORIBATID MITES
HYALE
HARPACTICOIDA
CIRRIPEDIA.

G. microps takes a higher proportion of oligochaetes in its diet. In salt marsh pools a high proportion of the food taken by *G. microps* consists of harpacticoid copepods and halacarid mites. Further out towards the sea in an area with numerous stones, the main food was found to be the cirri of barnacles. These were snapped off as the barnacles made its normal feeding movements, and only rarely was any other part of the barnacle's body taken.

In the Gulf of Finland *Gobius microps* spends the winter in deep water, but in May the mature males migrate inshore and occupy the shallow water. At the beginning of June they are joined by females. The territorial behaviour of the male *G. microps* is dependent on the availability of a suitable nesting site (Nyman, 1953). The favourite site is underneath an empty shell valve of the clam *Mya arenaria*. If the shell lies with the concave side uppermost the male turns it over by pushing with its head and making strong swimming movements. When the shell lies with the concave side downwards, the male creates a strong current near the tapering end of the shell and drives sand away from underneath it, so that the end of the shell sinks into the sand. The sand which has been thrown into suspension by the male then settles over the surface of the shell, so camouflaging it. The space under the concave side of the shell is then cleared of sand and the male takes up residence, with its head pointing outwards and the tail undulating gently to keep a flow of water through the nest.

When the male has possession of a nest it assumes its nuptual colours. The overall colour becomes much darker and the edges of the pectoral and ventral fins become blue. The iris of the eye also becomes blue and a conspicuous deep blue spot becomes apparent on the anterior dorsal fin. If a male is defeated and dispossessed of its nest by another male the

Figure 92 (opposite). Feeding habits of *Gobius microps* in the Gwendraeth Estuary. A, comparison with the minnow (*Phoxinus phoxinus*) in a fresh region. B, in salt marsh pools. C, in a stony area near the mouth of the estuary

nuptial colours are lost, and the normal dull coloration re-appears in about ten minutes.

When a male is in possession of a nest it will threaten any small fish that approaches by facing the intruder and then rotating rapidly about its longitudinal axis, alternately to the right and left. If the intruder retreats in front of this display it will be chased out of the territory, but if the intruder is a mature female of *G. microps*, a different sequence of events takes place. The female holds its tail in a curve upwards and then moves the pectoral fins so that the body is raised slowly from the substratum to display the swollen underside to the male. The male then turns and swims into the nest, followed in most instances by the female. If the female does not follow, the male turns and begins to revolve about his longitudinal axis again. The female then usually enters the nest with the male, who performs a series of trembling movements. The female turns her belly uppermost and presses the eggs as they emerge against the ceiling of the nest. After each deposition of a group of about 20 eggs the male also inverts and sheds sperm over the eggs. This process is repeated at intervals over the course of an hour until the ceiling of the nest is plastered with fertilised eggs. The female then leaves the nest and does not return.

The eggs are guarded by the male, who performs fanning movements with the tail and pectoral fins. If the male is removed soon after the eggs have been laid they do not develop, so that the current of aerated water which the male causes to pass over the eggs is essential to their development. Also, if the eggs are left unguarded, they are eaten by other males of *G. microps*, or by gammarid amphipods.

Development of the eggs takes about 10 days at 20°C, and when the young emerge they are about 2.5 mm in length. These small fish swim actively among vegetation in shallow water, but by the time they reach a length of 11 mm they become bottom dwellers like the adults.

ESTUARINE ELASMOBRANCHS

The concentration of salts in the blood of a marine elasmobranch is higher than that of a marine teleost and lower than sea water, but the blood also contains a considerable concentration of urea which brings the osmotic concentration up to or slightly above the concentration of sea water. In this way the elasmobranchs reduce the osmotic stress between internal and external environments and so do not face problems of the same magnitude as those faced by marine teleosts.

The rate of glomerular filtration by a marine elasmobranch may be as high as 80 ml/kg fish per day, but the volume of urine varies between 5 and 20 ml/kg fish per day (Smith 1936), so that over 75 per cent of the water passing through the glomeruli is resorbed. The kidney tubules are also efficient at resorbing monovalent ions and urea.

Sharks are usually considered as marine animals, but several species of the genus *Carcharhinus* enter estuaries and may be found many miles upstream. The Zambezi shark, *C. zambezensis*, has been captured 120 miles upstream from the mouth of the river, and another member of the genus, *C. gangeticus*, penetrates long distances into Asiatic rivers. The bull shark, *C. leucas*, occurs in estuaries in the southern states of the U.S.A. and in Lake Nicaragua, which is fresh. At one time the form found in the lake was regarded as a separate species (*C. nicaraguensis*), but it now seems that the shark can migrate from the sea to the lake *via* the Rio San Juan, and the lake form is not specifically distinct.

When the blood of *Carcharhinus leucas* from the sea is compared with the blood of specimens from Lake Nicaragua an interesting difference is found. The marine specimens have a blood urea content of 333 mM/1, but in the fresh-water specimens the figure falls to 132 mM/1. There is also a lowering of the sodium and chloride contents of the blood in fresh water, but this is by no means as marked as the drop in urea content. The sodium content of the blood falls from 223 mg/kg water when the shark is in the sea to 200 mg/kg water when in fresh water, while the corresponding figures for chloride are 236 and 181 mg/kg water (Urist, 1962). It seems that the major adjustment made by the shark on entering fresh water is to reduce the urea content of its blood so as to lower the osmotic difference between the internal and external environments.

Sawfish of the genus *Pristis* also enter estuaries. *Pristis pectinatus* is found in the Gulf of Mexico, and penetrates the lower Mississipi, while *P. perroteti* enters the Zambezi and other African rivers, and *P. cuspidatus* goes well beyond the tidal limits in asiatic rivers. Some of these species reach a length of 20 ft or more, with a saw up to 6 ft in length. This saw is said to be used for grubbing about in sand and mud to disturb prey. It may also be used to strike from side to side among a shoal of fish. The damaged fish can then be consumed in a leisurely manner. Mullet (*Mugil* spp.) are often among the fish that fall prey to *Pristis*.

The urea content of the blood of estuarine *Pristis* is lower than that of truly marine elasmobranchs (Smith, 1931, 1936). This is a point of similarity to the bull shark. Another feature which the estuarine sawfish shares with the estuarine *Carcharhinus* species is that both are viviparous. This is advantageous in that it relieves the embryos of any osmotic difficulties which might be encountered if eggs were laid.

CYCLOSTOMES

The cyclostomes fall naturally into two groups. The Myxini, or hagfishes, are stenohaline marine forms, but the Petromyzones, or lampreys, include several forms which migrate from fresh water to the sea and back again. In Britain there are three forms of lamprey: *Petromyzon marinus*, *Lampetra fluviatilis* and *L. planeri*. The last form spends the whole of its life in fresh water, but the other two migrate to the sea for a part of their life cycle. Adults of *P. marinus* enter estuaries and penetrate a short distance upstream to spawn; after spawning they die. The eggs give rise to eyeless larvae known as ammocoetes. These larvae bury themselves in mud, leaving only the head exposed, and feed by filtering fine particles from water taken into the pharynx. The particles are trapped in mucous secreted in the floor of the pharynx and passed upwards and backwards to the oesophagus. *Lampetra fluviatilis* spends about five years as an ammocoete before metamorphosing into an adult (Hardisty, 1961). The filter feeding mechanism is lost and the mouth becomes suctorial, with a series of teeth inside the sucker (Fig. 93). Eyes develop, the skin of the ventral surface

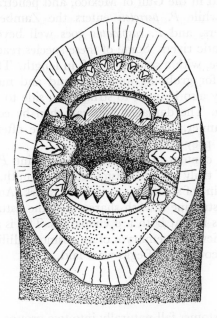

Figure 93. Mouth of *Lampetra fluviatilis*. The outer lips have been cut away to show the teeth and tongue.

becomes silver in colour (cf. the eel, p. 280) and the lampreys then migrate to sea where they attack fishes such as the shad (*Clupea alosa*) and salmon (*Salmo salar*). The sucking mouth is used to attach the lamprey to the side of the fish, and the teeth rasp a hole through the body wall. If the point of attachment to the prey is in a suitable position the lamprey may suck out some of the viscera. After about a year of feeding on fish the lampreys migrate back to fresh water to spawn and die.

When *Lampetra* migrates back into fresh water it loses the ability to osmoregulate in sea water. This loss of ability is gradual, because fresh-run lampreys are capable of withstanding higher salinities than mature individuals that have been in fresh water for some time (Morris, 1956). Some fresh-run individuals can osmoregulate in 50 per cent sea water (Δ —0.97°C) and can maintain the blood at Δ=-0.57°C. This regulation is achieved by a mechanism similar to that used by marine teleosts: the saline water is swallowed and chloride is secreted *via* an extra-renal route (Morris, 1958). The mature individuals can survive in 33 per cent sea water only by tolerating an increase in the osmotic pressure of the blood caused by a loss of water. If the water loss exceeds 14 to 19 per cent of the body weight the lamprey dies. Once the mature lamprey has lost the ability to osmoregulate in sea water it is not regained and it has not proved possible to acclimatise the mature forms to sea water (Hardisty, 1956).

The concentration of the blood serum of *Lampetra fluviatilis*, which is migratory, is somewhat higher than that of *L. planeri*, which spends the whole of its life in fresh water. Hardisty (1956) found that the serum of *L. planeri* was equivalent to 110 mM/1, while the corresponding figure for adult *L. fluviatilis* in fresh water was between 136 and 143 mM/1.

CHAPTER 11

ESTUARINE BIRDS

THE mass occurrence of invertebrates such as cockles, mussels, hydrobiids and nereids in estuaries provides a source of food for large numbers of birds. Open expanses of sand and mud, with an almost uniform fauna and little or no cover for the approach of predators, are ideal feeding grounds. In the winter large flocks of birds assemble in estuaries and feed on the dense populations of invertebrates.

Some of the birds which overwinter in estuaries change their habits completely in the summer. Many species of waders become inland birds, nesting in the tundra or on northern moorlands, and feeding on terrestrial foods such as insects and earthworms.

In addition to the winter assemblages there are also resident populations which feed on estuarine invertebrates during the summer, and there is usually a resident population of fish eaters such as cormorants, herons and kingfishers, although these may also be supplemented by migrants during the winter.

Estuarine birds can be considered under several groupings. First, the specialist fish eaters form an ecological group with examples from several different orders and families, while the next four groups are more homogeneous taxonomic units. The family Anatidae, including ducks, geese and swans, has a number of species which spend a considerable part of their time on estuaries. Flamingoes (family Phoenicopteridae) are highly specialised feeders, some species of which are associated with estuaries and coastal lagoons. The waders include a number of species that are notable for their mass occurrence in estuaries, and for their importance as predators of the sand and mud fauna. The gulls form a group of more generalised predators and scavengers which extend into estuaries from the sea, and are not so characteristically estuarine as the waders. A last group is formed by the terrestrial birds which are not normally associated with estuaries but which from time to time visit estuarine shores to feed.

Within the fish eaters there are several different techniques of hunting. The cormorants (family Phalacrocoracidae) and the mergansers (*Mergus* spp., family Anatidae) swim and dive from the surface. The herons and egrets (family Ardeidae) wade and use their long necks to reach their prey, while the kingfishers (family Alcedinidae) fly over the surface of the water and pick up fish which come near to the surface, often diving suddenly to do so. Pelicans show differing feeding habits according to the species.

PELICANS

There are six species of pelicans in the world. Five of these are basically dwellers on inland waters, which may winter on the coast or in estuaries, but the North American brown pelican (*Pelecanus occidentalis*) is a marine species inhabiting shallow coastal waters. This species catches fish by diving from the air. In Florida the main food is the menhaden (*Brevoortia* spp.) which is an important member of estuarine communities. Other estuarine fish are also taken, including mullet (*Mugil cephalus*), sheepshead (*Archosargus probatocephalus*) and the silversides of the genus *Menidia*.

The American white pelican (*Pelecanus erythrorhynchus*) feeds in estuaries during the winter. This species does not dive, but feeds from the surface, catching the fish by putting the head under water and scooping with the mouth open. In shallow water several birds may co-operate by forming a semicircle and then splashing and driving fish towards the shore, where they are more easily caught.

The brown pelican, in common with other marine birds, has the problem of maintaining its water balance when the only water available contains more salt than the kidney can cope with. The bird kidney in general is capable of producing urine with a salt concentration only about half that of sea water. If the bird drank sea water it would have to produce a volume of urine twice as great as the amount of sea water taken in. This impossible situation is relieved by the nasal glands, which lie in the anterior portions of the orbital cavities and open into the nasal cavities just behind the external nares. These glands can secrete a salt solution which is more concentrated than sea water, so that they provide the pelican with a mechanism whereby it can drink sea water, excrete the excess salts, and still retain some of the water (Schmidt-Nielsen and Fange, 1958). When feeding on fish the pelican probably has little need to drink sea water, but the nasal gland provides a safeguard to prevent excessive water loss when excreting salts.

The efficiency of the pelicans' salt gland is illustrated by the following experiment. A pelican was injected intravenously with 28 ml of 10 per cent sodium chloride solution. Within a minute the nasal gland started secreting, and continued to secrete for 2 hours. During this period 33·7 ml of secretion were produced. The concentration of salt in the secretion was equal to a 4 per cent sodium chloride solution, which is somewhat more concentrated than sea water (= 3·5 per cent sodium chloride). During the same period the kidneys produced 22 ml of urine, but the concentration was only equivalent to a 1·5 per cent solution of sodium chloride. Of the total amount of salt excreted during the 2 hours the nasal glands removed four-fifths and the kidneys one-fifth.

The size of the nasal glands varies in different species of birds. The largest glands are found in the most strictly marine species (Technau, 1936). The black-headed gull (*Larus ridibunda*), which often breeds near fresh water, has relatively smaller nasal glands than the common gull (*Larus canus*), which in turn has smaller nasal glands than the more marine Herring gull (*Larus argentatus*) and greater black-backed gull (*Larus marinus*). An experiment by Schildmacher (1932) demonstrated that ducks reared in salt water conditions developed larger nasal glands than individuals of the same species reared in fresh water.

CORMORANTS

Cormorants (family Phalacrocoracidae) are fairly closely related to pelicans, and share several features of their biology. The nasal glands are well developed, and all the species are fish eaters, although some of the smaller species such as the shag (*Phala-crocorax aristotelis*) include crustaceans in their diet. The cormorant, which has been studied in the greatest detail, is the common species in Europe (*Phalacrocorax carbo*) which is also widespread in other parts of the world, including Greenland, Iceland, Africa, Asia, Australia and New Zealand. This species occurs on the open coast, in estuaries and on inland waters.

The hunting cormorant swims on the surface of the water, and from time to time puts its head underwater to look for fish. At intervals the bird dives and remains underwater for about three-quarters of a minute. The depth of the dive does not usually exceed 3 metres, but up to 9·5 metres has been recorded. When swimming underwater the wings are held close to the body and the two webbed feet are used in unison. When swimming at the surface the two legs are used alternately.

Cormorants are remarkable among swimming and diving birds for the poor waterproofing of their feathers, which become wet. This contrasts with the ducks which have well-proofed feathers which repel water. The wetting of a cormorant's feathers may be an adaptation to rapid movement under water. If the feathers of such a large bird were water-repellent they would trap a substantial amount of air when the bird dived. The resulting buoyancy would impede rapid diving and cause the bird to exert more effort to remain submerged. When the cormorant has stopped fishing it leaves the water and usually stands with the wings extended as if hanging them out to dry. This characteristic posture may be held for an hour, with only slight changes in position.

There have been numerous studies of the food taken by *P. carbo* (Steven, 1933; Serventy, 1938; Falla and Stokall, 1945; Hartley, 1948; Madsen and Sparck, 1950, van Dobben, 1952; Mills, 1965—among others). The general impression gained from these studies is that the cormorant takes fish of a relatively large size, the detailed specific composition of the diet varying according to the habitat and geographical area. Fish with a length between 10 and 30 cm are most frequently taken, so that gobies and sand eels usually escape attention. Eels up to 60 cm long and cod 40 cm long have been recorded in the stomach of

P. carbo. When feeding in estuaries the diet includes a high proportion of flounders (*Platichthys flesus*).

The amount of food eaten by cormorants has been the subject of much speculation. Some early estimates were based upon the idea that the amount of food found in the stomach could be digested in four hours and that the cormorant ate three meals a day. This led to an estimate of 14 lb of fish per bird per day. Now the body weight of *P. carbo* is about 7 lb, so that these old estimates implied that each bird ate twice its own weight of fish per day! More realistic figures, based on captive cormorants which maintained or increased their body weight, indicate that the average daily requirement is a little over 1 lb of fish per day (Madsen and Sparck, 1950; du Plessis, 1957).

Nestling cormorants are fed with a fluid regurgitated from the parent stomach. The head of the young bird is placed in the lower bill of the parent; and fluid then pours down the nestling's throat. Adults may also bring water to the young. The mouth and gullet of the parent are filled with water which can then be squirted out as a fine jet.

HERONS, STORKS AND IBISES

Ardea cinerea, the common or grey heron of Europe, usually feeds around inland waters, but it also extends its activities into estuaries, and is often seen standing alone with its long legs partly immersed in water. The food taken by adult herons varies greatly with the locality, and usually reflects the local aquatic fauna, including insects, fish, amphibians, reptiles, young birds and mammals. The food brought to nestling *A. cinerea* in a heronry near the Thames Estuary has been studied by Owen (1960). Two species of stickleback, *Gasterosteus aculeatus* and *Pygosteus pungitius*, together with the brackish-water prawn *Palaemonetes varians* were the most numerous of the food items. The eel (*Anguilla anguilla*) also formed an important part of the diet, and small numbers of other estuarine fish, such as *Gobius* and the pipe fish *Syngnathus acus*, were also taken.

In North America there is a greater variety of herons than in Europe. The species vary in abundance, habitat preference and feeding habits. Most of the species are found in inland areas, but they also occur in coastal regions and feed in estuaries. Some species, such as the tricolored heron, *Hydranassa tricolor*, and the reddish egret, *Dichromanassa rufescens*, are predominantly coastal in habit. The latter species is often associated with the red mangroves (*Rhizophora*). The food of both these species includes small estuarine fish, but it may also include invertebrates such as crustaceans and grasshoppers. A tricolored heron killed in Florida had eaten about 200 grasshoppers.

The larger herons, such as the great blue heron, *Ardea herodias*, include the young of other birds in their diet. Palmer (1962) records the remains of young coots (*Fulica americana*), (avocets (*Recurvirostra americana*) and black-necked stilts (*Himantopus mexicanus*) in the pellets of *A. herodias* from the Bear River marshes in Utah. In addition a wide variety of fish, amphibians and mammals are eaten. When hunting on tidal flats and sand bars the great blue heron includes a high proportion of gar-fish (Belonidae) and mullet (Mugilidae) in its diet. The usual fishing technique of this species is to stand still until prey comes within striking distance.

The smaller herons and egrets often have a more active feeding technique, stepping briskly through the shallow water and striking rapidly as soon as they get near enough to their prey. A good example of one of the more active species is the snowy egret, *Egretta thula*, which may be seen 'running swiftly through the shallows, throwing up their wings' (Audubon, 1835). This species also uses its toes to disturb fish from among aquatic vegetation. In estuarine areas the diet of *E. thula* includes fiddler crabs of the genus *Uca*.

The storks (family Ciconiidae) are often more terrestrial in habit than the herons, but some species, such as *Mycteria americana*, the wood stork, are swamp dwellers that also feed in estuaries. The wood stork has been recorded as eating large numbers of small fish, such as *Mollienesia latipinna* and *Cyprinodon variegatus*, and includes crayfish and fiddler crabs in its diet.

Some of the ibises (family Threskiornithidae) may also be found in estuaries. The American white ibis, *Eudocimus albus*, is mainly coastal in habit and nests among mangroves. The long decurved bill is used to probe the mud between mangrove roots, and the diet includes fiddler crabs, fish and aquatic snakes.

Spoonbills belong to the same family as the ibises, but they have long, flattened bills which they sweep from side to side in muddy water and catch a variety of small animals. Feeding is often communal, with several birds working side by side. The diet of the roseate spoonbill (*Ajaia ajaja*) of Central and South America includes fish (*Cyprinodon, Fundulus, Menidia*), crabs (*Uca, Callinectes*), shrimps (*Palaemonetes, Penaeus, Hippolyte*) and molluscs (*Amnicola*). The European spoonbill *Platalea leucorodia* has been recorded feeding on the marine bivalve *Tellina*, but more usually it feeds in fresh water.

KINGFISHERS

The feeding techniques of kingfishers vary to some extent, but usually the food is taken by means of an aerial dive. This dive may originate from a perch, or the bird may hover over the water until it sights a fish and then drop suddenly. The common kingfisher of Europe, *Alcedo atthis*, uses both techniques, and has been seen diving from a high mud bank in the Gwendraeth Estuary. This species is widespread on inland waters, but also feeds in estuaries, particularly in autumn and winter. The belted kingfisher, *Megaceryle alcyon*, of N. America also has a similar tendency to be found more frequently in estuaries during the winter.

Europe and N. America each have a single species of kingfisher, but in the tropics several species may be found on a single estuary. Scott, Harrison and MacNae (1951) observed three species on the Klein River Estuary in South Africa. These species differed in size and in feeding techniques, so that they did not compete with one another. The largest of the three, *Megaceryle maxima*, or giant kingfisher, uses a perch from which it dives when a fish is sighted. The smallest, *Corythornis cristata*, the malachite kingfisher, is a low, fast flyer, inhabiting reedy areas. This species is small enough to perch on reeds and is capable of diving from this position to catch small fish; it also includes insects, such as dragonflies, in its diet. Hovering and diving are the main techniques employed by the third species, *Ceryle rudis*, the pied kingfisher, which is intermediate in size between the other two. This species often fishes in pairs or small groups, and takes a wide range of small fish, which rarely exceed 10 cm in length (Daget, 1954).

In the genus *Halcyon* there are about 38 species, most of which are found in Asia. Of these species several are found in mangrove swamps. On the eastern side of Africa *H. senegaloides* is known as the mangrove kingfisher, and feeds mainly on estuarine crabs. Perhaps the most widespread of these coastal kingfishers is the white-collared kingfisher, *H. chloris*, which inhabits mangroves from the Red Sea to Borneo. In New Zealand and Australia the sacred kingfisher, *H. sancta*, nests inland, but overwinters near the coast. Some individuals migrate northwards to the mangrove swamps around the coasts of S.E. Asia. This species captures its prey by diving down from a perch, then returning to the perch with the prey which it beats to death before swallowing (Falla, Sibson and Turbott, 1966).

Kingfishers are hole nesters; they dig tunnels into banks of rivers and creeks, using their bills and feet. Species living in estuarine areas must of necessity make their nests above the limits of the tides, so that although they may feed in the estuaries they usually nest further upstream.

FLAMINGOES

Flamingoes (family Phoenicopteridae) are found particularly in association with inland saline lakes, often on the borders of deserts. But some species, such as the greater flamingo *Phoenicopterus ruber*, are found around brackish lagoons at the mouths of rivers. In Europe *P. ruber* breeds at the mouth of the Rhône. Previously this species used to breed at the mouths of the Volga and Ural, but these colonies have been destroyed by man. The largest known colony nests in the Great Rann of Cutch in N.W. India where over 10,000 nests have been counted (Salim Ali, quoted in Voous, 1960).

The feeding mechanisms of flamingoes have been studied in detail by Jenkin (1957). The bill is modified to form a filter. Water is pumped in and out of the bill by action of the tongue, and particles are trapped on fringes of horny lamellae which line the bill. In *P. ruber* these lamellae are fairly widely spaced so that relatively large particles are retained. The smaller flamingoes, such as *Phoeniconaias minor*, have much finer filters which are capable of retaining diatoms and blue-green algae. There is also a difference in general feeding technique between the greater and lesser flamingoes: *P. minor* feeds by swinging the bill from side to side in the surface layers of water, but *Phoenicopterus ruber* feeds in bottom mud, dipping and raising the bill. Both species have been observed to stir up the bottom with their feet and then feed on the muddy suspension. The differences in feeding technique tend to reduce competition between the two species, and in some parts of Africa the two have been found feeding together.

The food taken by *Phoenicopterus ruber* includes a range of plant and animal materials. Small brackish-water snails are often the main item of diet. Species of *Paludestrina* are eaten in Europe, while *Cerithium* is the main food in America, and at Port Sudan *Tympanotomus fluviatile* was found in masses in the stomachs of *P. ruber*. These snails, which may reach a length of 12 mm, are ingested whole because the bill is too weak to break them. Chironomid larvae and Crustacea, such as *Sphaeroma* and *Artemia* are also eaten by *P. ruber*, as are plant seeds similar in size to these animals. When all else fails, mud is ingested. This mud will contain organic matter, bacteria, diatoms and a range of microscopic animals, and it is from these sources that the flamingo derives nutriment.

The nests of flamingoes are made of mud. If available a base of twigs or stones is laid in the ground, but this is covered with mud which dries to form a blunt cone with a shallow depression on top. Nesting colonies usually occur in remote areas near lakes or coastal lagoons. The nests are often very close to water, and heavy losses of eggs and chicks frequently occur when rainstorms cause the water level to rise suddenly.

GEESE AND DUCKS

The estuarine geese are mainly migratory vegetarians. In Britain the only wild resident populations occur in the north, but in winter the southern estuaries are invaded and used as feeding grounds. The greylag goose (*Anser anser*) breeds in Scotland, but winters as far south as the Mediterranean. The estuarine plants eaten by this species include *Scirpus* and *Zostera*. The latter plant also features in the diet of the brent goose (*Branta bernicla*), and when the unidentified epidemic of the 1930s destroyed much of the intertidal *Zostera* on the coasts of North America and Europe there was a marked diminution in the numbers of brent geese. Many of the geese that winter on estuaries migrate northwards to inland breeding grounds where they graze on terrestrial plants.

The smaller members of the Anatidae, the ducks, have a much more varied diet than the geese. It is not possible here to consider all the species that might be seen feeding on estuaries, but a few examples have been chosen to illustrate the range of their impact on the estuarine ecosystem.

The teal (*Anas crecca*) is a frequent visitor to salt marshes, as well as fresh-water localities. In Britain it is most numerous in winter, and when it feeds in salt marshes the main food is the seeds of *Salicornia* spp. A wide range of other salt marsh plant seeds are taken, and about a quarter or a third of the total volume of food is of animal origin. The most abundant animals in the teal's diet are hydrobiids, *Corophium volutator* and chironomid larvae (Olney, 1963).

The relative importance of each item in the diet appears to vary from one locality to another and from one season to another. Sometimes an individual bird is found which has been specialising on a particular item of diet. One teal, shot in October, in Kent, contained 1,250 specimens of *Hydrobia jenkinsi*; another shot in January contained over 800 seeds of *Atriplex patula*.

A curious feature of the diet of the teal was that the seeds of *Spartina townsendi* and *Halimione portulacoides* were rarely eaten even though abundant in the feeding area. This same feature was found in the diet of the mallard (*Anas platyrhynchos*) in the same area, but this duck seems to eat these two seeds a little more frequently than the teal. One exceptional mallard was found with 4,000 seeds of *H. portulacoides* in its stomach (Olney, 1964). The mallard is a seed eater, but takes a wide range of other foods as well. In some localities hydrobiids form an important part of the diet. In addition, *Palaemonetes varians*, *Crangon vulgaris*, *Corophium volutator* and *Carcinus maenas* are eaten when the mallard feeds in salt marshes and brackish-water ditches.

One of the most conspicuous estuarine ducks is the shelduck (*Tadorna tadorna*), which has a greenish-black head and neck contrasting with its general white plumage and broad chestnut band across the body. The primary feathers of the wings are black, so that a strong contrasting pattern is presented when the duck is in flight. The shelduck feeds on open mud flats and is visible from a considerable distance. The main food of this species on British estuaries is *Hydrobia ulvae*. Olney (1965) found that all the shelduck stomachs that he examined contained this snail, and he related the distribution of the bird and the time of feeding to the availability of *Hydrobia*.

The activities of *Hydrobia ulvae* have been described in Chapter 6. As the tide recedes most of the snails are found crawling on the mud surface, and the ebb tide is the time when the shelduck feeds most intensively. Other mud-dwelling animals are also included in the diet: *Corophium volutator* and *Nereis diversicolor* are the most frequent. Small amounts of the green alga *Enteromorpha* are also eaten.

The nesting site of the shelduck is normally away from the estuary, either in neighbouring sand dunes or some distance inland. After the breeding season the ducks migrate to moulting grounds, where huge assemblages moult and temporarily lose the ability to fly. The main moulting grounds in western Europe lie at the mouths of the Rhine and the Elbe. In Britain there are subsidiary moulting grounds at Bridgwater Bay.

WADERS

A notably variable feature among waders is the length of the beak (Table 24). These variations can be related to the predominant feeding habits of each species. The curlew and the godwits can use their long beaks to penetrate down to considerable depths in sand and mud to reach worms such as *Nereis* and *Arenicola*. But long beaks are not always related to the ability to probe sand. Avocets, with their upturned beaks, feed by wading in shallow water and swinging the head from side to side, or else they stir up the surface layers of mud and catch shrimps and worms as they are disturbed. The ratio of the length of the beak to the length of the leg is important in determining the potential feeding techniques of waders. The black-winged stilt has long legs, and normally feeds by wading and catching animals swimming near the water surface. On land the legs have to be bent to enable the tip of the beak to reach the ground.

TABLE 24

BILL LENGTH AND TARSUS LENGTH OF SOME EUROPEAN ESTUARINE WADERS*

Common name	Scientific name	Bill length (mm)	Tarsus length (mm)
Curlew	Numenius arquata	100–152	66–80
Whimbrel	Numenius phaeopus	76–99	50–61
Slender-billed Curlew	Numenius tenuirostris	67–94	57–65
Black-tailed Godwit	Limosa limosa	82–126	75–82
Bar-tailed Godwit	Limosa lapponica	72–106	46–51
Avocet	Recurvirostra avosetta	75–92	88–91
Black-winged Stilt	Himantopus himantopus	57–68	119–137
Spotted Redshank	Tringa erythropus	53–64	52–59
Greenshank	Tringa nebularia	50–59	55–61
Redshank	Tringa totanus	38–44	44–50
Common Sandpiper	Tringa hypoleucos	23–26	22–24
Curlew Sandpiper	Calidris ferruginea	33–42	27–32
Knot	Calidris canutus	30–38	27–31
Dunlin	Calidris alpina	25–34	21–25
Sanderling	Calidris alba	23–28	22–26
Purple Sandpiper	Calidris maritima	21–26	20–23
Little Stint	Calidris minuta	17–20	19–22
Grey Plover	Pluvialis squatarola	27–32	43–50
Golden Plover	Pluvialis apricaria	21–26	38–41
Turnstone	Arenaria interpres	20–24	23–26
Ringed Plover	Charadrius hiaticula	14–16	25–28
Kentish Plover	Charadrius alexandrinus	14–16	25–29
Little-ringed Plover	Charadrius dubius	12–14	22–25

*Data from Witherby et al. (1940)

The plovers, with their short bills, often pick up surface-dwelling animals, such as *Hydrobia*, or the small predatory beetle *Bembidion laterale* (see p. 267). Dunlin and knot often feed on the invertebrates which make comparatively shallow burrows, such as *Corophium* and some of the dipterous larvae.

Fig. 94 shows the relative length of the bills of some waders in relation to the burrows of some of the common mud-dwelling invertebrates.

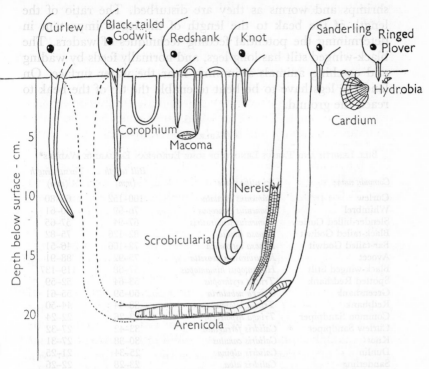

Figure 94. Diagram showing the length of bill in some common waders in relation to the burrows of some common mud dwelling invertebrates. Most waders can only catch *Arenicola* when it ascends the tails shaft in order to defaecate. Large specimens of *Scrobicularia* are beyond the reach of nearly all waders

Birds which feed in flocks tend to specialise in particular types of diet. The winter food of the oyster catcher (*Haematopus ostralegus*) sometimes consists entirely of *Cardium*, and sometimes exclusively of *Mytilus*. In contrast the birds which are more solitary, such as the redshank (*Tringa totanus*), have a more varied diet and use a wide range of techniques in feeding.

The members of the genus *Haematopus* are all fairly large waders with contrasting black and white patterns and straight powerful bills which can be used to open the shells of bivalved molluscs. In Britain *H. ostralegus*, the oyster catcher, forms large flocks in winter, particularly in areas where *Mytilus* or *Cardium* occur in abundance. The technique used to open these shells shows some variation in different localities. Norton-Griffiths (1966) compared two populations which were feeding mainly on *Mytilus*. One of the populations was feeding on *Mytilus*, which were very firmly attached and had thick shells. In this situation the oyster catchers stabbed the molluscs by jabbing the bill between the shell valves when they were slightly open in shallow water. The second population was feeding on *Mytilus* which were not so firmly attached, and could be torn away from their anchorage and then hammered by rapid blows of the beak. Drinnan (1958) has described this hammering in detail. The blows are delivered with the bill closed, and six or eight are usually enough to break into the shell. Once inside the shell the bill is opened and the mussel flesh seized, then separated from the shell by shaking. The whole process usually takes less than a minute.

When *H. ostralegus* feeds in an area where the mussels are small the number eaten is larger than in regions with large *Mytilus*. Drinnan (1958) found that small mussels were consumed at a rate of 82 per hour, or a total of 574 for a tidal period. Larger mussels were taken at a rate of 21 per hour. The wet weight of mussel flesh consumed daily was found to be approximately equal to the body weight of *H. ostralegus*. Similar quantities of *Cardium* are taken when this forms the main food. The rates of feeding observed by Drinnan (1957) in Morecambe Bay varied between 14 and 51 cockles per hour, with a daily consumption between 214 and 315 cockles per bird, varying with the size of the cockles.

In the breeding season many oyster catchers leave the shore and nest inland. The young are fed on earthworms and leather jackets (larvae of tipulid flies). On Skokholm Island, off the Welsh Coast, Safriel (1966) found that the nesting population included some pairs which fed their young mainly on the limpet *Patella*. These pairs were not very successful in rearing their young. On an average each pair of the limpet feeders reared 0·16 young, but the other birds which fed their young on earthworms and leatherjackets reared an average of 0·95 young per pair. This difference was not necessary related to any great lack of nutritive value of limpets, but appeared to be related to the fact that the limpet feeders left their young unattended more frequently than the others, and the young were killed by gulls. About 80 per cent of all the young which failed to fledge were killed by gulls.

The winter population of *Haematopus ostralegus* in Britain is augmented by populations which migrate southwards from breeding grounds in the Faroes and Iceland. These individuals overwinter mainly around the Irish Sea. There is also a general winter movement of British breeders southwards. Some of the first year birds migrate to the coasts of France and Spain, but most of the adults remain in Britain. Many oyster catchers return annually to winter on particular feeding grounds. The size of these flocks assembling in the winter feeding areas can be impressive. Drinnan (1957), using a photographic method, estimated that the flock of *H. ostralegus* on the Cartmel Wharf (Morecambe Bay) cockle beds in the winter of 1954–55 was about 30,000 strong. These birds were calculated to have eaten about 22 per cent of the total cockle population of the Cartmel Wharf area during the winter.

GULLS

Gulls are usually regarded as omnivores and scavengers. In general terms this is true, but there are differences in the ecology and diet of the different species, even those inhabiting a comparatively restricted geographical area. The black-headed gull (*Larus ridibunda*) spends a considerable part of its time away from the sea, often on inland waters or in marshy areas. A considerable population overwinters in the London parks. A large proportion of its diet consists of insects. Spärck (1950), for instance, goes so far as to say that both the common gull (*Larus canus*) and the black-headed gull are not marine or shore birds in their feeding habits, but land birds relying on a diet of insects, earthworms and plants. Nevertheless the black-headed gull does feed in estuaries and often adopts the technique of trampling in shallow pools, moving backwards and picking up the animals disturbed by its foot movements. This appears to be a particularly good technique for collecting *Corophium*. A similar technique is used by other gulls.

The larger gulls, such as the herring gull (*Larus argentatus*),

L

lesser black-back (*L. fuscus*) and greater black-back (*L. marinus*) eat almost anything that they can catch and kill. The last-named species is the largest of the three, and can kill the adults of some of the smaller sea birds. All three species eat eggs and young of other birds, and will even eat the young of their own species, including their own offspring, if they are encountered away from the nest. Crabs and mussels are eaten by these gulls on estuaries. Sometimes the prey is carried into the air and dropped in an endeavour to break it open, but this is often ineffective when the substratum is mud. I have seen a herring gull carry the same crab into the air six times and drop it ineffectively on to soft mud. The habit probably persists because it is effective when the substratum happens to be rock.

Most gulls are colonial nesters. This habit gives a certain amount of protection from predation. Black-headed gulls will react aggressively to predators entering their colony. The nests are usually spaced about a metre apart, so that any strange animal entering the colony will be noticed by neighbouring birds and a clamour of alarm can be raised.

In many areas the gull colonies are not static, but fluctuate in numbers, and one species may replace another. Gross (1950) describes a situation on Muskeget Island off the coast of Massachusetts where terns (*Sterna* spp.) were ousted by laughing gulls (*Larus atricilla*) and these in turn were replaced by herring gulls, which were later joined by great black-backed gulls. This last species tends to become most abundant near the nesting colonies of other species, where it can prey upon eggs and young or even adults.

TERRESTRIAL BIRDS

Many terrestrial birds from widely differing families visit estuaries to feed from time to time. Some of these visitors are more or less accidental, but some of the common terrestrial species are regular estuarine visitors. Carrion crows (*Corvus corone*), ravens (*Corvus corax*) and magpies (*Pica pica*) are often among the species scavenging on estuarine shores in Britain. Smaller birds, such as the pied wagtail (*Motacilla alba*) and wheatear (*Oenanthe oenanthe*) may also be seen. Apart from such visitors, which only spend a short time on the estuary, there are some basically terrestrial birds which have become restricted to the estuarine habitat. A particularly good example is the long-billed marsh wren (*Telmatodytes palustris griseus*) which inhabits the salt marshes of Georgia and has been the subject of a detailed study by Kale (1965).

In the summer the marsh wren spends all its time in the zone of *Spartina alterniflora*, bordering salt marsh creeks. The males occupy territories of about 60 or 100 sq metres, and the nests are built in the upper parts of the *Spartina* plants, which may grow as high as 2 metres above the surface of the marsh. The nests are beyond the reach of normal spring tides, but sometimes, if the wind reinforces a high spring tide, there may be as much as 21 per cent mortality by drowning among the nestlings. High tides also have an unexpected effect in that they tend to increase predation on nestlings by the rice rat (*Oryzomys palustris*). Normally the rice rat spends a lot of its time scavenging on the surface of the marsh and feeding on crabs and other marsh invertebrates. When the spring tides inundate the marsh, *Oryzomys* climbs up the stems of *Spartina*, and so is more likely to encounter the nests of the marsh wren. As well as eating the young wrens *Oryzomys* will also take over the nest and rear its own young. Another animal which will invade the nests if they are wetted by the tide is the marsh crab, *Sesarma cinereum*.

The long-billed marsh wren is an insect eater, and during the summer it feeds almost exclusively on the insects associated with *Spartina alterniflora*. Most of the insects that it takes are associated with the stems and leaves of the grass and are basically terrestrial forms. This makes an interesting comparison with the seaside sparrow (*Ammospiza maritima*) which occupies the same regions but feeds on the ground and so avoids competition with the marsh wren. As Kale (1965) puts it: 'I have never collected a marsh wren with muddy feet or a seaside sparrow with clean feet.'

A final group of basically terrestrial birds found around estuaries is constituted by the birds of prey. Some of these, such as the peregrine (*Falco peregrinus*), are widespread species which visit estuaries to prey on the birds that may be found there. In some parts of S.E. Asia the oriental hawk owl (*Ninox scutulata*), which normally inhabits forest clearings, has developed a taste for crabs, and may be seen feeding on open mud flats. Other birds of prey, such as the eagles of the genus *Haliaetus*, including the bald eagle of North America and the fish eagle of Africa, are more closely associated with water, and may become resident on estuaries. These eagles include a high proportion of fish in their diet, but will also take aquatic birds. Even closer associations are found in certain birds of prey, for instance the rufous crab hawk (*Buteogallus aequinoctialis*) is confined to the mangroves of tropical South America where it feeds on crabs.

Birds Associated with Mangroves

In the description of estuarine vegetation (Chapter 3) it was noted that in many tropical areas the salt marsh is replaced by mangrove swamps. Now some of the trees, particularly the *Rhizophora* species, grow to a considerable height and produce dense foliage suitable for roosting and nesting in. The upper stories of a mangrove constitute a terrestrial habitat over the top of an aquatic one, and a wide variety of birds have been recorded in this habitat.

In the Gambia, Cawkell (1964) found 10 species nesting in the mangroves. A further 9 were observed feeding in the mangroves, and another 26 species were seen in and around the swamps. An even richer bird fauna was found in the mangroves of Surinam by Haverschmidt (1965), who observed 43 nesting species, and another 39 species using the mangroves in other ways. Haverschmidt's list is remarkably similar to the list of birds in the mangroves of Trinidad, where Ffrench (1966) recorded 43 nesting species and another 51 species seen but not found nesting.

These lists from Africa and South America included many species associated with water, such as herons, ibises, cormorants, ducks, waders and rails, but there were other more terrestrial birds such as flycatchers, hawks, and even swallows.

CHAPTER 12

ESTUARINE PARASITES AND EPIBIONTS

THE problems encountered by animals living in estuaries are encountered to varying degrees by their parasites and epibionts. The latter term is used to denote organisms that attach themselves to the surfaces of animals. Such organisms will naturally be exposed to the same fluctuations in salinity as their hosts. This problem is not encountered to the same degree by internal parasites living in the body fluids of their host, which may be regulated with some degree or constancy. External parasites will encounter the same problems as epibionts. It is not always easy to distinguish between ectoparasites and epibionts. The basic difference lies in the method of obtaining food. An ectoparasite feeds on the tissues of the host, while an epibiont derives its food from other sources. A good example of an epibiont is provided by the colonial ciliate *Epistylis arenicolae*, which attaches to the surface of the gills of *Arenicola marina* and obtains its food by a ciliary mechanism which removes microorganisms from the water surrounding the worm in its burrow.

Epistylis belongs to the Peritrichida, a group of ciliates containing many species capable of attaching themselves to solid substrata and sometimes developing long stalks, which are contractile in some genera (Fig. 95B). Many peritrichs are epibionts.

A pioneer study of the epibionts in brackish water was made by Precht (1935), who studied the forms found in the Kiel Canal. Over 70 species of peritrichs were found, and a considerable number of these were host specific. For instance, the genus *Myoschiston* was found to contain the following species:

M. carcini on *Carcinus maenas*
M. duplicatum on *Carcinus* and *Idotea*
M. sphaeromae on *Sphaeroma rugicauda*

315

M. neomysidis on *Neomysis integer*
M. cypridicola on *Cyprideis litoralis*
M. balanorum on *Balanus improvisus*
M. centropagidarum on *Centropages* and *Eurytemora*

Thus 5 of the 7 species were found on a single host species. Similar examples could be given for other genera.

The ciliates associated with the Scandinavian species of *Gammarus* have been studied in detail by Fenchel (1965b).

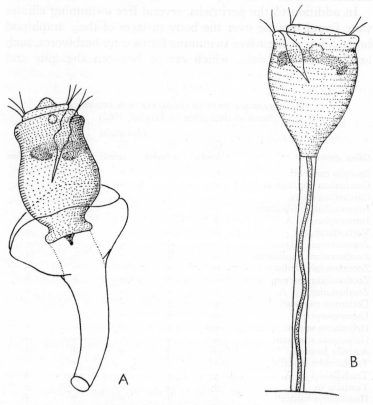

Figure 95. Epibiotic peritrichous ciliates

A, *Myoschiston sphaeromae* from *Sphaeroma rugicauda* in the Gwendraeth Estuary. This is a solitary individual of a species which forms small colonies. B, *Vorticella octava*, a common fresh-water species which occurs as an epibiont on *Daphnia magna* in rock pools on the shores of the Gulf of Finland, in salinities up to 8‰

Table 25 shows that *Gammarus duebeni* shares a number of species in common with the more marine members of its genus. If these species are compared with the ciliates recorded from two fresh-water species, *Gammarus pulex* and *Carinogammarus roeseli*, it is found that the former has 11 species and the latter has 14 species of peritrichs associated with it (Nenninger, 1948; Sommer, 1950). None of these species has as yet been recorded from *G. duebeni* in fresh water. It is clear from this that the ciliates associated with *G. duebeni* are derived from the marine habitat.

In addition to the peritrichs, several free swimming ciliates were found moving over the body surfaces of their amphipod hosts. Some of these free swimming forms were herbivores, such as *Trochiloides trivialis*, which creeps between the gills and

TABLE 25

CILIATES ASSOCIATED WITH GAMMARUS IN SCANDINAVIA
(Based on data given by Fenchel, 1965)

Ciliate species	Host species				
	G. locusta	G. oceanicus	G. salinus	G. zaddachi	G. duebeni
Epistylis gammari	–	+	+	+	–
Carchesium duplicatum	–	+	–	–	–
Carchesium sp. A.	–	–	–	–	+
Intranstylum duplicatum	–	–	–	–	+
Intranstylum sp. A.	+	–	–	–	–
Vorticella sp. A.	–	–	–	–	+
Zoothamnium hiketes	+	+	+	+	–
Zoothamnium duplicatum	–	+	–	–	–
Zoothamnium rigidum	–	+	–	–	–
Zoothamnium nanum	–	+	+	+	+
Zoothamnium sp. A.	+	–	–	–	–
Cothurnia gammari	+	+	–	–	+
Lagenophrys sp.	+	+	–	–	–
Heliochona sessilis	+	+	+	–	–
Heliochona scheuteni	+	+	+	+	+
Askoella janssoni	+	+	+	+	+
Trochilioides trivialis	+	+	+	+	–
Trochilioides sp. A.	–	–	–	–	+
Trochilia sp.	+	+	–	–	+
Hemiophrys baltica	–	+	+	+	+
Acineta foetida	+	+	+	+	+
Hypocoma parasitica	+	+	–	–	–
Gymnodinioides inkystans	+	+	+	+	–
Conidophrys pilisuctor	+	+	–	–	–
Silenella ovoidea	–	–	–	–	+

pleopods of *Gammarus* and feeds on blue-green algae which grow there. Others among the free swimmers were carnivores. *Hypocoma parasitica* was found attacking colonial peritrichs; the wall of the peritrich was penetrated and the cell contents sucked out.

Curious intermediates between parasites and epibionts are found in the ciliate genus *Gymnodinioides*. This genus contains a number of species that live on the surface of crustaceans, for instance *G. corophii* is found on species of *Corophium*, and *G. zonatum* is found on *Gammarus*. The life histories of these species are probably similar to that described for *G. inkystans* by Chatton and Lwoff (1935). They distinguished three main stages: phoront, trophont and tomont. The phoront is an encysted form found attached to the gills of the hermit crab *Eupagurus prideauxi*. When the host moults a trophont emerges from the phoront and feeds on the shed cuticle of the host. This feeding stage lasts for 6 or 8 hours, during which time the volume of the trophont may increase 60 times. The trophont then encysts on a suitable substratum and transforms into a

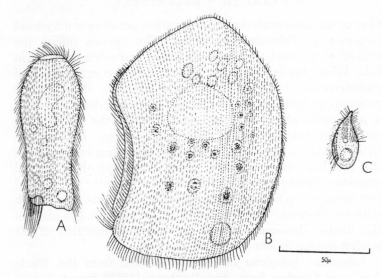

Figure 96. Ciliates associated with *Mytilus edulis*
A, *Ancistrum mytili;* B, *Peniculistoma mytili;* C, *Raabella helensis.* All are drawn to the same scale (after Fenchel, 1965a)

tomont. Within the cyst the tomont divides asexually to produce up to 64 tomites which eventually break out of the cyst and swim in search of a new host.

The mantle cavities of lamellibranch molluscs provide shelter for a variety of small organisms, and among these there are a large number of ciliates. From *Mytilus edulis* alone at least seven species have been recorded. By dwelling in the mantle cavity these ciliates will be subjected to variations in the salinity of the water taken in by the mollusc, but such variations may be considerably reduced compared with the external environment because the bivalve can close its shell and cease taking in water at low salinities.

Closely allied species of ciliates may inhabit bivalves with different salinity tolerances. Fenchel (1965a) found that *Ancistrum caudatum* from the mantle cavity of the marine horse mussel, *Modiolus modiolus*, was less resistant to dilution than *Ancistrum mytili* from the mantle cavity of *Mytilus edulis*.

PARASITIC COELENTERATA

One of the most remarkable of estuarine parasites is a hydroid coelenterate, *Polypodium hydriforme*. This species infests the ovaries of the sterlet (*Acipenser ruthenus*) and other acipenserid fish. When inside the ovary of the fish the ectoderm and endoderm of the hydroid are reversed from their normal positions so that the endoderm covers the ectoderm and is in a position to obtain nutritive material from the surrounding ovarian tissues. When the sterlet lays its eggs in May the hydroid stolons pass out with the eggs, and then evert, so that the ectoderm and endoderm come to occupy their normal positions. The hydroid then spends several months as a normal free-living polyp feeding on any small animals that it can catch with its tentacles. This free-living stage produces gonads, but the further development and the mode of infection of the sterlets ovaries do not seem to be known.

Polypodium hydriforme has been recorded from the Black, Aral and Caspian Seas, and Raykova (1958) found that in some parts of the Volga nearly 90 per cent of the females of *Acipenser ruthenus* were infected.

PLATYHELMINTHES

The free-living platyhelminths have been discussed in Chapters 6 and 7. All the non-parasitic form belong to the group Turbellaria, while the major groups of parasitic forms are the Trematoda and Cestoda. There are nevertheless a number of turbellarians which have adopted a parasitic mode of life. As examples one can take the two species of the rhabdocoele genus *Paravortex*, which are found in British estuaries. *Paravortex cardii* lives in the stomach of *Cardium edule*, and *P. scrobiculariae* lives in the intestine of *Scrobicularia plana* (Atkins, 1934; Freeman, 1957). Both species are viviparous. Egg capsules are produced by the ovaries and eventually lie free in the mesenchyme of the adult worm. The embryos develop inside the adult and when fully formed they escape through the body wall; this causes the death and disintegration of the parent. A single adult of *P. scrobiculariae* has been observed with 54 well-developed embryos lying free in the mesenchyme. After escaping from the maternal tissues the young rhabdocoeles pass out of the host's intestine and may spend a short period as free-living forms before they are taken in by the feeding current of a new host and so gain access to its alimentary tract.

Scrobicularia plana also serves as host to an adult digenetic trematode (Freeman and Llewellyn, 1958). This is remarkable because adult digenetic trematodes are typically parasites of vertebrates. The trematode *Proctoeces subtenuis* lives in the kidney of *Scrobicularia*. In this situation it is exposed to the variations in osmotic pressure that are undergone by the fluid produced by the kidney. Both the blood and the fluid produced by the kidney of *Scrobicularia* show variations in osmotic pressure which parallel variations in the external medium (Freeman and Rigler, 1957), so that the adult trematode does not gain much osmotic protection by living in this host, although the host can close its shell and reduce exchange with the external medium if the salinity falls to a low level.

In Britain the only localities where *Proctoeces subtenuis* has been found living in *Scrobicularia* are near the mouth of the Thames Estuary, in a region where there are no fish likely to prove suitable as final hosts (Freeman, 1962). Elsewhere in the world *P. subtenuis* has been recorded as a parasite of fish, notably labrids (wrasses) and sparids (sea bream). In North America various larval stages and adults have been found in the mussel, *Mytilus edulis* (Stunkard and Uzmann, 1959). It thus seems that *Proctoeces subtenuis* is capable of completing its life history in two bivalved molluscs, *Scrobicularia plana* and *Mytilus edulis*, without any necessity of infecting a fish, but it is also capable of infecting fish if the opportunity arises.

A more normal life history for a digenetic trematode is found in *Podocotyle atomon*. This species is parasitic in a wide range of marine fish, and in estuarine fish such as the flounder (*Platichthys flesus*), three-spined stickleback (*Gasterosteus aculeatus*), four-spined stickleback (*Apeltes quadracus*) and the common American eel (*Anguilla rostrata*). The early larval stages of this trematode are found in the winkle (*Littorina saxatilis*) and the cercariae which leave this host penetrate and encyst in the tissues of amphipods such as *Gammarus* and *Amphithoe*. The fish become infected when they eat the amphipods (Hunninen and Cable, 1943).

It is not possible to discuss here the life cycles of all the trematodes known to infect estuarine vertebrates. The number and variety of the species involved may be judged from the fact that Hutton (1952) was able to list nine distinct species of larval trematodes from the cockle (*Cardium edule*), and Rothschild (1938) recorded a similar number of distinct forms from *Hydrobia ulvae*. One of the forms with a larval stage in *Hydrobia* is *Cryptocotyle jejuna*. The egg of this trematode hatches when ingested by *Hydrobia*. The miracidium which emerges from the egg burrows into the tissues of the snail and then passes through a series of larval stages with a multiplication of numbers at each stage, culminating in the liberation of large numbers of cercaria larvae derived from a single egg. These larvae escape from the snail host and swim actively in water where they may encounter gobies (*Gobius* spp.). When a cercaria comes into contact with a goby it bores into the skin of the fish and encysts. When the fish is eaten by a bird such as a redshank (*Tringa totanus*) the cyst is digested and the larval trematode is liberated in the intestine of the bird.

Another species, *Cryptocotyle lingua*, uses *Littorina littorea* as the first host and a variety of inshore fish as the secondary host. The final host is usually a gull or other fish-eating bird. These two species of *Cryptocotyle* provide another example of ecological separation of allied species, with the added complication of the ecological preferences of the hosts. In using *Hydrobia ulvae* as its first host, *Crytocotyle jejunum* will enact most of its life history in a brackish environment, and will be most likely to infect a wading bird that has a varied diet including fish. In contrast *C. lingua* will have a wider range because *Littorina littorea* lives both in salt marshes and on open rocky shores. A wider range of final host will thus become available.

Most of the tapeworms (Cestoda) that parasitise estuarine animals also have complicated life cycles involving one or more intermediate hosts. *Bothriocephalus scorpii* is a good example. The early larval stages of this species are found in the planktonic copepod *Eurytemora hirundo*, and the secondary host is either *Gobius minutus* or *Gasterosteus aculeatus*. The final host is most frequently a turbot (*Rhombus maximus*) or a bullhead (*Myxocephalus scorpius*). The flounder (*Platichthys flesus*) also becomes infected with *Bothriocephalus scorpii*, but much less frequently than the turbot. This difference between the two flatfish is related to their feeding habits. The turbot includes a high proportion of *Gobius minutus* in its diet, but the flounder rarely eats fish (cf. p. 283).

One remarkable cestode, *Archigetes sieboldi*, seems to have eliminated its vertebrate host from the life cycle and continues to reinfect the oligochaete *Tubifex*, which is its only known host. The parasite is found in the coelom of the oligochaete, and while still basically a larval form the genital organs are developed. The eggs of Archigetes hatch only after they have been swallowed by an oligochaete. Conditions for the reinfection of *Tubifex* are often ideal in the upper reaches of estuaries where immense populations of these oligochaetes may be found in intertidal muds.

The life cycle of two of the tapeworms of the genus *Diphyllobothrium* can be completed in both fresh and brackish water. *Diphyllobothrium dendriticum* uses gulls (*Larus* spp.) as its final hosts, and has larval stages in copepods (*Cyclops* and *Diaptomus*) and sticklebacks (*Gasterosteus aculeatus*). The tapeworm eggs which pass out with the faeces of the gulls give rise to ciliated larvae known as coracidia. These larvae can swim actively, but survive for only twelve hours unless they are ingested by a copepod. Once in the gut of a copepod the coracidium sheds its outer ciliated coat and bores through the gut wall into the haemocoele of its host. The larva in the body cavity of the copepod becomes a procercoid, which is an elongated form bearing small hooklets at the posterior end which is separated from the rest of the body by a constriction. When an infected copepod is eaten by a stickleback the procercoid burrows through the gut wall of the fish and develops into a plerocercoid. This stage lacks the hooklets at the posterior end, and develops a scolex resembling the attachment apparatus of the adult tapeworm. The final host is reached when a gull eats an infected stickleback. Sometimes an infected stickleback is eaten by a larger fish such as a trout. In such a situation the larval tapeworm again burrows through the gut wall of the predator and becomes encysted among the viscera or in the muscles. If the large fish is too big to be eaten by a gull the larval tapeworms cannot develop further. If the stickleback is eaten by a piscivorous bird other than a gull the larval tapeworm fails to establish an infection.

The life history of the broad tapeworm, *Diphyllobothrium latum*, has many features in common with that of *D. dendriticum*, but the final host is man. The primary host is a copepod (most frequently *Diaptomus*), and the secondary host is often the stickleback (*Gasterosteus aculeatus*).

The stickleback forms an important item in the diet of several larger fish, and these accumulate the plerocercoids from their prey. A large pike (*Esox lucius*) may carry up to 1,000 plerocercoids embedded in its muscles and viscera. If one of these plerocercoids is eaten in uncooked fish by man it can develop into an adult tapeworm of impressive dimensions. A fully developed *D. latum* may produce 4,000 proglottids and reach a length of 40 ft.

In the Gulf of Finland *Diphyllobothrium latum* is very common: the eggs are found in the plankton of Helsinki Harbour (Valikangas, 1926), and near Petrodvorets about 90 per cent of the pike (*Esox lucius*) and burbot (*Lota lota*), as well as smaller percentages of perch (*Perca fluviatilis*) and eels (*Anguilla anguilla*) are infected (Petrushevski 1931).

The life cycle of *Diphyllobothrium latum* is basically of fresh-water origin, but it has been carried into brackish water by the adaptation of its intermediate hosts to the stable low salinities of the Baltic Sea. The life cycle of *D. dendriticum* is also capable of being completed in fresh water, but the maritime habits of the final hosts (*Larus* spp.) tend to bring the cycle into brackish water. The two fish which may be involved in the life cycle of *D. dendriticum* are both euryhaline forms capable of living over a considerable range of salinities.

PARASITIC NEMATODES

Among the nematodes it is a frequent feature to find members of a single genus differing in the class to which their final host belongs. For instance, among the forms likely to be found in estuaries, *Contracoecum bidentatum* uses the sterlet (*Acipenser ruthenus*) as a final host, while *C. spiculigerum* becomes adult in piscivorous birds. Several nematodes use estuarine inverte-brates as intermediate hosts. *Contracoecum aduncum* parasitises about 40 species of fish in the sea and in brackish water. The intermediate hosts of this nematode are the planktonic copepods *Acartia bifilosa* and *Eurytemora affinis*, both of which are wide-spread near the mouths of estuaries. *Capillaria tuberculata* is another parasite of the sterlet, and uses *Gammarus* species as intermediate hosts.

PARASITIC CRUSTACEA

The three major groups of parasitic Crustacea are the Copepoda, Cirripedia and Isopoda. Each of these groups has numerous parasitic members in the sea, but they also have representatives in brackish and fresh water. The Copepoda are the most numerous in terms of species, while the Cirripedia show the greatest structural modifications.

Copepods may be associated with other animals in a variety of ways. There are, for instance, a number of species known to inhabit the mantle cavities of bivalved molluscs; *Paranthesius rostratus* is commonly found in the mantle cavity of *Cardium edule*. The precise feeding habits of these copepods are not known. It is generally assumed that they are scavengers. Structurally they are not much different from free-living cyclopoid copepods.

Profound modifications of structure may be found in the copepods that have become internal parasites. To take but a single example, *Mytilicola intestinalis* (Fig. 97) is found in the

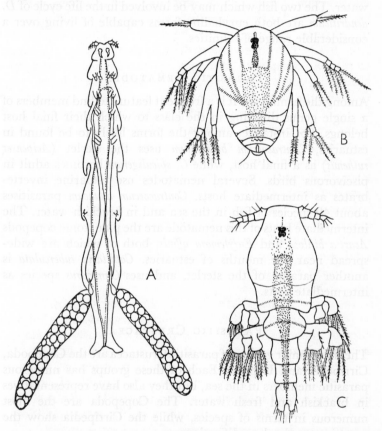

Figure 97. Mytilicola intestinalis (Crustacea: Copepoda) from the intestine of *Mytilus edulis*

A, Adult female, ventral view, actual length 7 mm (after Steuer). B, First metanaupliar stage, dorsal view, actual length 250μ (after Pesta). C, Copepodid stage, dorsal view, to same scale as B (after Pesta)

intestine of *Mytilus edulis* and certain other bivalves. The general form of *Mytilicola* is maggot-like; the legs are greatly reduced when compared with free-living forms, and the mouthparts of the adult also show reductions, including the loss of the mandibles (Hockley, 1951). The adult female lays her eggs in two elongated sacs, which together may contain up to 300 eggs.

The larva that emerges from the egg of *Mytilicola* is a metanauplius (Pesta, 1907), which is capable of swimming in the plankton. This soon moults to give rise to a second metanaupliar stage, and this then metamorphoses into a copepodid (Fig. 97C). The copepodid is the infective stage. It is possible to rear the copepodids from eggs without feeding the metanaupliar stages. It seems as if there is enough food in the egg to enable the infective stage to be reached without any necessity for feeding in the free-swimming stages. At a temperature of 14°C the copepodid stage is reached in about 40 hours (Grainger, 1951). The early development of *Mytilicola* is thus considerably telescoped when compared with free-living forms that pass through five naupliar stages over a much longer period of time.

It is not known exactly how the copepodid of *Mytilicola* enters the intestine of *Mytilus*, but once it has done so the next moult reveals a form much more like the adult, with a considerable reduction of the limbs. In warm weather *Mytilicola* may become sexually mature about 7 weeks after hatching (Korringa, 1951)

In parts of the North Sea the breeding of *Mytilicola* is suspended for part of the winter, but in regions with a milder climate, such as the Mediterranean and the coasts of Ireland, it is possible to find females with eggs throughout the year.

Mytilicola can tolerate a range of salinity at least as great as the range tolerated by its host, so that it is capable of enacting its life cycle well into estuarine conditions. The effect of this copepod on its host seems to vary with the local conditions. Where there is a plentiful supply of food for the mussels, they can tolerate several copepods in the gut without any serious loss in condition, but if there is any reduction in food supply the mussels lose condition. Meyer and Mann (1951) have shown that even a single parasite in the gut can reduce the filtering efficiency of *Mytilus*. With seven copepods in the gut the filtering efficiency of the mussel may be halved. The loss of mussel production caused by this parasite can be of commercial importance, and several investigations have been made with the control of this copepod in mind (e.g. Bolster, 1954; Hepper, 1955). It appears that mussels growing in fast-moving water tend to be less heavily infected than mussels growing in sluggish water, and mussels growing on pier piles or other raised structures are less liable to be infected than mussels lying on the floor of an estuary. The cultivation of mussels on fences or ropes seems to be a practical method of reducing the intensity of infection, but the ropes or fences should be in a position where they are submerged for 16 hours a day to provide enough time for the mussels to feed and grow at a fast rate.

Other estuarine copepods are parasitic on fish. Those found on the salmon are dealt with on p. 332. The life cycles of these parasites show considerable variation. *Thersitina gasterostei*, which attaches itself to the inside of the gill covers of stickle-backs (*Gasterosteus aculeatus* and *Pygosteus pungitius*), is parasitic only in the adult stage; the earlier stages are free swimming. In contrast *Lernaeocera branchialis* uses two different species during its life history. The early stages are parasitic on the flounder (*Platichthys flesus*) and the later stages parasitise cod or whiting (*Gadus* spp.). It is thus possible to find the early stages in estuaries, but the later stages tend to be more strictly marine.

Estuarine crustaceans may also be hosts to parasitic copepods. *Corophium volutator* sometimes carries a species of *Sphaeronella* in its brood pouch. This parasite has a rounded body with only two pairs of minute legs. The mouth is conical with small needle-like mandibles which pierce the wall of the brood pouch and enable the parasite to suck the hosts blood. A fully deve-loped parasite may produce enough eggs to completely displace those of its host.

The Cirripedia include the order Rhizocephala, which are parasites of other Crustacea. The best known example is *Sacculina carcini*, a parasite of the shore crab, *Carcinus maenas*. When fully developed the parasite appears as a sac-like bulge between the thorax and abdomen of the crab, pressing the forwardly directed abdomen away from its neat contact with the thorax. Before reaching this final stage *Sacculina* goes through a remarkable life history. The eggs are retained in the mantle cavity of the adult until they hatch as nauplii. These nauplii are liberated into the plankton and pass through four naupliar stages before metamorphosing into cypris larvae. The cypris larvae are of two sizes. The smaller ones attach to the base of a seta on a crab. The larval limbs are then cast off and the lava assumes a sac-like form. Within the sac a hollow dart-like structure is developed and through this tube the internal contents of the sac are injected into the body cavity of the crab. The small mass of cells migrates through the body of the crab and comes to rest underneath the intestine. In this position it grows and sends out root-like structures to all parts of the host body. The central body also increases in size and finally bursts through the host's cuticle between the thorax and abdomen.

The larger cypris larvae do not inject their body contents into a crab, but inject them into the mantle cavity of a juvenile *Sacculina* that has recently burst through the cuticle of its host. These large cypris larvae are in fact males, but such highly specialised males that when they have injected themselves into the mantle cavity of a female they take up residence in special pockets opening off the mantle cavity and resemble testes. For many years it was thought that *Sacculina* was a hermaphrodite, but thanks to the researches of Ichikawa and Yanagimachi (1960) and Yanagimachi (1961) it is now known that the apparent testes are the result of injection of the body contents of large cyprids.

A crab infected with *Sacculina* loses the capacity to reproduce, and male crabs lose their secondary sexual characteristics so that they resemble females. In particular the abdomen loses the characteristic narrow form of the male and becomes broad like that of the female. The change induced by the presence of *Sacculina* also extends to the behaviour of the animals. Rasmussen (1959) found that crabs with the parasite tended to behave like females with eggs and remained in deeper, more saline, water than the males. In estuarine conditions this results in the parasitised crabs being restricted to the mouth of the estuary.

A few of the Rhizocephala have penetrated into fresh water. These parasites produce larger eggs than the marine and estuarine forms, and hatch as cypris larvae, so that the naupliar stages are eliminated and the time spent free swimming is greatly reduced. One form, *Sacculina gregaria*, travels with its host, *Eriocheir japonicus*, into fresh water, but does not breed until its host returns to more saline water for its own breeding season. It may be suggested that in this case the hormones of the host influence the breeding of the parasite, but the details of this process have yet to be studied.

Paragnathia formica is one of the most typically estuarine of crustacean parasites. This isopod is, in its young stages, parasitic on fish, but as an adult it lives in cavities beneath the surface of salt marsh soils in Western Europe. The most frequent hosts of the young stages are *Gobius microps*, *Platichthys flesus* and the eel, *Anguilla anguilla*. The first larval stage that attaches itself to the external surface of a fish is a fairly typical looking isopod, but its mouthparts are modified for piercing and sucking. After gorging itself on the blood of the host the larva has the third, fourth and fifth segments of the thorax distended to about twice the diameter of the rest of the body. Before it has fed the segmented larva is known as a zuphea, but the distended larva is termed a praniza. Fixation to the host may last as little as 2 hours, or may be as long as 2 days. The larva then leaves the host and may spend about $2\frac{1}{2}$ months before its next meal. According to the findings of Stoll (1962) there are 3 larval stages, each lasting between 2 and 4 months.

The normal life cycle of *Paragnathia formica* in an estuary near Roscoff begins with the liberation of larvae in April or May. Metamorphosis into the adult does not occur until the following February. The males develop a broad head and large mandibles, but the females retain heads like those of the praniza, but with much smaller eyes. Once they have become mature the males and females do not feed, but remain in their chambers beneath the surface of the salt marsh. The females retain their eggs in internal pouches and the young are liberated as active larvae which leave the parental burrows to seek out new fish hosts.

OTHER PARASITIC ARTHROPODS

A group of parasites that has yet to be studied from the viewpoint of the estuarine biologist is formed by the insects and mites parasitic on birds. It is not known exactly what problems these parasites face when their hosts swim in waters of different salinities. The mites and lice on the feathers of ducks, for instance, probably do not get wet if they are close to the skin of their host, because the feathers of ducks trap air and present a hydrofuge surface to the water. But the lice on a cormorant may well get wet because the feathers of these birds are not well waterproofed. It is perhaps significant that the cormorants do not have head lice (Rothschild and Clay, 1952), for it is in this region of the body that the feathers are shortest and the lice would be most likely to come into contact with water.

The pelicans and cormorants of America have a genus of lice (*Piagetiella*) living inside their throat pouches. In this situation they are certain to get wet and are subjected to the additional hazard of abrasion by the scaly bodies of fish captured by the birds.

The feather mites of cormorants may also be confronted with special problems. Apart from the possibility of osmotic stress when the bird is submerged, they have to remain attached to the host while it is swimming. This is more of a problem for the males than the females, because the males have to be able to move about in search of females. It is a common feature of feather mites to find that the males have enlarged hind legs, but some of the species found on cormorants have developed a remarkable asymmetry. The male of *Dinalloptes anisopus* has the second leg on the right side greatly enlarged (Atyeo and Peterson, 1966). The left hind leg is also large, resembling the hind legs of other feather mites, but the right hind leg is very small. This means that the right hand side of the body shows a hypertrophy of the second leg and atrophy of the last leg. The significance of this modification is unknown, but it may be related in some way to the structure of cormorant feathers and the problems that the male encounters in moving about in search of females over the surfaces of the feathers when they are wet.

Mites may also be associated with estuarine invertebrates. *Phaulodinychus orchestiidarum* is found on the ventral surfaces of *Orchestia* and *Talitrus*, but it is not known if it obtains food from its host. A mite which does feed on its host is found on adult midges of the genus *Culicoides* living in salt marshes. The mite, *Evansiella culicoides*, attaches itself to the soft parts of the abdomen and thorax and feeds by sucking the body fluids of the host. In some instances the mites are attached in such a position that they hinder the flying of the host so that the midges are restricted to crawling movements over the vegetation.

PARASITIC FUNGI

The fungi infecting estuarine animals have not received a great deal of attention, but Atkins (1929, 1954, a, b, 1955) described three species which infect the eggs of the pea crab (*Pinnotheres pisum*), itself a parasite of the mussel, *Mytilus edulis*. Much of the material studied by Atkins came from estuaries in Devon and Cornwall. One of the fungus species, *Leptolegnia marina*, was also found growing on the gills of the lamellibranchs *Barnea candida* and *Cardium echinatum*, and infected the adult as well as the eggs of the pea crab. The fungus penetrates the body wall of the crab where the cuticle is very thin, often in the roof of the gill chamber. The hyphae spread throughout the gills and the crab dies a few days after infection has become evident. A second species, *Plectospira dubia*, was found to be capable of infecting the eggs of *Crangon vulgaris*, *Leander serratus* and several species of crabs. Unfortunately there are no data available concerning the salinity tolerances of these fungi, and it is not known how far into an estuary they can penetrate.

A truly brackish water fungus has been found infecting the rotifer *Synchaeta monopus* in the plankton of the Baltic. The fungus, *Synchaetophagus balticus*, produces robust hyphae which invade the body cavity of the rotifer and eventually kill it (Apstein, 1911). Another fungus was noted by Segerstråle (1957) as causing the deaths of millions of copepods (*Eurytemora hirundoides*) in the plankton of the Baltic in 1950. The identity of this fungus was not established.

PARASITES OF MIGRATORY FISH

Fish such as the salmon and eel pose interesting problems to the parasitologist. When a salmon leaves the river in which it hatched and migrates to the sea, does it carry a collection of fresh-water parasites with it, and do these survive during the marine phase of its life? From the work summarised by Dogiel (1961) it appears that the young salmon can aquire about 12 species of metazoan fresh-water parasites, including trematodes (e.g. *Crepidostomum farionis*), cestodes (e.g. *Triaenophorus nodulosus*), acanthocephalans (*Neoechinorhynchus rutilis*) and nematodes (e.g. *Rhabdochona denudata*). When the young salmon migrates to the sea all these parasites are lost. New parasites of marine origin are acquired, including the trematode *Derogenes varicus*, the cestode *Eubothrium crassum*, the acanthocephalan *Echinorhynchus gadi*, the nematode *Terranova decipiens*, and the copepod *Lepeophtheirus salmonis*.

When the salmon migrates back into fresh water from the sea there is a reduction in this fauna of marine parasites. The external parasites such as *Lepeophtheirus* are the most vulnerable to changes in osmotic pressure and generally disappear after the salmon has been in fresh water for a few weeks. The gut parasites also show a reduction in fresh water, this may not be an osmotic effect alone, but may also be connected with the fact that the salmon does not feed when it returns to breed in fresh water. The acanthocephalan *Echinorhynchus gadi* is one of the first of the intestinal parasites to disappear after the salmon has been in fresh water for a few weeks. Parasites of the body cavity are better protected against changes in the concentration of the external medium, and the larvae of *Terranova decipiens* appear to survive throughout the time that the salmon spends in fresh water.

Salmincola salmonea is a copepod which parasitises the gills of salmon in fresh water, but it also appears to be tolerant of sea water, so that when a salmon returns to the sea after spawning it may carry a number of these copepods with it, and they will survive until the salmon returns to the river again. As far as is known *Salmincola* breeds only in fresh water, and new infections of salmon occur only when they migrate into rivers to spawn. The salmon thus has two contrasting copepod parasites: *Lepeoptheirus salmonis* attaches itself to the anal region, breeds in the sea, and cannot survive permanently in fresh water; while *Salmincola salmonea* attaches to the gills, breeds in fresh water, but can also tolerate sea water.

The parasite fauna of the eel (*Anguilla anguilla*) provides a marked contrast with that of the salmon. The leptocephalus larva spends three years in the sea. During this time it migrates from the Sargasso Sea to the shores of Europe. Throughout this migration it appears to be quite free from parasites, or at least none seems to have been recorded. As soon as the young elvers begin to penetrate into brackish water some of them become infected with parasites that do not have intermediate hosts. Elvers in Italy have been found infected with the protozoan *Myxidium giardi* and the monogenetic trematode *Gyrodactylus*. Later the eels become infected with a wider range of parasites, including many with intermediate hosts. The eels that remain in the lower reaches of rivers tend to acquire the richest fauna of parasites, including some marine species such as *Podocotyle atomon* and *Deropristis inflata*. The latter species is known from Europe and North America. infecting both *Anguilla anguilla* and *A. rostrata*. When *A. rostrata* is the final host the primary host is the snail *Bittium alternata*, and the secondary host is *Nereis virens* (Cable and Hunninen, 1942). In the fresh parts of rivers several typical fresh-water parasites are found in eels. *Sphaerostoma bramae* and *Bucephalus polymorphus* are parasites of a wide range of fresh-water fish and infect the eel during its fresh-water feeding phase. The parasite fauna of the eel during its final marine phase does not seem to have been investigated.

PARASITES IN THE BRACKISH SEAS

More work has been done on the parasites of fish in the brackish seas than on those in estuaries. A large amount of information on this topic has been summarised by Dogiel, Petrushevski and

Polyanski (1961); only a few of the general points can be mentioned here.

In the Aral Sea most of the fish are of fresh-water origin, and so are their parasites. This situation results in a reduction of the prevalence of parasites in the more saline parts of the sea. In a low salinity area at the mouth of the Amu river a total of 66 species of fish parasites have been recorded, while in a more saline area near the island of Uzan Kuir, where the salinity was about 10‰ there were only 31 species of fish parasites.

As one penetrates into the Gulf of Finland the parasites of marine origin die out and are replaced by fresh-water forms. The disappearance of the marine forms is not always an effect of reduced salinity on the parasite or its final host, but may in some cases be due to the disappearance of an intermediate host. In parts of the Gulf of Finland it is possible to find fish of marine origin infected with parasites of fresh-water origin, and fish of fresh-water origin infected with parasites of marine origin. For instance the fresh-water pike–perch (*Lucioperca lucioperca*) has been found infected with a marine acanthocephalan, *Corynosoma strumosum*.

In the Black Sea about 190 species of fish parasites have so far been recorded. A high proportion of these are of marine origin. The Black Sea still maintains contact with the Mediterranean through the Sea of Marmara, and there are about 40 species of fish parasites common to the Black and Mediterranean Seas, but which have not been recorded elsewhere. In addition to these there are a few more widespread marine species, occurring for instance in the Atlantic Ocean as well as the Black Sea.

The parasite fauna of the Black Sea contrasts markedly with that of the Caspian Sea, where, of the 172 species so far recorded, only 10 are of marine origin. Some of these marine species are shared with the Black Sea, and are apparently ancient forms which entered the Caspian Sea early in its history when it was in contact with the Tethys Sea.

CHAPTER 13

ESTUARINE FOOD WEBS

THE food web of a single estuary is so complex that it is impossible to depict in a diagram. The feeding habits of many of the animals vary seasonally, and some of the fish vary their diet in different parts of the estuary. It follows from this that any accurate attempt to depict this state of affairs would have to take the form of a three-dimensional model, and would need some device to make allowance for seasonal variation. In this chapter the approach has been simplified by considering first the general groups of feeding habits and then giving diagrams of food webs in particular estuarine situations to give some details of the finer connections of the web.

Plants, by their photosynthetic activities, provide the primary source of food in estuaries. The assessment of the amount of plant material available to animals is not a simple matter. From the gross amount of primary production it is necessary to deduct the material utilised in respiration by the plants. A further complication is that many animals do not feed directly on the plant material, but only on the detritus formed from the breakdown of this material. This process involves the additional complication of bacterial growth.

Some plant material is utilised directly by herbivores. Geese, for instance, may graze on salt marsh plants, and at the other extreme of size the small Crustacea of the zooplankton may feed directly on the phytoplankton. But a high proportion of the estuarine vegetation is converted to detritus before it is utilised by animals. This detritus forms the substratum for bacterial growth, and the bacteria in turn provide an important source of food for filter feeders and detritus eaters.

The main source of estuarine detritus is from the larger rooted plants. *Zostera* beds near the mouth of estuaries and *Spartina* meadows in the salt marshes are probably the most

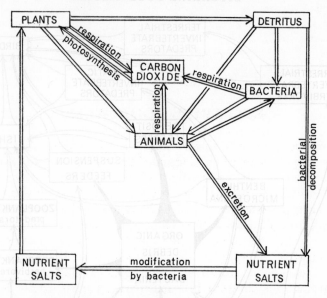

Figure 98. Primary diagram of the estuarine food web, showing basic relationships between nutrient salts, plants, bacteria and animals

important sources of detritus, although a great variety of other plants may also contribute.

Detritus and its coating of bacteria may be consumed while in suspension in the water. Suspension feeders, such as *Cardium* and *Mya*, are often dominant members of burrowing communities, and some members of the estuarine zooplankton, such as the mysids, have feeding mechanisms which are capable of collecting detritus from suspension. Once the detritus has settled it is eaten by a wide variety of animals. The lugworm *Arenicola marina* utilises detritus in the soil as its main source of food. But it can also utilise suspended material by driving water forwards through its burrow so that particles are trapped in the sand lying in front of the worm's head (Krüger, 1959). The relative importance of suspended material as food for *Arenicola* varies from one area to another. In intertidal areas it may form as little as 10 per cent of the total food, but in regions where the worm is permanently submerged the proportion may be greatly increased (Jacobsen, 1967).

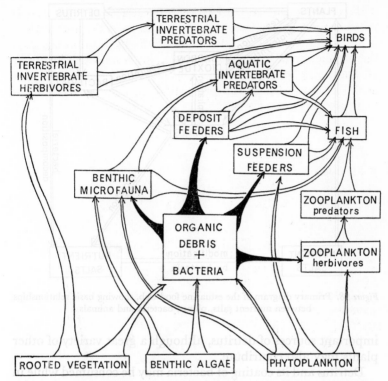

Figure 99. Diagram of the food relationships of the main ecological groups of estuarine animals. The arrows go from food to consumer

The faeces of some animals, such as *Hydrobia ulvae*, form substrata suitable for the further growth of bacteria (Newell, 1965). As many bacteria are capable of fixing nitrogen, the growth of these micro-organisms on the surface of detritus and animal faeces greatly increases the total nitrogenous material available to animals in estuaries.

The importance of bacteria as food for animals dwelling in mud has been emphasised by Zobell and Feltham (1942), who calculated that in a cubic foot of mud there would be about a gram dry weight of bacteria. This would represent the balance between the rate of bacterial reproduction and the rate of consumption of bacteria by other organisms. The rate of bacterial reproduction under natural conditions is not known with certainty, but on the basis of laboratory observations it is possible for a single bacterium to divide twenty times a day. If one assumes only half this rate under natural conditions then the production of bacteria in a cubic foot of mud will be about 10 grams dry weight per day. This gives an approximate idea of the amount of bacterial protoplasm, which has a relatively high protein content, available to the mud fauna.

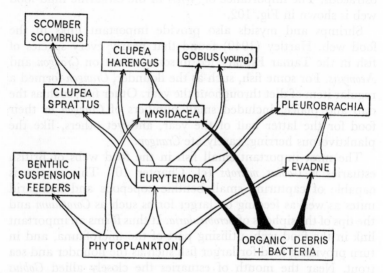

Figure 100. Simplified food web of the zooplankton in a British estuary

The bacteria in mud are eaten directly by many members of the microfauna. Ciliates, such as *Trachelocerca phoenicopterus* and *Uronema marinum,* collect bacteria by means of ciliary currents, and certain nematodes living in the black sulphide layer appear to feed exclusively on sulphide bacteria. Larger members of the mud fauna feed indirectly on the bacteria. For instance, *Scrobicularia plana* uses its inhalent siphon to suck up the surface layers of the mud and so ingests a mixture of silt, detritus, bacteria and diatoms, and probably includes a few members of the microfauna as well.

The microfauna of an estuarine mud is subjected to predation both by vertebrates and invertebrates. Some of the ciliates, such as *Lacrymaria coronata* and *Loxophyllum fasciolatum*, prey upon other ciliates, and some other members of the microfauna, such as turbellarians, also are predators. Young specimens of *Nereis diversicolor* exert an important effect on the microfaunal population. Perkins (1958) examined specimens of this polychaete ranging in length from 2 to 9 mm, and found them to be omnivorous. The food taken by the young worms included diatoms, foraminiferans, nematodes, copepods and ostracods. The importance of *Nereis* in the estuarine mud food web is shown in Fig. 102.

Shrimps and mysids also provide important links in the food web. Hartley (1940) found that nearly every species of fish in the Tamar Estuary fed to some extent on *Crangon* and *Neomysis*. For some fish, such as the flounder, *Crangon* formed a regular item of diet throughout the year. Other fish, such as the dab and plaice, included small numbers of *Crangon* in their food for the latter half of the year, and yet others, like the planktivorous herring, rarely ate *Crangon*.

The most important small fish in the food webs of British estuaries is *Gobius microps* (cf. Chapter 10). This species is capable of capturing small benthic copepods and halacarid mites as well as feeding on larger forms such as *Corophium* and the tips of the siphons of *Scrobicularia*. It thus forms an important link in the food web, utilising part of the microfauna, and in turn providing food for larger fish such as the flounder and sea trout. Near the mouth of estuaries the closely allied *Gobius minutus* feeds on small Crustacea, and is in turn eaten by fish such as the Brill (*Rhombus laevis*).

A convenient way of summarising the information available concerning feeding habits is to construct a trophic spectrum (Darnell, 1961). This method may be used to compare different age groups of one species, or two populations of one species, or a range of species in a particular habitat. Fig. 103 illustrates an application to three stages of one species. The food categories are listed in sections of equal height; this height represents 100 units. Within each age group of fish the percentage volume of each food is shown by a block which is proportional in height to the percentage of the total food formed by the particular food item. The sum of all the heights of the blocks for a group

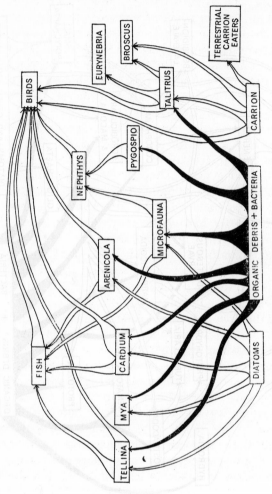

Figure 101. Simplified food web of a sandy shore in a British estuary

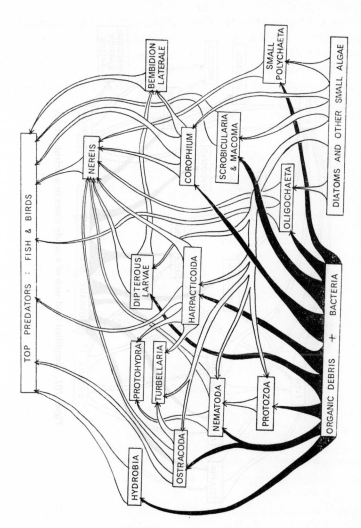

Figure 102. Simplified food web of a British estuarine mud flat

of fish adds up to 100 units. In this way the relative importance of each food item can easily be seen.

When the trophic spectrum technique is applied to a complete community of fishes it reveals a considerable amount of information about the structure of the food web. Fig. 104 shows Darnell's analysis of the feeding habits of the most important fish in Lake Pontchartrain, which is a large estuary on the Louisiana coast. Some important features of this trophic spectrum are as follows.

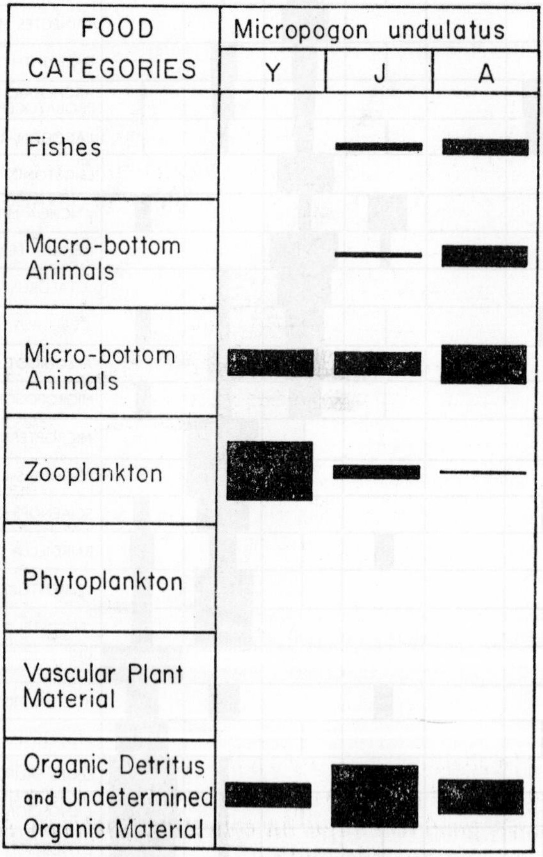

FOOD CATEGORIES	Micropogon undulatus		
	Y	J	A
Fishes		▬	▬
Macro-bottom Animals		▬	▬
Micro-bottom Animals	▬	▬	▬
Zooplankton	▬	▬	▬
Phytoplankton			
Vascular Plant Material			
Organic Detritus and Undetermined Organic Material	▬	▬	▬

Figure 103. Trophic spectrum of *Micropogon undulatus* at three stages of development. Y = young, J = juvenile, A = adult (after R. M. Darnell)

Organic detritus	Vascular plants	Phytoplankton	Zooplankton	Micro-benthos	Macro-benthos	Fishes	FOOD CATEGORIES / CONSUMER SPECIES
■	■						MUGIL CEPHALUS
■							BREVOORTIA PATRONUS
		■	■				DOROSOMA PETENENSE
■				■			DOROSOMA CEPEDIANUM
■					■		TRINECTES MACULATUS
■					■		GALEICHTHYS FELIS
	■				■		ARCHOSARGUS PROBATOCEPHALUS
■		■			■		LAGODON RHOMBOIDES
■				■			LEIOSTOMUS XANTHURURUS
			■	■		■	ANCHOA MITCHILLI
			■	■			MENIDIA BERYLLINA
■					■		ICTALURUS PUNCTATUS
■	■				■	■	ICTALURUS FURCATUS
			■		■		APLODINOTUS GRUNNIENS
■			■		■	■	MICROPOGON UNDULATUS
				■	■		MICROPTERUS SALMOIDES
					■		POGONIAS CHROMIS
■					■	■	SCIAENOPS OCELLATA
			■		■		BAIRDIELLA CHRYSURA
					■	■	LEPISOSTEUS OCULATUS
					■	■	LEPISOSTEUS SPATULA
						■	LEPISOSTEUS OSSEUS
			■		■	■	ROCCUS MISSISSIPPIENSIS
■						■	STRONGYLURA MARINA
						■	ELOPS SAURUS
						■	CYNOSCION NEBULOSUS
						■	CYNOSCION ARENARIUS
						■	PARALICHTHYS LETHOSTIGMA
						■	CARCHARHINUS LEUCAS
						■	CARANX HIPPOS

M

1. Phytoplankton and vascular plants were utilised directly by few of the fish species. Blue-green algae were eaten by the striped mullet (*Mugil cephalus*), but otherwise the utilisation of algae was negligible. The main vascular plants eaten by fish were *Ruppia maritima* and *Vallisneria spiralis*. A few fish, such as the sheepshead (*Archosargus probatocephalus*) included a high proportion (up to 54 per cent) of these plants in their diet, but in most fish the proportion was much lower.

2. Organic detritus and its associated bacteria formed an appreciable fraction of the diet of a wide range of species.

3. Zooplankton was eaten mainly by the smaller fish, and by the immature stages of the larger fish. The dominant zooplankter was the calanoid copepod *Acartia tonsa*, which formed a high percentage of the food of the bay anchovey (*Anchoa mitchilli*), the most abundant fish in the estuary.

4. The microbenthos was consumed by a wide range of fish. The spot (*Leistomus xanthurum*) specialised in feeding on ostracods and harpacticoid copepods, but included other microbenthos as well. The tide water silverside (*Menidia beryllina*) fed to a large extent on amphipods, isopods and chironomid larvae.

5. The three most important species of larger benthic invertebrates were the clam, *Rangia cuneata*, the mud crab, *Rithropanopeus harissii*, and the blue crab, *Callinectes sapidus*. Large clams were eaten by only two species of fish: the freshwater drum (*Aplodinotus grunniens*) and the black drum (*Pogonias cromis*). But the crabs were eaten by a much wider range of fish. The young of the blue crab were much more liable to predation than the adults, but even the adults were subjected to predation by the alligator gar (*Lepisosteus spatula*) which was found foraging in the marshy creeks.

6. A considerable number of fish species fed on other fish. The most intense predation fell on three species: the large-scale menhaden (*Brevoortia patronus*), bay anchovey (*Anchoa mitchilli*), and the young of the Atlantic croaker (*Micropogon undulatus*). These species were present in large populations and were widely distributed so that they were accessible to a range of predators.

An aspect of estuarine ecology that has received little attention can be considered under the term 'terrestrial export'. By this term is meant the removal of material from the estuarine environment by terrestrial animals which are not normally a part of the estuarine fanua. Rabbits and voles (*Microtus* spp.) often enter salt marshes to feed on the vegetation, and man sets his cattle and sheep to graze on the upper parts of salt marshes. A wide variety of terrestrial birds may also be seen feeding on estuarine shores. Crows, rooks and jackdaws scavenge on the shore; wheatears sometimes feed on *Talitrus*, and starlings may descend upon the shore in large numbers when their terrestrial feeding grounds are covered by snow (Lord, 1867).

Invertebrates may also play a part in removing material from the estuary, but their role is generally smaller than that of the yertebrates. A large number of species of carabid beetles may be found in grassland at the margins of estuaries. Some of these, which may be regarded as a true part of the estuarine fauna, have been discussed in Chapter 9. But many others are not found permanently on the estuarine shore, and only occasionally maraud down the shore or into the salt marshes. Some of the ground-dwelling spiders have similar habits. The numerous flies and bees which visit flowers in salt marshes also remove a certain amount of material in the form of pollen and nectar, but from a quantitative point of view this is probably negligible.

The food webs of British salt marshes have not been studied in detail, although Nicol (1935) has provided some indications of the general features which may be found. Such information as is available from Nicol, Sutcliffe (1961b) and my own observations is summarised in Fig. 105.

In North America a detailed quantitative survey of a salt marsh in Georgia has been made by Teal (1962). The most important plant on the marsh was *Spartina alterniflora*. When calculated in terms of energy the gross primary production by this plant amounted to 34,580 k cal/m²/yr, of which 28,000 k cal were used in respiration, leaving a net production of 6,580 k cal/m²/yr. Algae contributed a further net production of 1,620 k cal/m²/yr.

Some insects were found which fed directly on the living *Spartina*. The grasshopper, *Orchelimum fidicinum*, consumed the tissues, while the leaf hopper, *Prokelisia marginata*, sucked the juices. Between them these two insects assimilated

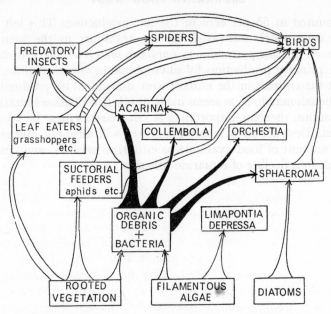

Figure 105. Simplified food web of the surface of a British salt marsh

304 k cal/m²/yr, which amounted to approximately 4.6 per cent of the net production of *Spartina.*

A slightly smaller amount of material was assimilated by the crabs which fed on detritus and algae. The dominant members of this group were *Uca pugnax, U. pugilator* and *Sesarma reticulatum.*

Nematodes assimilated 85 k cal/m²/yr, annelids 35, and molluscs 132 k cal/m²/yr. Bacterial decomposition of the dead *Spartina* was a most important factor in the salt marsh, accounting for 3,890 k cal/m²/yr. This activity of the bacteria was not entirely deleterious. Laboratory experiments showed that during the decomposition of *Spartina* over a period of 16 weeks the protein content was twice as high at the end of the experiment as at the beginning. Thus the bacterial coating on the *Spartina* debris increased the percentage protein available to the detritus eaters.

Altogether the total energy dissipated by the consumers on the salt marsh, including predatory birds and raccoons, amounted to 55 per cent of the net production. This left 45 per cent which was available for transport to the nearby estuary. In this particular estuary the water was very turbid, and primary production by phytoplankton was negligible, so that the animals in the estuary were dependent on production in the salt marsh. If, as seems likely, this is true of other estuarine situations, then the importance of salt marshes must be recognised. Destruction of coastal marshes will strongly influence the amount of food available to estuarine animals, and could result in a decline of estuarine fisheries.

BIBLIOGRAPHY

ANCONA, U. D' (1927) "Ricerche sull' ingrandimento dell' occhio dell' anguilla in rapporto alla maturità sessuale e considerazioni sul suo significato biologico', Atti Accad. naz. Lincei Re. (6).

ANDRÉ, M. (1904) 'Thalassarctus marina', Faune de France, vol. 49, pp. 152.

ANNANDALE, N. (1907) 'The fauna of brackish ponds at Port Canning, Lower Bengal, Part III. An isolated race of the actinian Metridium schillerianum (Stoliczka)', Rec. Indian Mus. 4: 47–74.

ARNDT, C. (1917) 'Syndesmophagie bei(?), ein Symbiose lebender Pilz', Wus. Mitteilung, Abh. N.F. 12: 163–6.

ARNDT, C. H. (1914) 'Some insects of the between tides zone', Proc. Indiana Acad. Sci. 1914: 323–30.

ARUDPRAGASAM, K. D. and NAYLOR, E. (1964a) 'Gill ventilation and the role of reversed respiratory currents in Carcinus maenas (L.)', J. exp. Biol. 41: 299–307.

ARUDPRAGASAM, K. D. and NAYLOR, E. (1964b) 'Gill ventilation volumes, oxygen consumption, and respiratory rhythms in Carcinus maenas (L.)', J. exp. Biol. 41: 309–21.

ASCROFT, R. L. (1900) 'Notes on the white flake, or flounder', Trans. Lancs. Nat. Soc. 1: 174.

ATKINS, D. (1929) 'On a fungus allied to the Saprolegniaceae found in the pea crab Pinnotheres', J. mar. biol. Ass. U.K. 16: 203–19.

ATKINS, D. (1934) 'Two parasites of the common cockle, Cardium edule: a rhabdocoele Paravortex cardii Hallez and a copepod Paranthessius rostratus (Canu)', J. mar. biol. Ass. U.K. 19: 669–76.

ATKINS, D. (1936–8) 'On the ciliary mechanisms and interrelationships of lamellibranchs, Parts I–VII, Quart. J. micr. Sci. 79: 181–308, 339–445; 80: 321–436.

ATKINS, D. (1954a) 'Further notes on a marine member of the Saprolegniaceae, Leptolegnia marina n. sp., infecting certain invertebrates', J. mar. biol. Ass. U.K. 33: 613–25.

ATKINS, D. (1954b) 'A marine fungus, Plectospira dubia n. sp. (Saprolegniaceae) infecting crustacean eggs and small Crustacea', J. mar. biol. Ass. U.K. 33: 721–??.

ATKINS, D. (1954c) 'Leg disposition in the brachyuran megalopa when swimming', J. mar. biol. Ass. U.K. 33: 627–36.

ATKINS, D. (1955) 'Pelogenia inclusions n.sp. infecting the egg mass of the pea crab Pinnotheres pisum', Trans. Brit. mycol. Soc. 38: 29–46.

ATYEO, W. T. and PETRAKSON, F. C. (1886) 'The feather mite genus Dinalloptes (Acarina, Proctophyllodidae)', Acarologia 8: 470–4.

AUDOUIN (1835) Quoted in Palmer (1962).

BIBLIOGRAPHY

ANCONA, U. D' (1927) 'Ricerche sull' ingrandimento dell' ochio dell' anguilla in rapporto alla maturita sessuale e considerazioni sul suo significato biologico', *Atti Accad. naz. Lincei Rc.* (6).

ANDRE, M. (1964) 'Halacariens marins' *Faune de France*, vol. 46, pp. 152.

ANNANDALE, N. (1907) 'The fauna of brackish ponds at Port Canning, Lower Bengal. Part III. An isolated race of the actinian *Metridium schillerianum* (Stoliczka)' *Rec. Indian Mus.*, 1: 47–74.

APSTEIN, C. (1911) '*Synchaetophagus balticus*, ein *Synchaeta* lebender Pilz', *Wiss. Meeresunters, Kiel*, N.F. 12: 163–6.

ARNDT, C. H. (1914) 'Some insects of the between tides zone', *Proc. Indiana Acad. Sci.*, 1914: 323–36.

ARUDPRAGASAM, K. D. and NAYLOR, E. (1964a) 'Gill ventilation and the role of reversed respiratory currents in *Carcinus maenas* (L.)', *J. exp. Biol.*, 41: 299–307

ARUDPRAGASAM, K. D. and Naylor, E. (1964b) 'Gill ventilation volumes, oxygen consumption, and respiratory rhythms in *Carcinus maenas* (L.), *J. exp. Biol.*, 41: 309–21.

ASCROFT, R. L. (1900) 'Notes on the white fluke, or flounder', *Trans. Liverp. biol. Soc.*, 14: 174.

ATKINS, D. (1929) 'On a fungus allied to the Saprolegniaceae found in the pea crab *Pinnotheres*', *J. mar. biol. Ass. U.K.*, 16: 203–19.

ATKINS, D. (1934) 'Two parasites of the common cockle, *Cardium edule*; a rhabdocoele *Paravortex cardii* Hallez and a copepod *Paranthessius rostratus* (Canu)', *J. mar. biol. Ass. U.K.*, 19: 669–76.

ATKINS, D. (1936–8) 'On the ciliary mechanisms and interrelationships of lamellibranchs'. Parts I–VII, *Quart. J. micr. Sci.*, 79: 181–308, 339–445; 80: 321–436.

ATKINS, D. (1954a) 'Further notes on a marine member of the Saprolegniaceae, *Leptolegnia marina* n. sp. infecting certain invertebrates', *J. mar. biol. Ass. U.K.*, 33: 613–25.

ATKINS, D. (1954b) 'A marine fungus, *Plectospira dubia* n. sp. (Saprolegniaceae) infecting crustacean eggs and small Crustacea', *J. mar. biol. Ass. U.K.*, 33: 721–32.

ATKINS, D. (1954c) 'Leg disposition in the brachyuran megalopa when swimming', *J. mar. biol. Ass. U.K.*, 33: 627–36.

ATKINS, D. (1955) '*Pythium thalassium* n.sp. infecting the egg mass of the pea crab *Pinnotheres pisum*', *Trans. Brit. mycol. Soc.*, 38: 29–46.

ATYEO, W. T. and PETERSON, P. C. (1966) 'The feather mite genus *Dinalloptes* (Acarina, Proctophyllodidae)', *Acarologia*, 8: 470–4.

AUDUBON, (1835) Quoted in Palmer (1962).

350 BIBLIOGRAPHY

AYERS, J. C. (1956) 'Population dynamics of the marine clam *Mya arenaria*', *Limnol. Oceanogr.*, 1: 26–34.

BAAS BECKING, L. G. M. and WOOD, E. J. F. (1955) 'Biological processes in the estuarine environment', *Proc, K. ned. Akad. Wet.*, B 58: 160–81.

BACKLUND, H. O. (1945) 'Wrack fauna of Sweden and Finland. Ecology and chorology', *Opusc. Ent. Lund Suppl.*, 1945, pp. 236.

BALFOUR-BROWNE, F. (1940) *British Water Beetles*, Vol. I, London: Ray Society.

BALFOUR-BROWNE, F. (1950) *British Water Beetles*, Vol. II, London: Ray Society.

BALFOUR-BROWN, F. (1958) *British Water Beetles*, Vol. III, London: Ray Society.

BARKER, D. (1963) 'Size in relation to salinity in fossil and recent euryhaline ostracods', *J. mar. biol. Ass. U.K.*, 43: 785–95.

BARNES, H. and BARNES, M. (1958) 'A note on the opening response of *Balanus balanoides* (L.) in relation to salinity and certain inorganic ions', *Veröff. Inst. Meeresforsch. Bremer*, 5: 160–4.

BARNES, H. FINLAYSON, D. M. and PIATIGORSKY, J. (1963) 'The effect of desiccation and anaerobic conditions on the behaviour, survival and general metabolism of three common cirripedes', *J. Anim. Ecol*, 32: 233–52.

BARNETT, P. R. O. (1966) 'The comparative development of two species of *Platychelipus* Brady (Harpacticoida)', *Some Contemporary Studies in Marine Science*, pp. 113–27, London: Allen and Unwin.

BASSINDALE, R. (1938) 'The intertidal fauna of the Mersey Estuary', *J. mar. biol Ass. U.K.*, 23: 83–98.

BASSINDALE, R. (1942) 'The distribution of amphipods in the Severn Estuary and Bristol Channel', *J. Anim. Ecol.*, 11: 131–44.

BASSINDALE, R. (1943) 'A comparison of the varying salinity conditions of the Tees and Severn estuaries', *J. Anim. Ecol.*, 12: 1–10.

BAYLY, I. A. E. (1961) 'A revision of the inland water genus *Calamoecia* (Copepoda: Calanoida)', *Austrl. J. mar. freshw. Res.*, 12: 54–91.

BAYLY, I. A. E. (1962) 'Additions to the inland water genus *Calamoecia* (Copepoda: Calanoida)', *Austral. J. mar. freshw. Res.*, 13: 252–64.

BAYLY, I. A. E. (1963) 'A revision of the coastal water genus *Gladioferens* (Copepoda: Calanoida)', *Austral. J. mar. freshw. Res.*, 14: 194–217.

BAYLY, A. I. E. (1964) 'A new species of *Isias* (Copepoda: Calanoida) from the Brisbane River Estuary and a comparison of the Australasian centropagid genera', *Austral. J. mar. freshw. Res.*, 15: 239–47.

BAYLY, A. I. E. (1965) 'Ecological studies on the planktonic Copepoda of the Brisbane River Estuary with special reference to *Gladioferens pectinatus* (Brady) (Calanoida)', *Austral. J. mar. freshw. Res.*, 16: 315–50.

BAYNE, B. L. (1965) 'Growth and the delay of metamorphosis of the larvae of *Mytilus edulis*', *Ophelia*, 2: 1–47.

BEADLE, L. C. (1931) 'The effect of salinity changes on the water content and respiration of marine invertebrates', *J. exp. Biol.*, 8: 211–27.

BEADLE, L. C. (1934) 'Osmotic regulation in *Gunda ulvae*', *J. exp. Biol.*, 11, 382–96.

BEADLE, L. C. (1937) 'Adaptation to changes of salinity in polychaetes 1. Control of body volume and body fluid concentration in *Nereis diversicolor*', *J. exp. Biol.*, 14: 56–70.

BEADLE, L. C. (1939) 'Regulation of the haemolymph in the saline water mosquito larva *Aedes detritus*', *J. exp. Biol.*, 16: 346–62.

BEANLAND, F. (1940) 'Sand and mud communities of the Dovey Estuary', *J. mar. biol. Ass. U.K.*, 24, 589–611.

BECKER, P. (1958) 'Some parasites and predators of biting midges, *Culicoides* Latreille (Dipt., Ceratopogonidae)', *Ent. mon. Mag.*, 94: 186–9.

BECKER, P. (1961) 'Observations on the life cycle and immature stages of *Culicoides circumscriptus* Kieff. (Diptera, Ceratopogonidae)', *Proc. Roy. Soc. Edin.*, 67: 363–86.

BERTIN, L. (1925) 'Recherches bionomiques, biochimiques et systematiques sur les Épinoches', *Ann. Inst. Oceanogr, Monaco*, 2: 1–204.

BEYER, A. (1939) 'Morphologische, okologische und physiologische Studien an den Larven der Fliegen *Ephydra riparia, E. micans* und *Caenia fumosa*', *Kieler Meeresforsch*, 3: 265–320.

BHATNAGER, K. M. and CRISP, D. J. (1965) 'The salinity tolerance of nauplius larvae of cirripedes', *J. Anim. Ecol.*, 34: 419–28.

BLEGVAD, H. (1932) 'On the flounder (*Pleuronectes flesus*) and the Danish flounder fishing in the Baltic', *Cons. int. Exp. Mer. Rapp. Proc. verb.*, 78: 1–28.

BLOCK, J. W. DE, and GEELEN, H. J. (1958) 'The substratum required for the settling of mussels (*Mytilus edulis*)', *Arch. néerl. Zool.*, 13 (suppl. 1): 446–60.

BOCQUET, C. (1953) 'Recherches sur le polymorphisme natural des *Jaera marina* (Fabr.) (Isopodes asellotes)', *Archs Zool. exp. gen.*, 90: 187–450.

BOLSTER, G. C. (1954) 'The biology and dispersal of *Mytilicola intestinalis* Steuer. A copepod parasite of mussels', *Fish. Invest. Lond.*, (2) 18 (6): 1–30.

BOOKHOUT, C. G. and HORN, E. C. (1949) 'The development of *Axiothella mucosa* (Andrew)', *J. Morph.*, 84: 145–83.

BORDEN, M. A. (1931) 'A study of the respiration and of the function of haemoglobin in *Planorbis corneus* and *Arenicola marina*', *J. mar. biol. Ass. U.K.*, 17: 709–38.

BOUCHER-FIRLEY, S. (1935) 'Recherches biochimiques sur les Teleostéens Apodes', *Ann. Inst. Oceanogr. Monaco*, 15: 1–327.

BOWMAN, T. E. (1965) '*Cyathura specus*, a new cave isopod from Cuba', *Stud. Faun. Curacao Caribb.*, 22: 88–97.

BRADSHAW, J. S. (1955) 'Preliminary laboratory experiments on the ecology of foraminiferal populations', *Micropalaeontology*, 1: 351–8.

BRADSHAW, J. S. (1957) 'Laboratory studies on the rate of growth of the foraminifer *Streblus beccarii* (Linné) var *tepida* (Cushman)', *J. Palaeont.*, 31: 1138–47.

BRADSHAW, J. S. (1961) 'Laboratory experiments on the ecology of Foraminifera', *Contr. Cushman Fedn.*, 12: 87–106.

BRADY, G. S. and ROBERTSON, D. (1870) 'Ostracoda and Foraminifera of tidal rivers', *Ann. Mag. nat. Hist.* (4) 6: 1–33. 'Foraminifera', by H. B. Brady, pp. 273–309.

BRAFIELD, A. E. (1964) 'The oxygen content of interstitial water in sandy shores', *J. Anim. Ecol.*, 33: 97–116.

BRAFIELD, A. E. and NEWELL, G. E. (1961) 'The behaviour of *Macoma balthica*', *J. mar. biol. Ass. U.K.*, 41: 81–7.

BRATTSTROM, H. (1954) 'Notes on *Victorella pavida* Kent', *Lunds Univ. Arssk.*, N.F. 50 (9): 1–29.

BREGNBALLE, F. (1961) 'Plaice and flounder as consumers of the microscopic bottom fauna', *Medd. Danm. Fisk. Havund. Kbh.*, 3: 133–82.

BRESCIANI, J. (1960) 'Some features of the larval development of *Stenhelia (Delavalia) palustris* Brady, 1868 (Copepoda Harpacticoida)', *Vidensk. Medd. Dansk. naturh. Foren.*, 123: 237–47.

BRICTEUX-GRÉGOIRE, S., DUCHÂTEAU-BOSSON, G., JEUNIAUX, C. and FLORKIN, M. (1962) Constituants osmotiquement actifs des muscles du crabe chinois *Eriocheir sinensis* adapté a l'eau douce ou a l'eau de la mer', *Arch. int. Physiol. Biochim.*, 70: 273–86.

BRINKHURST, R. O. (1963) 'Notes on the brackish-water and marine species of Tubificidae (Annelida, Oligochaeta)', *J. mar. biol. Ass. U.K.*, 43: 709–15.

BRINKHURST, R. O. (1964) 'Observations on the biology of the marine oligochaete *Tubifex costatus*', *J. mar. biol. Ass. U.K.*, 44: 11–16.

BRISTOWE, W. S. (1923) 'A British semi-marine spider', *Ann. Mag. nat. Hist.* (9) 12: 154–6.

BRISTOWE, W. S. (1958) *The world of spiders*, London: Collins, 304 pp.

BROEKHUYSEN, G. J. (1935) 'The extremes in the percentage of dissolved oxygen to which the fauna of a *Zostera* field in the tide zone at Niewdiep can be exposed', *Arch. néerl. Zool.*, 1: 339–46.

BROEKHUYSEN, G. J. (1936) 'On the development, growth and distribution of *Carcinides maenas* (L.)', *Arch. néerl. Zool.* 2: 257–399.

BRO-LARSEN, E. (1936) 'Biologische Studien über die tunnelgrabenden Käfer auf Skallingen', *Vidensk. Medd. Dansk. naturh. Foren.*, 100: 1–231.

BRO-LARSEN, E. (1952) 'On sub social beetles from the salt marsh, their care of progeny and adaptation to salt and tide', *Trans. 9th Int. Cong. Ent.*, 1: 502–6.

BROWN, E. S. (1948) 'The ecology of Saldidae (Hemiptera-Heteroptera) inhabiting a salt marsh, with observations on the evolution of aquatic habits in insects', *J. Anim. Ecol.*, 17: 180–8.

BRUCE, J. R. (1928) 'Physical factors on the sandy beach', *J. mar. biol. Ass. U.K.*, 15: 535–65.

BRYAN, G. W. (1961) 'The accumulation of radioactive caesium in crabs', *J. mar. biol. Ass. U.K.*, 41: 551–76.

BULL, H. O. (1931) 'Resistance of *Eurytemora hirundoides* Nordquist, a brackish water copepod to oxygen depletion', *Nature, Lond.*, 127: 406.

BURBANCK, W. D. (1959) 'The distribution of the estuarine isopod *Cyathura* sp. along the eastern coast of the United States', *Ecology* 40: 507–11.

BURBANCK, W. D. (1962) 'An ecological study of the distribution of the isopod *Cyathura polita* (Stimpson) from the brackish waters of Cape Cod, Massachusetts', *Amer. Mid. Nat.* 67: 449–76.

BURCKHARDT, G. (1913) 'Zooplankton aus ost und sud-asiatischen Binnengewässern', *Zool. Jahrb. (syst.)* 34: 341–472.

BUTLER, P. M. and POPHAM, E. J. (1958) 'The effects of the floods of 1953 on the aquatic insect fauna of Spurn (Yorkshire)', *Proc. Roy. Ent. Soc. Lond.*, A. 33: 149–58.

CABLE, R. M. and HUNNINEN, A. V. (1942) 'Studies on *Deropristis inflata* (Molin), its life history and affinities to trematodes of the family Acanthocolpidae', *Biol. Bull. mar. biol. Lab. Woods Hole*, 82: 292–312.

CALDWELL, J. M. (1955) 'Tidal currents at inlets in the United States', *Proc. Amer. Soc. civil Engin.*, 81 (716): 1–12.

CANNON, H. G. and MANTON, S. M. (1927) 'On the feeding mechanism of a mysid crustacean, *Hemimysis lamornae*', *Trans. Roy. Soc. Edin.*, 55: 219–53.

CAPSTICK, C. K. (1957) 'The salinity characteristics of the middle and upper reaches of the River Blyth Estuary', *J. Anim. Ecol.*, 26: 295–315.

CAPSTICK, C. K. (1959) 'The distribution of free-living nematodes in relation to salinity in the middle and upper reaches of the River Blyth Estuary', *J. Anim. Ecol.*, 28: 189–210.

CARLISLE, D. B. (1957) 'On the hormonal inhibition of moulting in decapod Crustacea II. The terminal anecydsis in crabs', *J. mar. biol. Ass. U.K.* 36: 291–307.

CARLISLE, D. B. and Denton, E. J. (1959) 'On the metamorphosis of the visual pigments of *Anguilla anguilla* (L.)', *J. mar. biol. Ass. U.K.*, 38: 97–102.

CARPENTER, S. J., MIDDLEKAUFF, W. W. and CHAMBERLIN, R. W. (1946) 'The mosquitoes of the southern United States east of Oklahoma and Texas', *Amer. Midl. Nat. Monogr.*, 3: 1–292.

CARTER, N. (1932–3) 'A comparative study of the algal flora of two salt marshes', *J. Ecol.*, 20: 341–70; 21: 128–208; 385–403.

CAWKELL, E. M. (1964) 'The utilisation of mangroves by African birds', *Ibis* 106: 251–3.

CHAPMAN, G. (1949) 'The thixotrophy and dilatancy of a marine soil', *J. mar. biol. Ass. U.K.*, 28: 123–40.

CHAPMAN, G. and NEWELL, G. E. (1956) 'The role of the body fluid in the movement of soft-bodied invertebrates II. The extension of the siphons of *Mya arenaria* L. and *Scrobicularia plana* (da Costa)', *Proc. roy. Soc. B.*, 145: 564–80.

CHAPMAN, V. J. (1960) *Salt marshes and salt deserts of the World*, London: Hill, 392 pp.

CHATER, E. H. and JONES, H. (1957) 'Some observations on *Spartina townsendii*, H. and J. Groves in the Dovey Estuary', *J. Ecol.*, 45: 157–67.

CHATTON, E. and LWOFF, A. (1935) 'Les Ciliés apostomes: morphologie, cytologie, ethologie, evolution, systematique. I. Aperçu historique et général. Etude mongraphique des genres et des espéces', *Arch. Zool. exp. gen.*, 77: 1–453.

CHIPPERFIELD, P. N. J. (1953) 'Observations on the breeding and settlement of *Mytilus edulis* (L.) in British waters', *J. mar. biol. Ass. U.K.*, 32: 449–76.

CLARK, R. B. and CLARK, M. E. (1960) 'The ligamentary system and the segmental musculature of *Nephtys*', *Quart. J. micr. Sci.*, 101: 149–176.

CLAUS, A. (1937) 'Vergleichend-physiologische Untersuchungen zur Okologie der Waserwanzen, mit besonderer Berücksichtigung der Brackwasserwanze *Sigara lugubris* Fieb', *Zool. Jahrb.*, 58: 364–432.

COE, W. R. (1943) 'Biology of the nemerteans of the Atlantic Coast of North America', *Trans. Connect. Acad. Arts Sci.*, 35: 129–328.

COMFORT, A. (1957) 'The duration of life in molluscs', *Proc. malacol. Soc. Lond.*, 32: 219–41.

COOPER, L. N. H. and MILNE, A. (1939) 'The ecology of the Tamar Estuary V. Under-water illumination. Revision of data for red light.' *J. mar. biol. Ass. U.K.*, 23: 391–6.

COSTLOW, J. D. and BOOKHOUT, C. G. (1957) 'Larval development of *Balanus eburneus* in the laboratory', *Biol. Bull. mar. biol. Lab. Woods Hole*, 112: 313–24.

COSTLOW, J. D. and BOOKHOUT, C. G. (1959) 'The larval development of *Callinectes sapidus* Rathbun reared in the laboratory', *Biol. Bull. mar. biol. Lab. Woods Hole*, 116: 373–96.

CRAIG-BENNET, A. (1931) 'The reproductive cycle of the three spined stickleback, *Gasterosteus aculeatus* Linn', *Phil. Trans.*, B 219: 197–279.

CRAWFORD G. I. (1937) 'A review of the amphipod genus *Corophium*, with notes on the British species', *J. mar. biol. Ass. U.K.*, 21: 589–630.

CRISP, D. J. and MEADOWS, P. S. (1962) 'The chemical basis of gregariousness in cirripedes', *Proc. Roy. Soc. Lond.*, B 156: 500–20.

CRISP, D. J. (1958) 'The spread of *Elminius modestus* Darwin in north west Europe', *J. mar. biol. Ass. U.K.*, 37: 483–520.

CRONIN, L., E. DAIBER, J. C. and HULBERT, E. M. (1962) 'Quantitative seasonal aspects of zooplankton in the Delaware River Estuary', *Chesapeake Sci.*, 3: 63–93.

CROWELL, S. and DARNELL, R. M. (1955) 'Occurrence and ecology of the hydroid *Bimeria franciscana* in Lake Pontchartrain, Louisiana', *Ecology*, 36: 516–18.

DAGET, J. (1954) 'Note sur le régime alimentaire du Céryle pie', *Notes Afr.*, 62: 55–7.

DAHL, I. (1960) 'The oligochaetes of 3 Danish brackish water areas. (Taxonomic and biological observations)', *Medd. Danm. Fisk. Havund.*, N.S. 2 (26) 1–20.

DALES, R. P. (1950) 'The reproduction and larval development of *Nereis diversicolor*, O. F. Muller', *J. mar. biol. Ass. U.K.*, 29: 321–60.

DALES, R. P. (1958) 'Survival of anaerobic periods by two intertidal polychaetes, *Arenicola marina* (L.) and *Owenia fusiformis* Delle Chiaje', *J. mar. biol. Ass. U.K.*, 37: 521–9.

DARNELL, R. M. (1959) 'Studies on the life history of the blue crab (*Callinectes sapidus* Rathbun) in Louisiana waters', *Trans. Amer. fish. Soc.*, 88: 294–304.

DARNELL, R. M. (1961) 'Trophic spectrum of an estuarine community based on studies of Lake Pontchartrain, Louisiana', *Ecology* 42: 553–68.

DAVIES, M. J. (1953) 'The crop contents of some British carabid beetles', *Ent. mon. Mag.*, 89: 18–23.

DAVIES, J. H. (1940) 'The ecology and geological role of mangroves in Florida', *Publ. Carnegie Inst. Wash.*, 517: 303–412.

DAVIS, L. V. and GRAY, I. E. (1966) 'Zonal and seasonal distribution of insects in North Carolina salt marshes', *Ecol. monogr.*, 36: 275–95.

DAY, J. H. (1951) 'The ecology of South African estuaries. 1. A review of estuarine conditions in general', *Trans. Roy. Soc. S. Afr.*, 33: 53–91.

DAY, J. H., MILLARD, N. A. H. and HARRISON, A. D. (1952) 'The ecology of South African estuaries III. Knysna, a clear open estuary'. *Trans. Roy. Soc. S. Afr.*, 33: 367–414.

DAY, J. H. MILLARD, N. A. H. and BROEKHUYSEN, G. J. (1953) 'The ecology of South African estuaries Part IV: The St Lucia system', *Trans. Roy. Soc. S. Afr.*, 34: 129–56.

DICKINSON, C. H. (1965) 'The mycoflora associated with *Halimione portulacoides*. III. Fungi on green and moribund leaves', *Trans. Brit. mycol. Soc.*, 48: 603–10.

DICKINSON, C. H. and PUGH, G. J. F. (1965) 'The mycoflora associated with *Halimione portulacoides* 1. The establishment of the root surface flora of mature plants', *Trans. Brit. mycol. Soc.*, 48: 381–90.

DOGIEL, V. A., PETRUSHEVSKI, G. K. and POLYANSKI, Y. I. (1961) *Parasitology of fishes*, Edinburgh: Oliver and Boyd, 384 pp.

DRINNAN, R. E. (1957) 'The winter feeding of the oyster catcher (*Haematopus ostralegus*) on the edible cockle (*Cardium edule*)', *J. Anim. Ecol.*, 26: 441–69.

DRINNAN, R. E. (1958) 'The winter feeding of the oystercatcher (*Haematopus ostralegus*) on the edible mussel (*Mytilus edulis*) in the Conway Estuary, North Wales', *Fish. Invest. Lond.* (II), 22 (4): 1–15.

DUCHÂTEAU-BOSSON, G., JEUNIAUX, C. and FLORKIN, M. (1961) 'Role de la variation de la composante amino-acide intracellulaire dans l'euryhalinité d'*Arenicola marina* L', *Arch. int. Physiol. Biochim.*, 69: 30–5.

DUNCAN, A. (1960) 'The spawning of Arenicola marina (L.) in the British Isles', *Proc. zool. Soc. Lond.*, 134: 137–56.

DU PLESSIS, S. S. (1957) 'Growth and daily food intake of the white-breasted cormorant in captivity', *Ostrich* 28: 197–201.

DUVAL, M. (1925) 'Recherches physico-chemique et physiologiques sur le milieu interieur des animaux aquatiques. Modifications sous l'influence du milieu exterieur', *Ann. Inst. Oceanogr.*, 2: 232–407.

DYTE, C. E. (1959) 'Some interesting habitats of larval Dolichopodidae (Diptera)', *Ent. mon. Mag.*, 95: 139–43.

EDNEY, E. B. (1957) *The water relations of terrestrial arthropods*, Cambridge: University Press, 109 pp.

EKMAN, S. (1919) 'Studien uber die marinen Relikte der nord-europäischen Binnengewässer. VI Die morphologischen Folgen des Relickwerdens', *Int. Rev. Hydrobiol.*, 8: 477–528.

ELLIS, W. G. (1937) 'The water and electrolyte exchange of *Nereis diversicolor* (Müller)', *J. exp. Biol.*, 14: 340–50.

FALLA, R. A., SIBSON, R. B. and TURBOTT, E. G. (1966) *A field guide to the birds of New Zealand*, London: Collins.

FALLA, R. A. and STOKELL, G. (1945) 'Investigations of the stomach contents of the New Zealand freshwater shags', *Trans. Roy. Soc. N.Z.*, 74: 320–31.

FAURÉ-FREMIET, E. (1950) 'Écologie des Ciliés psammophiles littoraux', *Bull. biol.*, 84: 35–75.

FAURÉ-FREMIET, E. (1950) 'The marine sand-dwelling ciliates of Cape Cod shores', *Biol. Bull. mar. biol. Lab. Woods Hole*, 99: 349–50.

FAURÉ-FREMIET, E. (1951) 'The marine sand-swelling ciliates of Cape Cod', *Biol. Bull. mar. biol. Lab. Woods Hole*, 100: 59–70.

FENCHEL, T. (1965a) 'Ciliates from Scandinavian molluscs. *Ophelia*', 2: 71–174.

FENCHEL, T. (1965b) 'On the ciliate fauna associated with the marine species of the amphipod genus *Gammarus* J. G. Fabricius', *Ophelia*, 2: 281–303.

FENCHEL, T. and JANSSON, B. O. (1966) 'On the vertical distribution of the microfauna in the sediments of a brackish-water beach', *Ophelia*, 3: 161–77.

FFRENCH, R. P. (1966) 'The utilisation of mangroves by birds in Trinidad', *Ibis*, 108: 423–4.

FLORKIN, M. and SCHOFFENIELS, E. (1965) 'Euryhalinity and the concept of physiological radiation', in *Studies in Comparative Biochemistry*, Oxford: Pergammon Press.

FOX, H. M. (1952) 'Anal and oral intake of water by Crustacea', *J. exp. Biol.*, 29: 583–99.

FRANKENBERG, G. D. and BURBANCK, W. D. (1963) 'A comparison of the physiology and ecology of the estuarine isopod *Cyathura polita* in Massachusetts and Georgia', *Biol. Bull. mar. biol. Lab. Woods Hole*, 125: 81–95.

FRANZINI, J. B. (1951) 'Porosity factor for case of laminar flow through granular media', *Amer. geophys. Union*, 32: 443–6.

FRASER, J. H. (1932) 'Observations on the fauna and constituents of an estuarine mud in a polluted area', *J. mar. biol. Ass. U.K.*, 18: 69–85.

FRASER, J. H. (1936) 'The occurrence, ecology and life history of *Tigriopus fulvus* (Fischer)', *J. mar. biol. Ass. U.K.*, 20: 523–6.

FRASER, J. H. (1938) 'The fauna of fixed and floating structures in the Mersey Estuary and Liverpool Bay', *Proc. Trans. Liverp. biol. Soc.*, 51: 1–21.

FRAZER, H. J. (1935) 'Experimental study of porosity and permeability of clastic sediments', *J. Geol.*, 43: 910–1010.

FREEMAN, R. F. H. (1957) '*Paravortex scrobiculariae* (Graff) in Great Britain', *Nature, Lond.*, 180: 1213–4.

FREEMAN, R. F. H. (1963) 'Experimental infection of two species of wrasse with the digenean *Proctoeces subtenuis*', *J. mar. biol. Ass. U.K.*, 43: 113–23.

FREEMAN, R. F. H. and LLEWELLYN, J. (1958) 'An adult digenetic trematode from an invertebrate host: *Proctoeces subtenuis* (Linton) from the lamellibranch *Scrobicularia plana* (Da Costa)', *J. mar. biol. Ass. U.K.*, 37: 435–57.

FREEMAN, R. F. H. and RIGLER, (1957) 'The responses of *Scrobicularia plana* (Da Costa) to osmotic pressure changes', *J. mar. biol. Ass. U.K.*, 36: 553–67.

FRETTER, V. (1955) 'Uptake of radioactive sodium (^{24}Na) by *Nereis diversicolor* (Müller) and *Perinereis cultrifera* (Grube)', *J. mar. biol. Ass. U.K.*, 34: 151–60.

GAITHER, A. (1953) 'A study of porosity and grain relationships in experimental sands', *J. sediment. Petrol.*, 23: 180–95.

GALLIEN, L. (1929) 'Étude de deux mollusques opisthobranches d'eau saumâtre', *Bull. Soc. Linn. Normand. Caen*, (8), 1: 162–89.

GALLIFORD, A. L. (1956) 'Notes on the ecology of pools in the salt marshes of the Dee Estuary', *Proc. Liverp. Fld Nat.Cl.*, 1955: 15–19.

GAMESON, A. L. H. (1959) 'Some aspects of the carbon, nitrogen and sulphur cycles in the Thames Estuary. II. Influence on the oxygen balance, pp. 51–9; in 'The effects of pollution on living material', *8th Symp. Inst. Biol. Lond.*

GASCOIGNE, T. (1956) 'Feeding and reproduction in the Limapontiidae', *Trans. Roy. Soc. Edin.*, 63: 129–51.

GEE, J. M. (1961) 'Ecological studies in South Benfleet Creek, with special reference to the amphipod genus *Corophium*'. *Essex Nat.*, 1961: 291–309.

GIBB, D. C. (1957) 'The free living forms of *Ascophyllum nodosum* (L.) Le Jol', *J. Ecol.*, 45: 49–83.

GILBERT, A. B. (1959) 'The composition of the blood of the shore crab, *Carcinus maenas* Pennant, in relation to sex and body size. II. Blood chloride and sulphate', *J. exp. Biol.*, 36: 356–62.

GOLDACRE, R. J. (1949) 'Surface films on natural bodies of water', *J. Anim. Ecol.*, 18: 36–9.

GOMPEL, M. and LEGENDRE, R. (1928) 'Limites de temperature et de salure supportés par *Convoluta roscoffensis*', *C. r. Seanc. Soc. biol. Paris*, 98: 572–3.

GOODHART, C. B. (1941) 'The ecology of the Amphipoda in a small estuary in Hampshire', *J. Anim. Ecol.*, 10: 306–22.

GRAINGER, J. N. R. (1951) 'Notes on the biology of the copepod *Mytilicola intestinalis* Steuer', *Parasitology*, 41: 135–42.

GRATON, L. C. and FRAZER, H. J. (1935) 'Systematic packing of spheres with particular relation to porosity and permeability', *J. Geol.*, 43: 785–909.

GRAY, I. E. (1957) 'A comparative study of the gill area of crabs', *Biol. Bull. mar. biol, Lab. Woods Hole*, 112: 34–42.

GREEN, J. (1954) 'The food, predators and a parasite of *Bembidion laterale* (Samouelle) (Col, Carabidae)', *Ent. mon. Mag.*, 90: 226–7.

GREEN, J. (1956) 'The mouthparts of *Eurynebria complanata* (L.) and *Bembidion laterale* (Sam.) (Col., Carabidae)', *Ent. mon. Mag.*, 92: 110–3.

GREEN, J. (1957a) 'The growth of *Scrobicularia plana* (Da Costa) in the Gwendraeth Estuary', *J. mar. biol. Ass. U.K.*, 36: 41–7.

GREEN, J. (1957b) 'The feeding mechanism of *Mesidotea entomon* (Linn.) (Crustacea: Isopoda)', *Proc. zool. Soc. Lond.*, 129: 245–54.

GREEN, J. (1957c) 'Osmoregulatory function of the contractile vesicle of *Asplanchna*', *Nature, Lond.*, 179: 432.

GREEN, J. (1959) 'Pigmentation of an ostracod, *Heterocypris incongruens*', *J. exp. Biol.*, 36: 575–82.

GREEN, J. (1967) 'Activities of the siphons of *Scrobicularia plana* (Da Costa)', *Proc. malacol. Soc. Lond.*, 37: 339–41.

GRINDLEY, J. R. (1964) 'Effect of low-salinity water on the vertical migration of estuarine plankton', *Nature, Lond.*, 203: 781–2.

GROSS, A. O. (1950) 'The herring gull–cormorant control project', *Proc. Xth Int. Congr. Ornithol. Uppsala*, pp. 532–6.

GUNTER, G. (1956) 'Some relations of faunal distributions to salinity in estuarine waters', *Ecology*, 37: 616–9.

GUNTER, G. (1961) 'Some relations of estuarine organisms to salinity', *Limnol. Oceanogr.*, 6: 182–90.

GURNEY, R. (1924) 'The larval development of some British prawns (Palaemonidae) 1. *Palaemonetes varians*', *Proc. zool. Soc. Lond.*, 1924: 297–328.

GURNEY, R. (1931) *British freshwater Copepoda*, Vol I, 238 pp., London: Ray Soc.

GURNEY, R. (1933) *British freshwater Copepoda*, Vol III, 384 pp., London: Ray Soc.

HAAHTELA, I. and NAYLOR, E. (1965) '*Jaera hopeana*, an intertidal isopod new to the British fauna', *J. mar. biol. Ass. U.K.*, 45: 367–71.

HALL, D. N. F. (1962) 'Observations on the taxonomy and biology of some Indo-West-Pacific Penaeidae (Crustacea: Decapoda)', *Colonl. Fish. Publ. Lond.*, 17: 1–229.

HANCOCK, D. A. (1960) 'The ecology of the molluscan enemies of the edible mollusc', *Proc. malacol. Soc. Lond.*, 34: 123–43.

HAND, C. and STEINBERG, J. (1955) 'On the occurrence of the nudibranch *Alderia modesta* (Lovén, 1844) on the central Californian coast', *Nautilus*, 69: 22–8.

HARDISTY, M. W. (1956) 'Some aspects of osmotic regulation in lampreys', *J. exp. Biol.*, 33: 431–47.

HARDISTY, M. W. (1961) 'The growth of larval lamprey', *J. Anim. Ecol.*, 30: 357–71.

HARLEY, M. B. (1950) 'Occurrence of a filter feeding mechanism in the polychaete *Nereis diversicolor*', *Nature, Lond.*, 165: 734–5.

HARTLEY, P. H. T. (1940) 'The Saltash tuck-net fishery and the ecology of some estuarine fishes', *J. mar. biol. Ass. U.K.*, 24: 1–68.

HARTOG, C. DEN (1958) 'Nieuwe gegevens over de Kwelderroodwieren van Terschelling', *Levend, Natuur*, 61 (10): 231–5.

HARVEY, H. W. (1960) *The chemistry and fertility of sea waters*, Cambridge.

HAVERSCHMIDT, F. (1965) 'The utilisation of mangroves by South American birds', *Ibis*, 107: 540–2.

HAVINGA, B. (1929) 'Krebse und Weichtiere', *Handbuch der Seefischerie Nordeuropas*, 3: 101–2.

HECHT, F. (1932) 'Der chemische Einfluss organischer Zersetzungsstoffe auf das Benthos, dargelegt an Untersuchungen mit marinen Polychaeten, insbesondere *Arenicola marina* L.', *Senkenbergiana* 14: 199–220.

HEDGEPETH, J. W. (1947) 'The Laguna Madre of Texas', *Trans. 12th North Amer. Wildl. Conf.*, pp. 364–80.

HENSCHEL, L. (1936) 'Wasserhaushalt und osmoregulation von Scholle und Flunder', *Wiss. Meeresunters*, 22: 89–121.

HEPPER, B. T. (1955) 'Environmental factors governing the infection of mussels, *Mytilus edulis*, by *Mytilicola intestinalis*', *Fish. Invest. Lond.*, (2) 20 (3): 1–21.

HERLANT-MEEWIS, H. (1958) 'La reproduction asexuée chez les Annélides', *Ann. Biol.*, 34: 133–66.

HEUTS, M. J. (1947) 'Experimental studies on adaptive evolution in *Gasterosteus aculeatus*', *Evolution*, 1: 89–102.

HJULSTRÖM, F. (1935) 'Studies of the morphological activity of rivers as illustrated by the River Fyris', *Bull. geol. Inst. Uppsala*, 25: 221–527.

HOCKLEY, A. R. (1951) 'On the biology of *Mytilicola intestinalis* (Steuer)', *J. mar. biol. Ass. U.K.*, 30: 223–32.

HODGE, D. (1963) 'The distribution and ecology of the mysids in the Brisbane River', *Univ. Queensl. Pap. Zool.*, 2: 91–104.

HODGE, D. (1964) 'A redescription of *Tenagomysis chiltoni* (Crustacea: Mysidacea) from a freshwater coastal lake in New Zealand', *New Zeal. J. Sci.*, 7: 387–95.

HOLLIDAY, F. G. T. and BLAXTER, J. H. S. (1960) 'The effects of salinity on the developing eggs and larvae of the herring', *J. mar. biol. Ass. U.K.*, 39: 591–603.

HOLLIDAY, F. G. T. and BLAXTER, J. H. S. (1961) 'The effects of salinity on herring after metamorphosis', *J. mar. biol. Ass. U.K.*, 41: 37–48.

HOLME, N. A. (1949) 'The fauna of sand and mud banks near the mouth of the Exe Estuary', *J. mar. biol. Ass. U.K.*, 28: 189–237.

HOLME, N. A. (1950) 'Population dispersion in *Tellina tenuis* Da Costa', *J. mar. biol. Ass. U.K.*, 29: 267–80.

HOLME, N. A. (1961) 'Notes on the mode of life of the Tellinidae (Lamellibranchia)', *J. mar. biol. Ass. U.K.*, 41: 699–703.

HOLMES, A. (1944) *Principles of physical geology*, London: Nelson.

HOLMES, R. M. (1961) 'Kidney function in migrating salmonids', *Rep. Challenger Soc.*, 3: 23.

HOLTHUIS, L. B. (1950) 'The Decapoda of the Siboga Expedition. Part X. The Palaemonidae. Subfamily Palaemoninae', *Siboga Expedit.*, 39a: 1–268.

HOPKINS, J. T. (1963) 'A study of the diatoms of the Ouse estuary, Sussex. 1. The movement of the mud flat diatoms in response to some chemical and physical changes', *J. mar. biol. Ass. U.K.*, 43: 653–63.

HOPKINS, J. T. (1964a) 'A study of the diatoms of the Ouse estuary, Sussex. II. The ecology of the mud flat diatom flora', *J. mar. biol. Ass. U.K.*, 44: 333–41.

HOPKINS, J. T. (1964b) 'A study of the diatoms of the Ouse estuary, Sussex. III. The seasonal variation in the littoral epiphyte flora and the shore plankton', *J. mar. biol. Ass. U.K.*, 44: 613–44.

HORA, S. L. (1934) 'Wanderings of the Bombay Duck, *Harpodon nehereus* (Ham. Buck.) in Indian waters', *J. Bombay nat. Hist. Soc.*, 37: 640–54.

HOWES, N. H. (1939) 'The ecology of a saline lagoon in south east Essex', *J. Linn. Soc. Zool.*, 40: 383–445.

HOWIE, D. I. D. (1959) 'The spawning of *Arenicola marina* (L.) 1. The breeding season', *J. mar. biol. Ass. U.K.*, 38: 395–406.

HUNNINEN, A. V. and CABLE, R. M. (1943) 'The life history of *Podocotyle atomon* (Rudolphi) (Trematoda: Opecoelidae)', *Trans. Amer. micr. Soc.*, 62: 57–68.

HUNTER, W. R. and HUNTER, M. R. (1962) 'On a population of *Hydrobia ulvae* in the Clyde estuary', *Glasg. Nat.*, 18: 198–205.

HUSTEDT, F. (1939) 'Die Diatomeenflora des Kustengebietes der Nordsee vom Dollart biz zur Elbemundung. 1. Die Diatomeenflora in den Sedimenten der Leybucht, des Memmert und bei der Insel Juist', *Abh. nat. Ver. Bremen*, 31: 572–677.

HUTTON, R. F. (1952) 'Studies on the parasites of *Cardium edule* L. *Cercaria fulbrighti* n. sp., a *Gymnophallus* larva with a forked tail', *J. mar. biol. Ass. U.K.*, 31: 317–26.

HYNES, H. B. N. (1950) 'The food of freshwater sticklebacks (*Gasterosteus aculeatus* and *Pygosteus pungitius*) with a review of methods used in studies of the food of fishes', *J. Anim. Ecol.*, 19: 36–80.

HYNES, H. B. N. (1954) 'The ecology of *Gammarus duebeni* Lilljeborg and its occurrence in fresh water in Western Britain', *J. Anim. Ecol.*, 23: 38–84.

HYNES, H. B. N. (1955) 'The reproductive cycle of some British freshwater Gammaridae', *J. Anim. Ecol.*, 24: 352–87.

ICHIKAWA, A. and YANAGIMACHI, R. (1960) 'Studies on the sexual organisation of the Rhizocephala II. The reproductive function of the larval (cypris) males of *Peltogaster* and *Sacculina*', *Annot. Zool. Japon.*, 33: 42–56.

INGLIS, C. C. and ALLEN, F. H. (1957) 'The regimen of the Thames Estuary as affected by currents, salinities and river flow', *Proc. Inst. civil Engin.*, 7: 827–68.

INGLIS, C. C. and KESTNER, F. J. T. (1958) The long term effects of training walls, reclamation and dredging on estuaries. *Proc. Inst. civil Engin.* 9: 193–216.

INMAN, D. L. (1952) 'Measures for describing the size distribution of sediments', *J. sediment. Petrol.*, 22: 125–45.

JACOB, F. H. (1956) 'A new British species of *Lipaphis* Mordvilko 1928 (Homoptera: Aphididae)', *Proc. Roy. ent. Soc. Lond.*, B. 25: 85–9.

JACQUIN, A. (1956) 'Recherches biologiques sur *Ochthebius quadricollis* Mulsant (Coléoptère Hydrophilide)', *Bull. Soc. Hist. nat. Afr. nord*, 47: 270–90.

JEFFRIES, H. P. (1962) 'Succession of two *Acartia* species in estuaries', *Limnol. Oceanogr.*, 7: 354–64.

JENKIN, P. M. (1957) 'The filter-feeding and food of flamingoes (Phoenicopteri)', *Phil. Trans. Roy. Soc.*, B. 240: 401–93.

JENSEN, J. P. (1955) 'Biological observations on the isopod *Sphaeroma hookeri* Leach', *Vidensk. Medd. Dansk. Naturh. Foren.*, 117: 305–39.

JESPERSON, P. (1942) 'Indo-Pacific leptocephalids of the genus *Anguilla*. Systematic and biological studies', *Dana Rept.*

JEUNIAUX, C. and FLORKIN, M. (1961) 'Modification de l'excretion azotée du Crabe chinois au cours de l'adaptation osmotique', *Arch. int. physiol.*, 69: 385–6.

JOHNSON, D. S. (1965) 'A review of the brackish water prawns of Malaya', *Bull. nat. Mus. Singapore*, 33 (2): 7–11.

JOHNSTON, R. (1964) 'Recent advances in the estimation of salinity', *Oceanogr. mar. Biol. ann. Rev.*, 2: 97–120.

JONES, N. S. (1948) 'The ecology of the Amphipoda of the south of the Isle of Man', *J. mar. biol. Ass. U.K.*, 27: 400–39.

JONES, J. W. and HYNES, H. B. N. (1950) 'The age and growth of *Gasterosteus aculeatus, Pygosteus pungitius*, and *Spinachia vulgaris* as shown by their otoliths', *J. Anim. Ecol.*, 19: 59–75.

JONES, D. and WILLS, M. S. (1956) 'The attenuation of light in sea and estuarine waters in relation to the concentration of suspended solid matter', *J. mar. biol. Ass. U.K.*, 35: 431–44.

JØRGENSEN, C. B. (1946) 'Lamellibranch larvae, in Thorson, G. Reproduction and larval development of Danish marine bottom invertebrates with special reference to the planktonic larvae in the Sound (Øresund)', *Medd. Komm. Danm. Fisk. Havunders*, ser. Plankton 4: 277–311.

JØRGENSON, C. B. (1949) 'The rate of feeding by *Mytilus* in different kinds of suspension', *J. mar. biol. Ass. U.K.*, 28: 333–44.

JØRGENSEN, C. B. and DALES, R. P. (1957) 'The regulation of volume and osmotic regulation in some nereid polychaetes'. *Physiol. comp. Oecol.*, 4: 357–74.

JØRGENSEN, O. M. (1929) 'The plankton of the Tyne Estuary', *Proc. Univ. Durham Phil. Soc.*, 8: 41–54.

KALE, H. W. (1965) 'Ecology and bioenergetics of the long-billed marsh wren *Telmatodytes palustris griseus* (Brewster) in Georgia salt marshes', *Publ. Nuttall ornith. Cl.*, 5: 1–142.

KAWAMURA, T. (1966) 'Distribution of phytoplankton populations in Sandy Hook Bay and adjacent areas in relation to hydrographic conditions in June 1962', *Tech. Pap. Bur. sport Fish. Wildl. Wash.*, 1: 1–37.

KING, F. H. (1898) 'Principles and conditions of the movements of ground water. *U.S. geol. Surv. Ann. Rep*', 19: 208–18.

KINNE, O. (1952) 'Zur Biologie und Physiologie von *Gammarus duebeni* Lillj. III. Zahlenverhaltnis der Geschlechter und Geschlechtsbestimmung', *Kiel. Meeresforsh*, 9: 126–33.

KINNE, O. (1953) 'Zur Biologie und Physiologie von *Gammarus duebeni* Lillj. VII. Uber die Temperaturabhängigkeit der Geschlechtsbestimmung', *Biol. Zbl.*, 72: 260–70.

KINNE, O. (1954) 'Eidonomie, Anatomie und Lebenszyklus von *Sphaeroma hookeri* Leach (Isopoda)', *Kiel. Meeresforsh*, 10: 100–20.

KINNE, O. (1955) '*Neomysis vulgaris* Thompson eine autokölogisch-biologische Studie', *Biol. Zbl.*, 74: 160–202.

KINNE, O. (1956) 'Über den Einfluss des Salzgehaltes und der Temperatur auf Wachstum, Form und Vermehrung bei dem Hydroidpolypen *Cordylophora caspia* (Pallas), Thecata Clavidae', *Zool. Jahrb.* (*Physiol.*), 66: 565–638.

KINNE, O. (1957) 'Über den Einfluss von Temperatur und Salzgehalt auf die Kopfchenform des Brackwasserpolypen *Cordylophora*', *Verh. Deutsch. zool. Geselsch.*, 1956: 445–9.

KINNE, O. (1960) '*Gammarus salinus*—einige Daten über den Umwelteinfluss auf Wachstum, Hautungsfolge, Herzfrequenz und Eientwicklungsdauer', *Crustaceana*, 1: 208–17.

KINNE, O. (1961a) 'Growth, molting frequency, heart beat, number of eggs, and incubation time in *Gammarus zaddachi* exposed to different environments', *Crustaceana*, 2: 26–36.

KINNE, O. (1961b) 'Die Geschlechtsbestimmung des Flohkrebses *Gammarus duebeni* Lillj (Amphipoda) ist temperaturabhängig—eine Entegnung', *Crustaceana*, 3: 56–69.

KITCHING, J. A. (1938) 'Contractile vacuoles', *Biol. Rev.*, 13: 403–44.

KITCHING, J. A. (1952) 'Contractile vacuoles', *Symp. Soc. exp. Biol.*, 6: 145–65.

KITCHING, J. A. (1954) 'Osmoregulation and ionic regulation in animals without kidneys', *Symp. Soc. exp. Biol.*, 8: 63–75.

KLIE, W. (1938) 'Krebstiere oder Crustacea III: Ostracoda, Muschelkrebse', *Tierwelt Deutschl.*, 34: 1–230

KOCH, H. J., EVANS, J. and SCHICKS (1954) 'The active absorption of ions by the isolated gills of the crab *Eriocheir sinensis* (M. Edw.)', *Meded. vlaamsche Acad. Kl. Wet.*, 16: 3–16.

KOLBE, R. W. (1927) 'Zur Oekologie, Morphologie und Systematik der Brackwasserdiatomeen', *Pflanzenforschung*, 7: 1–46.

KORNICKER, L. S. and WISE, C. D. (1960) 'Some environmental boundaries of a marine ostracode', *Micropalaeontology*, 6: 393–8.

KORRINGA, P. (1951) 'Le *Mytilicola intestinalis* Steuer (Copepoda parasitica) menace l'industrie moulière en Zélande', *Rev. Trav. Off. Pêches marit.*, 17: 9–13.

KORRINGA, P. (1957) 'Water temperature and breeding throughout the geographical range of *Ostrea edulis*', *Ann. biol.*, 33: 109–16.

KRISTENSEN, I. (1957) 'Differences in density and growth in a cockle population in the Dutch Wadden Sea', *Arch. néerl. Zool.*, 12: 351–453.

KROGH, A. (1937) 'Osmotic regulation in freshwater fishes by active absorption of chloride ions', *Z. vergl. Physiol.*, 24: 656–66.

KROGH, A. (1939) *Osmotic regulation in aquatic animals*, Cambridge: University Press, 242 pp.

KROMHOUT, G. (1943) 'A comparison of the protonephridia of fresh-water, brackish-water and marine specimens of *Gyratrix hermaphroditus*', *J. Morphol.*, 72: 167–79.

KRÜGER, F. (1959) 'Zur Ernährungsphysiologie von *Arenicola marina* L.' *Zool. Anz.*, 22 suppl., Bd: 115–20.

KRUMBEIN, W. C. (1932) 'The mechanical analysis of fine-grained sediments', *J. sediment. Petrol.*, 2: 140–49.

KRUMBEIN, W. C. (1936) 'Application of logarithmic moments to size frequency distribution of sediments', *J. sediment. Petrol.*, 6: 35–47.

LAFON, M., DURCHON, M. and SAUDRAY, Y. (1955) 'Recherches sur les cycles saisonniers du plankton', *Ann. Inst. Oceanogr.* 31 (3): 126–230.

LAGERSPETZ, K. (1963) 'Humidity reactions of three aquatic amphipods, *Gammarus duebeni*, *G. oceanicus* and *Pontoporeia affinis* in the air', *J. exp. Biol.*, 40: 105–10.

LANCE, J. (1962) 'Effects of water of reduced salinity on the vertical migration of zooplankton', *J. mar. biol. Ass. U.K.*, 42: 131–54.

LANCE, J. (1963) 'The salinity tolerance of some estuarine planktonic copepods', *Limnol. Oceanogr.*, 8: 440–9.

LANG, K. (1948) *Monographie der Harpacticiden*, Lund: Ohlsson, 1,682 pp.

LANG, K. (1965) 'Copepoda Harpacticoidea from the Californian Pacific Coast', *Kgl. svenska Vetensk-Akad Handl.*, (4) 10: 1–560.

LARSEN, K. (1936) 'The distribution of the invertebrates of the Dybsø Fjord, their biology and their importance as fish food', *Rep. Dan. biol. Stat.*, 61: 3–36.

LEE, D. J. and WOODHILL, A. R. (1944) 'The anopheline mosquitoes of the Australasian region', *Publ. Univ. Sydney Dept. Zool. Monogr.*, 2: 1–209.

LEGRAND, J. J. and JUCHAULT, P. (1963) 'Mise en evidence d'hermaphrodisme protogynique fonctionnel chez l'Isopode Anthuridé *Cyathura carinata* (Kroyer) et etude du mechanisme d'inversion sexuelle', *C. r. Acad. Sci. Paris*, 256: 2931–3.

LINDBERG, H. (1944) 'Okologisch-geographische Untersuchungen zur Insektenfauna der Felsentümpel an den Küsten Finnlands', *Acta. zool. Fennica* 41: 1–178.

LINDBERG, H. (1948) 'Zur Kenntnis der Insektenfauna im Brackwasser des Baltischen Meeres', *Comm. biol. Soc. sci. Fenn.* 10 (9): 1–206.

LINDQUIST, A. (1959) 'Studien über das Zooplankton der Bottensee II. Zur Verbreitung und Zusammensetzung des Zooplanktons', *Rep. Inst. mar. Res. Lysekil* (Biol.), 11: 1–136.

LINKE, O. (1939) 'Die Biota des Jadebusenwattes', *Helgol. wiss. Meeresunters*, 1: 201–348.

LLOYD, A. J. and YONGE, C. M. (1947) 'The biology of *Crangon vulgaris* L. in the Bristol Channel and Severn Estuary', *J. mar. biol. Ass. U.K.*, 26: 626–61.

LOCKWOOD, A. P. M. (1959a) 'The osmotic and ionic regulation of *Asellus aquaticus* (L.)', *J. exp. Biol.*, 36: 546–55.

LOCKWOOD, A. P. M. (1959b) 'The regulation of the internal sodium concentration of *Asellus aquaticus* in the absence of sodium chloride in the medium', *J. exp. Biol.*, 36: 556–61.

LOCKWOOD, A. P. M. (1960) 'Some effects of temperature and concentration of the medium on the ionic regulation of the isopod *Asellus aquaticus* (L)', *J. exp. Biol.*, 37: 614–30.

LOCKWOOD, A. P. M. (1961) 'The urine of *Gammarus duebeni* and *G. pulex*', *J. exp. Biol.*, 38: 647–58.

LOCKWOOD, A. P. M. (1964) 'Activation of the sodium uptake system at high blood concentration in the amphipod, *Gammarus duebeni*', *J. exp. Biol.*, 41: 447–58.

LOCKWOOD, A. P. M. (1965) 'The relative losses of sodium in the urine and across the body surface in the amphipod *Gammarus duebeni*', *J. exp. Biol.*, 42: 59–69.

LOCKWOOD, A. P. M. and CROGHAN, P. C. (1957) 'The chloride regulation of the brackish and fresh-water races of *Mesidotea entomon* (L.)', *J. exp. Biol.*, 34: 253–8.

LOOSANOFF, V. L. (1942) 'Shell movements of the edible mussel *Mytilus edulis* (L.) in relation to temperature', *Ecology*, 23: 231–4.

LORD, W. B. (1867) *Crab, shrimp and lobster lore*, London, 122 pp.

LUBBERT, H. and EHRENBAUM, E. (1936) *Pleuronectes flesus*, Handbuch der Seefischerei Nordeuropas, Bd II: 216–9.

LUBET, P. (1957) 'Cycle sexual de *Mytilus edulis* L. et de *Mytilus galloprovincialis* Lmk. dans le basin d'Arcachon (Gironde)', *Ann. biol.*, 33: 19–29.

LUXTON, M. (1964) 'Some aspects of the biology of salt marsh Acarina', *Proc. Ist Int. Congr. Acarol.*, pp. 172–82.

LYSTER, I. H. J. (1965) 'The salinity tolerance of polychaete larvae', *J. Animal Ecol.*, 34: 517–27.

MACFADYEN, A. (1952) 'The small arthropods of a *Molinia* fen at Cothill', *J. Anim. Ecol.*, 21: 87–117.

MACGINITIE, G. E. (1935) 'Ecological aspects of a California marine estuary,' *Amer. Midl. Nat.*, 16: 629–765.

MACGINITIE, G. E. (1941) 'On the method of feeding of four pelecypods', *Biol. Bull. mar. biol. Lab. Woods Hole*, 80: 18–25.

MCMILLAN, N. F. (1948) 'Possible biological races in *Hydrobia ulvae* (Pennant) and their varying resistance to lowered salinity', *J. Conch.*, 23: 14–16.

MADSEN, F. J. and SPÄRCK, R. (1950) 'On the feeding habits of the southern cormorant (*Phalacrocorax carbo sinensis* Shaw) in Denmark', *Dan. Rev. Game Biol.*, 1: 45–76.

MARSHALL, J. F. (1938) *British Mosquitoes*, London: British Museum, 341 pp.

MARSHALL, N. and WHEELER, B. M. (1965) 'Role of the coastal and upper estuarine waters contributing phytoplankton to the shoals of the Niantic Estuary', *Ecology* 46: 665–73.

MATHIESEN, H. and NIELSEN, J. (1956) 'Botaniske underøgelser i Randers Fjord og Grund Fjord', *Bot. Tidsskr.*, 53: 1–34.

MATTHIESSEN, G. C. (1960) 'Observations on the ecology of the soft clam, *Mya arenaria*, in a salt pond', *Limnol. Oceanogr.*, 5: 291–300.

MEADOWS, P. S. (1964a) 'Experiments on substrate selection by *Corophium* species: films and bacteria on sand particles', *J. exp. Biol.*, 41: 499–511.

MEADOWS, P. S. (1964b) 'Experiments on substrate selection by *Corophium volutator* (Pallas): depth selection and population density', *J. exp. Biol.*, 41: 677–87.

MEYER, P. F. and MANN, H. (1951) 'Recherches allemandes relatives au "Mytilicola", copépode parasite de la moule existant dans les watten allemandes 1950–51', *Rev. Trav. Off. Pêches marit.*, 17: 63–74.

MILLS, D. H. (1965) 'The distribution and food of the cormorant in Scottish inland waters', *Freshw. Salm. Fish. Res.*, 35: 1–16.

MILNE, A. (1938) 'The ecology of the Tamar Estuary III. Salinity and temperature conditions in the lower estuary', *J. mar. biol. Ass. U.K.*, 22: 529–42.

MILNE, A. (1940) 'The ecology of the Tamar Estuary. IV. The distribution of the fauna and flora on buoys', *J. mar. biol. Ass. U.K.*, 24: 69–87.

MORDUKHAI-BOLTOVSKOI, P. D. (1964) 'New species of *Apagis* and *Cercopagis* (Cladocera, Polyphemidae) from the Caspian Sea', *Crustaceana*, 7: 21–6.

MORRIS, R. (1956) 'The osmoregulatory ability of the lampern (*Lampetra fluviatilis* L.) in sea water during the course of its spawning migration', *J. exp. Biol.*, 33: 235–48.

MORRIS, R. (1958) 'The mechanism of marine osmoregulation in the lampern (*Lampetra fluviatilis* L.) and the causes of its breakdown during the spawning migration', *J. exp. Biol.*, 35: 649–65.

MOTAIS, R. (1961) 'Sodium exchange in the euryhaline teleost *Platichthys flesus flesus*', *Endocrinology*, 70: 724–6.

MOYSE, J. (1963) 'A comparison of the value of various flagellates and diatoms as food for barnacle larvae', *J. Cons. int. Expl. Mer.*, 28: 175–87.

MURIE, J. (1903) 'Report on the sea fisheries and fishing industries of the Thames Estuary', *Kent Essex Sea Fish. Comm. Lond.*

MURRAY, J. W. (1963) 'Ecological experiments on Foraminiferida', *J. mar. biol. Ass. U.K.*, 43: 621–42.

MURRAY, J. W. (1965) 'On the Foraminiferida of the Plymouth region', *J. mar. biol. Ass. U.K.*, 45: 481–505.

MUUS, B. J. (1963) 'Some Danish Hydrobiidae with the description of a new species, *Hydrobia neglecta*', *Proc. malacol. Soc. Lond.*, 35: 131–8.

MUUS, K. (1966) 'Notes on the biology of *Protohydra leuckarti* Greef (Hyroidea, Protohydridae)'. *Ophelia*, 3: 141–50.

NAGEL, H. (1934) 'Die Aufgaben der Excretionsorgane und der Kiemen bei der Osmoregulation von *Carcinus maenas*', *Z. vergl. Physiol.*, 21: 468–91.

NAYLOR, E. (1958) 'Tidal and diurnal rhythms of locomotory activity in *Carcinus maenas* (L.)', *J. exp. Biol.*, 35: 602–10.

NAYLOR, E. (1960) 'Locomotory rhythms in *Carcinus maenas* from non-tidal conditions', *J. exp. Biol.*, 37: 481–8.

NAYLOR, E. (1962) 'Seasonal changes in a population of *Carcinus maenas* (L.) in the littoral zone', *J. Anim. Ecol.* 31: 601–9.

NAYLOR, E. and HAAHTELA, I. (1966) 'Habitat preferences and interspersions of species within the superspecies *Jaera albifrons* Leach (Crustacea: Isopoda)', *J. Anim. Ecol.*, 35: 209–16.

NAYLOR, E., SLINN, D. J. and SPOONER, G. M. (1960) 'Observations on the British species of *Jaera* (Isopoda: Asellota)', *J. mar. biol. Ass. U.K.*, 41: 817–28.

NEALE, J. W. (1965) 'Some factors influencing the distribution of recent British Ostracoda', *Publ. Staz. zool. Napol.*, 33 suppl.: 247–307.

NEMENZ, H. (1960) 'On the osmotic regulation of the larvae of *Ephydra cinerea*', *J. Insect Physiol.*, 4: 38–44.

NENNINGER, U. (1948) 'Die Peritrichen der Umgebung von Erlangen mit besonderer Berücksichtigung ihrer Wirtspezifität', *Zool. Jahrb. (Syst.)*, 77: 163–281.

NEWELL, G. E. (1948) 'A contribution to our knowledge of the life history of *Arenicola marina* L.', *J. mar. biol. Ass. U.K.*, 27: 554–80.

NEWELL, G. E. (1951) 'The life history of *Clymenella torquata* Leidy (Polychaeta)', *Proc. zool. Soc. Lond.*, 121: 561–86.

NEWELL, I. M. (1947) 'A systematic and ecological study of the Halacaridae of eastern North America,' *Bull. Binghm oceanogr. Coll.*, 10 (3): 1–232.

NEWELL, R. (1962) 'Behavioural aspects of the ecology of *Peringia* (= *Hydrobia*) *ulvae* (Pennant) (Gastropoda, Prosobranchia)', *Proc. zool. Soc. Lond.*, 138: 49–75.

NEWELL, R. (1964) 'Some factors controlling the upstream distribution of *Hydrobia ulvae* (Pennant) (Gastropoda, Prosobranchia)', *Proc. zool. Soc. Lond.*, 142: 85–106.

NEWELL, R. (1965) 'The role of detritus in the nutrition of two marine deposit feeders, the prosobranch *Hydrobia ulvae* and the bivalve *Macoma balthica*', *Proc. zool. Soc. Lond.*, 144: 25–45.

NICHOLLS, A. G. (1935) 'The larval stages of *Longipedia coronata* Claus, *L. scotti* G. O. Sars, and *L. minor* T. and A. Scott, with a description of the male of *L. scotti*', *J. mar. biol. Ass. U.K.*, 20: 29–45.

NICLOUX, M. (1938) 'L'eau des tissues. Nouvelle contribution a l'étude d'une eau lee aux proteides', *Bull. Soc. Chim. biol.*, 20: 981–1032.

NICOL, E. A. T. (1935) 'The ecology of a salt marsh', *J. mar. biol. Ass. U.K.*, 20: 203–61.

NOODT, W. (1957) 'Zur Ökologie der Harpacticoidea (Crust. Cop.) des Eulitorals der Deutschen Meeresküste und der angrenzenden Brackgewässer'. *Z. Morph. Okol. Tiere*, 46: 149–242.

NORTON-GRIFFITHS, M. (1966) 'Oystercatchers and mussels', *Ibis*, 108: 455–6.

NYHOLM, K. G. (1951) 'A contribution to the study of the sexual phase of *Protohydra leuckarti*', *Ark. Zool.*, 2: 529–30.

NYMAN, K. J. (1953) 'Observations on the behaviour of *Gobius microps*', *Acta Soc. Faun. Flor. Fenn.*, 69 (5): 1–11.

OGLESBY, L. C. (1965a) 'Steady-state parameters of water and chloride regulation in estuarine nereid polychaetes', *Comp. Biochem. Physiol.*, 14: 621–40.

OGLESBY, L. C. (1965b) 'Water and chloride fluxes in estuarine nereid polychaetes', *Comp. Biochem. Physiol.*, 16: 437–55.

OLNEY, P. J. S. (1963) 'The food and feeding habits of the teal *Anas creca creca* L.', *Proc. zool. Soc. Lond.*, 140: 169–210.

OLNEY, P. J. S. (1964) 'The food of mallard *Anas platyrhynchos platyrhynchos* collected from coastal and estuarine areas', *Proc. zool. Soc. Lond.* 142: 397–418.

OLNEY, P. J. S. (1965) 'The food and feeding habits of shelduck *Tadorna tadorna*', *Ibis*, 107: 527–32.

OSBURN, R. C. (1944) 'A survey of the Bryozoa of Chesapeake Bay', *Chesap. biol. Lab. Publ.*, 63: 1–59.

OWEN, D. F. (1960) 'The nesting success of the heron *Ardea cinerea* in relation to the availability of food', *Proc. zool. Soc. Lond.*, 133: 597–617.

PALMEN, E. (1954) 'Effect of soil moisture upon temperature preferendum in *Dyschirius thoracicus* Rossi (Col., Carabidae)', *Suom. Hyont. Aikak.*, 20(1): 1–13.

PALMER, J. D. and ROUND, F. E. (1965) 'Persistent vertical-migration rhythms in benthic microflora. I. The effect of light and temperature on the rhythmic behaviour of *Euglena obtusa*', *J. mar. biol. Ass. U.K.*, 45: 567–82.

PALMER, R. S. (1962) *Handbook of North American Birds. Vol. 1. Loons through flamingoes*, Yale Univ. Press, 567 pp.

PANIKKAR, N. K. (1941) 'Osmoregulation in some palaemonid prawns' *J. mar. biol. Ass. U.K.*, 25: 317–59.

PANTIN, C. F. A. (1931a) 'The adaptation of *Gunda ulvae* to salinity I. The environment', *J. exp. Biol.*, 8: 63–72.

PANTIN, C. F. A. (1931b) 'The adaptation of *Gunda ulvae* to salinity III. The electrolyte exchange', *J. exp. Biol.*, 8: 82–94.

PARRY, G. (1953) 'Osmotic and ionic regulation in the isopod crustacean *Ligia oceanica* L.', *J. exp. Biol.*, 30: 567–74.

PARRY, G. (1955) 'Urine production by the antennal glands of *Palaemonetes varians* (Leach)', *J. exp. Biol.*, 32: 408–22.

PARRY, G. (1957) 'Osmoregulation in some freshwater prawns', *J. exp. Biol.*, 34: 417–23.

PARRY, G. (1961a) 'Osmoregulation of the freshwater prawn *Palaemonetes antennarius*', *Mem. Ist. Ital. Idrobiol.*, 13: 139–49.

PARRY, G. (1961b) 'Osmotic and ionic changes in the blood and muscle of migrating salmonids', *J. exp. Biol.*, 38: 411–28.

PARRY, G. (1966) 'Osmotic adaptation in fishes', *Biol. Rev.* 41: 392–444.

PATEL, B. and CRISP, D. J. (1960) 'Rates of development of the embryos of several species of barnacles', *Physiol. Zool.* 33: 104–19.

PATTERSON, A. H. (1904) *Notes of an east coast naturalist*, London.

PATTERSON, A. H. (1906) 'Some fish notes from Great Yarmouth for 1906', *Zoologist* (4), 10: 453–8.

PATTERSON, A. H. (1907) *Wild life in a Norfolk Estuary*, London.

PELKWIJK, J. J. and TINBERGEN, N. (1937) 'Eine reizbiologische Analyse einiger Verhaltensweisen von *Gasterosteus aculeatus* L.', *Z. Tierpsychol.*, 1: 193–200.

PERCIVAL, E. (1929) 'A report on the fauna of the estuaries of the River Tamar and the River Lynher', *J. mar. biol. Ass. U.K.*, 16: 81–108.

PERKINS, E. J. (1956) 'The fauna of a sand-bank in the mouth of the Dee Estuary', *Ann. Mag. nat. Hist.*, (12), 9: 112–28.

PERKINS, E. J. (1957) 'The black sulphide containing layer of the shore with particular reference to Whitstable, Kent', *Ann. Mag. nat. Hist.*, (12), 10: 25–35.

PERKINS, E. J. (1958a) 'The hardness of the soil of the shore at Whitstable, Kent', *J. Ecol.*, 46: 71–81.

PERKINS, E. J. (1958b) 'The food relationships of the microbenthos, with particular reference to that found at Whitstable, Kent', *Ann. Mag. nat. Hist.* (13), 1: 64–77.

PERKINS, E. J. (1960) 'The diurnal rhythm of the littoral diatoms of the River Eden Estuary, Fife', *J. Ecol.*, 48: 725–8.

PERTTUNEN, V. (1951) 'The humidity preferences of various carabid species (Col., Carabidae) of wet and dry habitats', *Ann. Entom. Fenn.*, 17: 72–84.

PESTA, O. (1907) 'Die Metamorphose von *Mytilicola intestinalis* Steuer', *Z. wiss. Zool.*, 88: 78–98.

PESTA, O. (1935) 'Ein Mysidaceen-Nachweis auf der Insel Korfu (Griechenland)', *Zool. Anz.*, 111: 332–3.

PETERSON, G. H. (1958) 'Notes on the growth and biology of the different *Cardium* species in Danish brackish water areas', *Medd. Danm. Fisk. Havunders.*, N.S. 2 (22): 1–31.

PETRUSHEVSKI, G. K. (1931) Über die Verbreitung der Plerocercoide von *Diphyllobothrium latum* in den Fischen von Nevabucht', *Zool. Anz.*, 94: 58.

PFITZENMEYER, H. T. (1962) 'Periods of spawning and setting of the soft shelled clam, *Mya arenaria* at Solomons Maryland', *Chesapeake Sci.*, 3: 114–20.

PHLEGER, F. B. (1960) *Ecology and distribution of recent Foraminifera*, John Hopkins Press, 297 pp.

PHLEGER, F. B. (1965) 'Patterns of marsh Foraminifera, Galveston Bay, Texas', *Limnol. Oceanogr. Suppl.* R.: 169–84.

PILLAY, T. V. R. (1953) 'Studies on the food, feeding habits and alimentary tract of the grey mullet, *Mugil tade* Forskal', *Proc. Nat. Inst. Sci. India*, 19: 777–827.

PONTIN, R. M. (1964) 'A comparative account of the protonephridia of *Asplanchna* (Rotifera) with special reference to the flame bulbs', *Proc. zool. Soc. Lond.*, 142: 511–25.

PONTIN, R. M. (1966) 'The osmoregulatory function of the vibratile flames and the contractile vesicle of *Asplanchna* (Rotifera)', *Comp. Biochem. Physiol.*, 17: 1111–26.

POPHAM, E. J. (1966) 'The littoral fauna of the Ribble Estuary, Lancashire, England', *Oikos* 17: 19–32.

PORTIER, P. and DUVAL, M. (1922) 'Variation de la pression osmotique du sang de l'Anguille en fonction des modifications de salinité du milieu exterieur', *C. r. Seanc. Biol.*, 175: 324–6.

POTTS, W. T. W. and PARRY, G. (1964) *Osmotic and ionic regulation in animals*, London: Pergamon.

PRECHT, H. (1935) 'Epizoen der Kieler Bucht', *Nova Acta Leopold*, 3: 405–74.

PRITCHARD, D. W. (1955) 'Estuarine circulation patterns', *Proc. Amer. Soc. civil Engin.*, 81 (717): 1–11.

PUGH, G. J. F. (1960) 'The fungal flora of tidal mud flats', pp. 202–8 in *Ecology of Soil Fungi*. Liverpool: University Press.

PUGH, G. J. F. (1961) 'Fungal colonisation of a developing salt marsh', *Nature, Lond.*, 190: 1032–3.

PUGH, G. J. F. (1962a) 'Studies on fungi in coastal soils 1. *Cercospora salina* Sutherland', *Trans. Brit. mycol. Soc.*, 45: 255–60.

PUGH, G. J. F. (1962b) 'Studies on fungi in coastal soils II. Fungal ecology in a developing salt marsh', *Trans. Brit. mycol Soc.*, 45: 560–6.

RADFORTH, I. (1940) 'The food of the grayling (*Thymallus thymallus*), Flounder (*Platichthys flesus*), roach (*Rutilus rutilus*) and gudgeon (*Gobio fluviatilis*) with special reference to the Tweed watershed', *J. Anim. Ecol.*, 9: 302–18.

RAMSAY, J. A. (1949) 'The osmotic relations of the earthworm', *J. exp. Biol.*, 26: 46–56.

RAMSAY, J. A. (1950) 'Osmotic regulation in mosquito larvae', *J. exp. Biol.*, 27: 145–57.

RANADE, M. R. (1957) 'Observations on the resistance of *Tigriopus fulvus* (Fischer) to changes in temperature and salinity', *J. mar. biol. Ass. U.K.*, 36: 115–19.

RANWELL, D. S. (1961) '*Spartina* salt marshes in southern England 1. The effect of sheep grazing at the upper limits of *Spartina* marsh in Bridgwater Bay', *J. Ecol.*, 49: 325–40.

RASMUSSEN, E. (1951) 'Faunistic and biological notes on marine invertebrates II. The eggs and larvae of some Danish marine gastropods', *Vidensk. Medd. Dansk. naturh. Foren.*, 113: 201–49.

RASMUSSEN, E. (1956) 'Faunistic and biological notes on marine invertebrates III. The reproduction and larval development of some polychaetes from the Isefjord, with some faunistic notes', *Biol. Medd. Kongl. Dan. Vid. Selsk.*, 23: 1–84.

RASMUSSEN, E. (1959) 'Behaviour of sacculinized shore crabs (*Carcinus maenas* Pennant)', *Nature, Lond.* 183: 479–80.

RAYKOVA, E. V. (1958) 'On the life cycle of *Polypodium hydriforme* Ussov (Coelenterata)', *Zool. Zh.*, 37: 3.

REES, C. B. (1940) 'A preliminary study of the ecology of a mud flat', *J. mar. biol. Ass. U.K.*, 24: 185–99.

REID, D. M. (1930) 'Salinity interchange between sea water in sand and overflowing fresh-water at low tide', *J. mar. biol. Ass. U.K.*, 16: 609–14.

REID, D. M. (1932) 'Salinity interchange between salt water in sand and overflowing fresh-water at low tide II', *J. mar. biol. Ass. U.K.*, 18: 299–306.

REMANE, A. (1958) 'Ökologie des Brackwassers', pp. 1–213 in *Die Biologie des Brackwassers*, Stuttgart: Schweizerbartsche Verlag.

REUTER, M. (1961) 'Untersuchungen uber Rassenbildung bei *Gyratrix hermaphroditus* (Turbellaria Neorhabdocoela)', *Acta zool. Fenn.*, 100: 1–32.

RITCHIE, J. (1927) 'Report on the prevention of growth of mussels in submarine shafts and tunnels at Westbank electricity station, Portobello', *Trans. Roy. Scot. Soc. Arts*, 19: 1.

ROBERTSON, J. D. (1960) 'Ionic regulation in the crab *Carcinus maenas* (L.) in relation to the moulting cycle', *Comp. Biochem. Physiol.*, 1: 183–212.

ROBINSON, A. R. and ROHWER, C. (1955) 'Measurement of canal seepage', *Proc. Amer. Soc. civil Engin.*, 81 (728): 1–20.

ROBSON, E. A. (1957) 'A sea anemone from brackish water', *Nature, Lond.*, 179: 787–8.

ROCH, F. (1924) 'Experimentelle Untersuchungen an *Cordylophora caspia* (Pallas) (= *lacustris* Allman) über die Abhängigkeit ihrer geographischen Verbreitung und ihrer Wuchsform von den physikalisch-chemischen Bedingung des umgehenden Mediums', *Z. Morph. Okol. Tiere*, 2: 350–426.

ROCHFORD, D. J. (1951) 'Studies in Australian estuarine hydrology 1. Introductory and comparative features', *Austral. J. mar. freshw. Res.*, 2: 1–116.

ROGERS. H. M. (1940) 'Occurrence and retention of plankton within the estuary', *J. Fish. Res. Bd. Can.*, 5: 164–71.

ROTHSCHILD, M. (1938) 'Further observations on the effect of trematode parasites on *Peringia ulvae* (Pennant, 1777)', *Novit. Zool.*, 41: 84–102.

ROTHSCHILD, M. and CLAY, T. (1952) *Fleas, flukes and Cuckoos. A study of bird parasites*, London: Collins, 304 pp.

ROUND, F. E. (1960) 'Diatom flora of a salt marsh on the River Dee', *New Phytol.*, 59: 332–48.

ROUND, F. E. and PALMER, J. D. (1966) 'Persistent vertical-migration rhythms in benthic microflora II. Field and laboratory studies on diatoms from the banks of the River Avon', *J. mar. biol. Ass. U.K.*, 46: 191–214.

ROWBOTHAM, F. (1964) *The Severn Bore*, London: MacDonald.

RUEBUSH, T. K. (1940) 'Morphology, encapsulation and osmoregulation of *Dinophilus gardineri* Moore', *Trans. Amer. micr. Soc.*, 59: 205–23.

RUSSELL, F. S. (1937) 'The seasonal abundance and distribution of the pelagic young of teleostean fishes caught in the ring trawl in offshore waters in the Plymouth area. Part IV. The year 1936, with notes on the conditions as shown by the occurrence of plankton indicators', *J. mar. biol. Ass. U.K.*, 21: 679–86.

RUTTNER, F. (1963) *Fundamentals of limnology*, 3rd Ed., Toronto: Univ. Press.

SAFRIEL, U. (1966) 'Food and survival of oystercatcher chicks on Skokholm in 1965', *Ibis* 108: 455.

SANDISON, E. E. (1966a) 'The effect of salinity fluctuations on the life cycle of *Balanus pallidus stutsburi* Darwin in Lagos Harbour, Nigeria', *J. Anim. Ecol.*, 35: 363–78.

SANDISON, E. E. (1966a) 'The effect of salinity fluctuations on the life cycle of *Gryphaea gasar* ((Adanson) Dautzenberg) in Lagos Harbour, Nigeria', *J. Anim. Ecol.*, 35: 379–89.

SANDISON, E. E. and HILL, M. B. (1966) 'The distribution of *Balanus pallidus stutsburi* Darwin, *Gryphaea gasar* ((Adanson) Dautzenberg), *Mercierella enigmatica* Fauvel and *Hydroides uncinata* (Phillippi) in relation to salinity in Lagos Harbour and adjacent creeks', *J. Anim. Ecol.*, 35: 235–50.

SANDON, H. (1957) 'Neglected animals—the Foraminifera', *New Biol.*, 24: 7–32.

SARS, G. O. (1897) 'Pelagic Entomostraca of the Caspian Sea', *Ann. Mus. zool. Acad. Sci. St. Petersb.*, 2: 1–78.

SARS, G. O. (1902) 'On the Polyphemidae of the Caspian Sea', *Ann. Mus. zool. Acad. Sci. St. Petersb.* 7: 31–54.

SAVAGE, R. E. (1956) 'The great spatfall of mussels (*Mytilus edulis* L.) in the River Conway Estuary in spring 1940', *Fish. Invest. Lond.*, (2), 20 (7): 1–22.

SCHILDMACHER, H. (1932) 'Ueber den Einfluss des Salzwassers auf die Entwicklung der Nasendrüsen', *J. Ornithol.* 80: 293–9.

SCHLIEPER, C. (1929) 'Über die Einwirkung niederer Salzkonzentrationen auf marine Organismen', *Z. vergl. Physiol.* 9: 478–514.

SCHMIDT, J. (1909) 'Remarks on the metamorphosis and distribution of the larvae of the eel', *Medd. Komm. Hav. Fisk.*, 3 (3): 1–17.

SCHMIDT, J. (1924) 'The breeding places of the eel', *Smithson. Rept.*, 1924: 279–316.

SCHMIDT-NIELSEN, K. and FANGE, R. (1958) 'The function of the salt gland in the brown pelican', *Aul*, 75: 282–9.

SCOTT, A. (1895) 'Examination of food in fishes' stomachs', *Lancs. Sea Fish. Rept.*, 1895: 6–12.

SCOTT, K. M. F., Harrison, A. D. and MACNAE, W. (1950) 'The ecology of South African Estuaries II. The Klein River Estuary, Hermanus Cape', *Trans. Roy. Soc. S. Afr.*, 33: 283–331.

SEGERSTRALE, S. G. (1957) 'Baltic Sea. In: Treatise on marine ecology and palaeoecology', *Geol. Soc. Amer. Mem.*, 67, vol. I: 751–800.

SERVENTY, D. L. (1938) 'The feeding of cormorants in south-western Australia', *Emu* 38: 293–316.

SEVENSTER-BOL, A. C. A. (1962) 'On the causation of drive reduction after a consummatory act', *Arch. néerl. Zool.*, 15: 175–236.

SEXTON, E. W. and SPOONER, G. M. (1940) 'An account of *Marinogammarus* (Schellenberg) gen. nov. (Amphipoda), with a description of a new species. *M. pirloti*', *J. mar. biol. Ass. U.K.*, 24: 633–82.

SHAW, J. (1955a) 'Ionic regulation in the muscle fibres of *Carcinus maenas* I. The electrolyte composition of single fibres', *J. exp. Biol.*, 32: 383–96.

SHAW, J. (1955b) 'Ionic regulation in the muscle fibres of *Carcinus maenas* II. The effect of reduced blood concentration', *J. exp. Biol.*, 32: 664–80.

SHAW. J. (1958) 'Further studies on ionic regulation in the muscle fibres of *Carcinus maenas*', *J. exp. Biol.*, 35: 902–19.

SHAW, J. (1961) 'Studies on ionic regulation in *Carcinus maenas* (L.) I. Sodium balance', *J. exp. Biol.*, 38: 135–52.

SHAW, J. and SUTCLIFFE, D. W. (1961) 'Studies on sodium balance in *Gammarus duebeni* Lilljeborg and *G. pulex pulex* (L.)', *J. exp. Biol.*, 38:1–15.

SIMMONS, E. G. (1957) 'An ecological study of the upper Laguna Madre of Texas', *Publ. Inst. mar. Sci. Texas*, 4: 156–200.

SMIDT, E. L. B. (1944) 'Biological studies of the invertebrate fauna of the harbour of Copenhagen', *Vidensk. Medd. Dansk. naturh. Foren.*, 107: 235–316.

SMIDT, E. L. B. (1951) 'Animal production in the Danish waddensee', *Medd. Komm. Hav. Fisk.*, 11: 1–151.

SMITH, E. A. (1889) 'Notes on British Hydrobiidae with a description of a supposed new species', *J. Conch.*, 6: 142–5.

SMITH, H. W. (1931) 'The absorption and excretion of water and salts by the elasmobranch fishes. 1. Fresh water elasmobranchs', *Amer. J. Physiol.*, 98: 279–95.

SMITH, H. W. (1936) 'The retention and physiological role of urea in the Elasmobranchii', *Biol. Rev.*, 11: 49–82.

SMITH, J. E. and NEWELL, G. E. (1955) 'The dynamics of the zonation of the common periwinkle (*Littorina littorea* (L.)) on a stony beach', *J. Anim. Ecol.*, 24: 35–56.

SMITH, R. I. (1955a) 'On the distribution of *Nereis diversicolor* in relation to salinity in the vicinity of Tvärminne, Finland, and the Isefjord, Denmark', *Biol. Bull. mar. biol. Lab. Woods Hole*, 108: 326–45.

SMITH, R. I. (1955b) 'Salinity variation in interstitial water at Kames Bay, Millport, with reference to the distribution of *Nereis diversicolor*', *J. mar. biol. Ass. U.K.*, 34: 33–46.

SMITH, R. I. (1956) 'The ecology of the Tamar Estuary VII. Observations on the interstitial salinity of intertidal muds in the estuarine habitat of *Nereis diversicolor*', *J. mar. biol. Ass. U.K.*, 35: 81–104.

SMITH, R. I. (1964) 'On the early development of *Nereis diversicolor* in different salinities', *J. Morph.*, 114: 437–64.

SNELLING, B. (1959) 'The distribution of intertidal crabs in the Brisbane River', *Austral. J. mar. freshw. Res.*, 10: 67–83.

SOIKA, A. G. (1955) 'Éthologie, Écologie, Systématique et Biogeographie des *Eurydice* s. str. (Isop. Cirolanides)', *Vie et Milieu*, 6: 38–52.

SOMMER, G. (1950) 'Die peritrichen Ciliaten des Grossen Plöner Sees', *Arch. Hydrobiol.*, 44: 349–440.

SOUTHWOOD, T. R. E. and LESTON, D. (1957) 'Notes on the nomenclature and zonal occurrence of the *Orthotylus* species (Hem. Miridae) of British salt marshes', *Ent. mon. Mag.* 93: 166–8.

SPÄRCK, R. (1950) 'The food of the North European gulls', *Proc. Xth Int. ornith. Congr. Uppsala*, pp. 588–91.

SPENCER, R. (1956) 'Studies in Australian estuarine hydrology II. The Swan River', *Austral. J. mar. freshw. Res.*, 7: 193–253.

SPOONER, G. M. 'The distribution of *Gammarus* species in estuaries', *J. mar. biol. Ass. U.K.*, 27: 1–52.

SPOONER, G. M. and MOORE, H. B. (1940) 'The ecology of the Tamar Estuary. VI. An account of the macrofauna of the intertidal muds', *J. mar. biol. Ass. U.K.*, 24: 283–330.

STADEL, O. (1936) 'Nahrungsuntersuchungen an Elbfischen', *Z. Fisch. Hilfwiss.*, 34: 45–61.

STEEL, W. O. (1955) 'Notes on the habits of some British *Bledius* species (Col., Staphylinidae)', *Ent. mon. Mag.*, 91: 240.

STEPHEN, A. C. (1929) 'Studies on the Scottish marine fauna: the fauna of the sandy and muddy areas of the tidal zone', *Trans. Roy. Soc. Edin.*, 56: 291–306.

STEPHENSON, T. A. (1935) *British sea anemones*. Vol II, London: Ray Society.

STEVEN, G. A. (1933) 'The food consumed by shags and cormorants around the shores of Cornwall (England)', *J. mar. biol. Ass. U.K.*, 19: 277–92.

STEVEN, G. A. (1949) 'Contributions to the biology of the mackerel *Scomber scomber* L. II. A study of the fishery in the south west of England with special reference to spawning, feeding and "fishermen's signs"', *J. mar. biol. Ass. U.K.*, 28: 555–81.

STOCK, J. H. (1952) 'Some notes on the taxonomy, the distribution and the ecology of four species of the Amphipod genus *Corophium* (Crustacea: Malacostraca)', *Beaufortia*, 21: 1–10.

STOLL, C. (1962) 'Cycle évolutif de *Paragnathia formica* (Hesse) (Isopode-Gnathiidae)', *Cah. Biol. mar.*, 3 (4): 401–16.

STOPFORD, S. C. (1951) 'An ecological survey of the Cheshire foreshore of the Dee Estuary', *J. Anim. Ecol.*, 20: 103–22.

STUNKARD, H. W. and UZMANN, J. R. (1959) 'The life cycle of the digenetic trematode *Proctoeces maculatus* (Looss 1901), Odhner 1911 (syn. *P. subtenuis* (Linton) Hanson 1950) and a description of *Cercaria adranocerca* n. sp.', *Biol. Bull. mar. biol. lab. Woods Hole*, 116: 184–93.

SUTCLIFFE, D. W. (1960a) 'Observations on the salinity tolerance and habits of a euryhaline caddis larva, *Limnephilus affinis* Curtis (Trichoptera: Limnephilidae)', *Proc. Roy. ent. Soc. Lond.*, A. 35: 156–62.

SUTCLIFFE, D. W. (1960b) 'Osmotic regulation in the larvae of some euryhaline Diptera', *Nature, Lond.*, 187: 331–2.

SUTCLIFFE, D. W. (1961a) 'Studies on salt and water balance in caddis larvae (Trichoptera): 1. Osmotic and ionic regulation of body fluids in *Limnephilus affinis* Curtis', *J. exp. Biol.*, 38: 501–19.

SUTCLIFFE, D. W. (1961b) 'Salinity fluctuations and the fauna in a salt marsh with special reference to aquatic insects', *Trans. nat. Hist. Soc. Northumb. Durh.*, 14: 37–56.

TAIT, J. (1927) 'Experiments and observations on Crustacea. Part VII. Some structural and physiological features of the valviferous isopod *Chiridotea*', *Proc. Roy. Soc. Edin.*, 46: 334–8.

TATTERSAL, O. S. (1957) 'Report on a small collection of Mysidacea from the Sierra Leone Estuary together with a survey of the genus *Rhopalophthalmus* Illig and a description of a new species of *Tenagomysis* from Lagos, Nigeria', *Proc. zool. Soc. Lond.*, 129: 81–128.

TATTERSALL, W. M. and TATTERSALL, O. S. (1951) *The British Mysidacea*, London: Ray Society, 460 pp.

TEAL, J. M. (1962) 'Energy flow in the salt marsh ecosystem of Georgia', *Ecology*, 43: 614–24.

TEBBLE, N. (1966) *British bivalve seashells*, London: British Museum (Natural History), 212 pp.

TECHNAU, G. (1936) 'Die Nasendrüse der Vögel', *J. Ornithol.*, 84: 511–617.

THAMDRUP, H. M. (1935) 'Beiträge zur Okologie der Wattenfauna auf experimenteller Grundlage', *Medd. Komm. Hav. Fisk.*, 10 (2): 1–125.

THEISEN, B. F. (1966) 'The life history of seven species of ostracods from a Danish brackish-water locality', *Medd. Danm. Fisk. Havunders.*, N.S.4:215–70.

THOMPSON, T. E. (1959) 'Feeding in nudibranch larvae', *J. mar. biol. Ass. U.K.*, 38: 239–48.

THOMSON, J. M. (1966) 'The grey mullets', *Oceanogr. mar. Biol. Ann. Rev.*, 4: 301–35.

THORSON, G. (1946) 'Reproduction and larval development of Danish marine bottom invertebrates, with special reference to the planktonic larvae in the Sound (Oresund)', *Medd. Komm. Hav. Plankt.*, 4 (1): 1–523.

TICKELL, F. G. and HIATT, W. N. (1938) 'Effect of angularity of grain on porosity and permeability of unconsolidated sands', *Amer. Ass. Petrol. Geol. Bull.*, 22: 1272–74.

TINBERGEN, N. (1951) *The study of instinct*, Oxford.

TODD, M. E. (1964) 'Osmotic balance in *Hydrobia ulvae* and *Potamopyrgus jenkinsi* (Gastropoda: Hydrobiidae)', *J. exp. Biol.*, 41: 665–77.

TRUEMAN, E. R. (1966) 'Observations on the burrowing of *Arenicola marina* (L.)', *J. exp. Biol.*, 44: 93–118.

TUCKER, D. W. (1959) 'A new solution to the Atlantic eel problem', *Nature, Lond.*, 183: 495–501.

URIST, M. R. (1962) 'Calcium and other ions in the blood and skeleton of the Nicaraguan freshwater shark', *Science* 137: 985–6.

VÄLIKANGAS, I. (1926) 'Planktologische Untersuchungen im Hafengebiet von Helsingfors I. Über das Plankton insbesondere das Netzzooplankton des Sommerhalbjahres', *Acta zool. Fennica* 1: 1–298.

VALKANOV, A. (1938) 'Ubersicht der Hydrozoenfamilie Moerisidae', *Annu. Univ. Sofia, Fac. Phys.-Math.*, 34: 251–320.

VALKANOV, A. (1953) 'Revision der Hydrozoenfamilie Moerisiidae', *Arb. biol. Meerest. Varna*, 18: 33–47.

VAN DOBBEN, W. H. (1952) 'The food of the cormorant in the Netherlands', *Ardea*, 42: 1–61.

VOOUS, K. H. (1960) *Atlas of European birds*, London: Nelson.

WATKIN, E. E. (1941) 'The yearly life cycle of the amphipod *Corophium volutator*', *J. Anim. Ecol.*, 10: 77–93.

WEBB D. A. (1940) 'Ionic regulation in *Carcinus maenas*', *Proc. Roy. Soc. B.*, 129': 107–36.

WEBB, J. E. (1958) 'The ecology of Lagos Lagoon V. Some physical properties of lagoon deposits', *Phil. Trans. Roy. Soc.*, B. 241: 393–419.

WEBB, M. G. (1956) 'An ecological study of brackish water ciliates', *J. Anim. Ecol.*, 25: 148–75.

WELLS, G. P. (1937) 'Studies on the physiology of *Arenicola marina* L. I. The pacemaker role of the oesophagus and the action of adrenaline and acetylcholine', *J. exp. Biol.*, 15: 117–57.

WELLS, G. P. (1945) 'The mode of life of *Arenicola marina* L.', *J. mar. biol. Ass. U.K.*, 26: 170–207.

WELLS, G. P. (1949a) 'Respiratory movements of *Arenicola marina* L.; intermittent irrigation of the tube, and intermittent aerial respiration', *J. mar. biol. Ass. U.K.*, 28: 447–64.

WELLS, G. P. (1949b) 'The behaviour of *Arenicola marina* L. in sand, and the role of spontaneous activity cycles', *J. mar. biol. Ass. U.K.*, 28: 465–78.

WELLS, G. P. (1950) 'The anatomy of the body wall and appendages in *Arenicola marina* L., *Arenicola claparedii* Levinsen and *Arenicola ecaudata* Johnston', *J. mar. biol. Ass. U.K.*, 29: 1–44.

WELLS, G. P. (1952) 'The proboscis apparatus of *Arenicola*', *J. mar. biol. Ass. U.K.*, 31: 1–28.

WELLS, G. P. (1957) 'Variation in *Arenicola marina* (L.) and the status of *Arenicola glacialis* Murdoch (Polychaeta)', *Proc. zool. Soc. Lond.*, 129: 397–419.

WELLS, G. P. (1959) 'The genera of Arenicolidae (Polychaeta)', *Proc. zool. Soc. Lond.*, 133: 301–14.

WELLS, G. P. (1962) 'The warm water lugworms of the world', *Proc. zool. Soc. Lond.*, 138: 331–55.

WELLS, G. P. and LEDINGHAM, I. C. (1940) 'Physiological effects of a hypotonic environment I. The action of hypotonic salines on isolated rhythmic preparations from polychaete worms (*Arenicola marina, Nereis diversicolor, Perinereis cultrifera*)', *J. exp. Biol.*, 17: 337–63.

WELLS, J. B. J. (1963) 'Copepoda from the littoral region of the estuary of the River Exe (Devon, England)', *Crustaceana*, 5: 10–26.

WERNSTEDT, C. (1942) 'Studies on the food of *Macoma baltica* and *Cardium edule*', *Vid. Medd. Dansk. naturh. Foren.*, 106: 241–52.

WERNTZ, H. O. (1962) 'Osmotic regulation in marine and freshwater gammarids (Amphipoda)', *Biol. Bull. mar. biol. Lab. Woods Hole*, 124: 225–39.

WHEATLAND, A. B. (1959) 'Some aspects of the carbon, nitrogen and sulphur cycles in the Thames Estuary. I. Photosynthesis, denitrification and sulphate reduction, pp. 33–50 in: The effects of pollution on living material', *8th Symp. Inst. Biol. Lond.*

WIESER, W. (1952) 'Die Beziehung zwischen Mundhohlengestalt, Ernahrungsweise und Vorkommen bei freilehenden marinen Nematoden', *Ark. Zool.* (2) 4: 439–84.

WIKGREN, B. (1953) 'Osmotic regulation in some animals with special reference to temperature', *Acta zool. Fenn.*, 71: 1–102.

WILLIAMS, R. B. (1964) 'Division rates of salt marsh diatoms in relation to salinity and cell size', *Ecology*, 45: 877–80.

WILSON, D. P. (1953) 'Notes from the Plymouth aquarium II. An alga disliked by *Mugil chelo* Cuvier', *J. mar. biol. Ass. U.K.*, 32: 203.

WOOD, E. J. F. (1964) 'Studies in microbial ecology of the Australasian Region V. Microbiology of some Australian estuaries', *Nova Hedwig.*, 8: 461–527.

YANAGIMACHI, R. (1962) 'The life cycle of *Peltogasterella*', *Crustaceana*, 2: 183–6.

YONGE, C. M. (1949) 'On the structure and adaptations of the Tellinacea, deposit feeding Eulamellibranchiata', *Phil. Trans. Roy. Soc.*, B. 234: 29–76.

ZENKEVICH, L. A. (1957) 'Caspian and Aral Seas. In Treatise on marine ecology and palaeoecology', *Geol. Soc. Amer. Mem.*, 67, vol. 1: 891–916.

ZENKEVITCH, L. (1963) *Biology of the seas of the U.S.S.R.*, London: Allen and Unwin, 955 pp.

ZOBELL, C. E. (1938) 'Studies on the bacterial flora of marine bottom sediments', *J. sedim. Petrol.*, 8: 10–18.

ZOBELL, C. E. and ANDERSON, D. Q. (1936) 'Observations on the multiplication of bacteria in different volumes of stored sea water and the influence of oxygen tension and solid surfaces', *Biol. Bull. mar. biol. Lab. Woods Hole*, 71: 324–42.

ZOBELL, C. E. and FELTHAM, C. B. (1942) 'The bacterial flora of a marine mud-flat as an ecological factor', *Ecology*, 23: 69–77.

N

ACKNOWLEDGEMENT TO SOURCES

THE following list is of authors and publishers who have granted permission for the reproduction of the figures noted.

Figs. 5 and 6 by Inglis and Allen (1957) from the *Proceedings of the Institute of Civil Engineering.*

Fig. 7 by Inglis and Kestner (1958) from the *Proceedings of the Institute of Civil Engineering.*

Fig. 8 by King (1898) from the *U.S. Geological Survey Annual Report.*

Fig. 11 is based on two separate figures by Reid (1931, 1932) from the *Journal of the Marine Biological Association of the U.K.*

Fig. 13 by Webb (1958) from the *Philosophical Transactions of the Royal Society.*

Fig. 16 by Mathiesen and Nielsen (1956) from the *Botaniske Tidsskrifter.*

Fig. 19 is based on a figure by Sandison and Hill (1966) from the *Journal of Animal Ecology.*

Figs. 21, 22, 49, 51, 54, 70 and 72 by G. O. Sars from *The Crustacea of Norway*, by permission of the University of Bergen.

Fig. 27 by G. O. Sars (1897) from the *Annuaire du Musee Zoologiques de l'Academie Imperiale des Sciences de St. Petersbourg.*

Fig. 28 by G. O. Sars (1894) from *Mélanges Biologiques.*

Fig. 31 by Annandale (1907) from the *Records of the Indian Museum.*

Fig. 34A by R. P. Dales (1950) from the *Journal of the Marine Biological Association of the U.K.*

Figs. 34B and 39B by Thorson (1946) from *Meddelelser fra Kommissionen for Danmarks Fiskeri-og Havundersogelser. Serie Plankton.*

Fig. 34C by Rasmussen (1956) from the *Biologiske Meddelelser det Kongelige Danske Videnskabernes Selskab.*

Figs. 36 and 61B by Atkins (1954 and 1955) from the *Journal of the Marine Biological Association of the U.K.*

Fig. 42 by Bayne (1965) from *Ophelia.*

Fig. 48 by Sandison (1966) from the *Journal of Animal Ecology.*

Fig. 73 by Bresciani (1960) from *Videnskabelige Meddelelser fra Dansk naturhistorisk Forening i Kjøbenhavn.*

Fig. 96 by Fenchel (1965) from *Ophelia.*

Fig. 97A by Steuer (1905) from *Arbeiten aus den Zoologischen Instituten der Universität, Wein.*

Figs. 97B and C by Pesta (1907) from the *Zeitschrifte für wissenschaftliche Zoologie.*

Figs. 103 and 104 by Darnell (1961) from *Ecology.*

INDEX

379